The Gas Situation in the ECE Region Around the Year 1990

United Nations Economic Commission for Europe

Some other titles in this series from Pergamon Press

HOUSING FOR SPECIAL GROUPS

BEHAVIOUR OF WOOD PRODUCTS IN FIRE

PROTEIN AND NON-PROTEIN NITROGEN FOR RUMINANTS

FROZEN AND QUICK-FROZEN FOOD

FACTORS OF GROWTH AND INVESTMENT POLICIES

COAL: 1985 AND BEYOND

NON-WASTE TECHNOLOGY AND PRODUCTION

THE GAS INDUSTRY AND THE ENVIRONMENT

BUILDING RESEARCH POLICIES

STATISTICAL SERVICES IN TEN YEARS' TIME

HUMAN SETTLEMENTS AND ENERGY

ELECTRICAL LOAD-CURVE COVERAGE

ENGINEERING EQUIPMENT FOR FOUNDRIES

INTEGRATED STATISTICAL INFORMATION SYSTEMS 1977

FINE PARTICULATE POLLUTION

PROBLEMS OF THE AGRICULTURAL DEVELOPMENT OF LESS FAVOURED AREAS

SELECTED WATER PROBLEMS IN ISLANDS AND COASTAL AREAS

The Gas Situation in the ECE Region Around the Year 1990

Proceedings of an international symposium of the Committee on Gas of the Economic Commission for Europe, held in Evian, France, at the invitation of the Government of France, 2-5 October 1978

Published for the
UNITED NATIONS
by
PERGAMON PRESS
OXFORD · NEW YORK · TORONTO · SYDNEY · PARIS · FRANKFURT

U.K.	Pergamon Press Ltd., Headington Hill Hall, Oxford OX3 0BW, England
U.S.A.	Pergamon Press Inc., Maxwell House, Fairview Park, Elmsford, New York 10523, U.S.A.
CANADA	Pergamon of Canada, Suite 104, 150 Consumers Road, Willowdale, Ontario M2J 1P9, Canada
AUSTRALIA	Pergamon Press (Aust.) Pty. Ltd., P.O. Box 544, Potts Point, N.S.W. 2011, Australia
FRANCE	Pergamon Press SARL, 24 rue des Ecoles, 75240 Paris, Cedex 05, France
FEDERAL REPUBLIC OF GERMANY	Pergamon Press GmbH, 6242 Kronberg-Taunus, Pferdstrasse 1, Federal Republic of Germany

First edition 1979

British Library Cataloguing in Publication Data

Symposium on the Gas Situation in the ECE Region Around the Year 1990, *Evian, 1978*
The gas situation in the ECE region around the year 1990.
1. Gas industry - Europe - Congresses
2. Gas, Natural - Europe - Congresses
3. Gas industry - North America - Congresses
4. Gas, Natural - North America - Congresses
I. Title II. Economic Commission for Europe. *Committee on Gas*
339.4'8'6657094 HD9581.E82 79-40031

ISBN 0-08-024465-3

In order to make this volume available as economically and as rapidly as possible the typescript has been reproduced in its original form. This method unfortunately has its typographical limitations but it is hoped that they in no way distract the reader.

Printed in Great Britain by Page Bros (Norwich) Ltd, Norwich and London

Contents

Introduction

The future of gas in the energy requirements of Europe and North America — the region encompassed by the Economic Commission for Europe (ECE) — is the subject of attention in this perspective study of natural gas markets, natural gas availabilities and import possibilities.

In recent years intensified efforts have been made to meet existing or foreseen energy needs with gas. New techniques of production and application are being explored; evaluations of energy requirements, supplies, distribution, storage, and financing are taking place with a view to facilitating informed projections and optimizing resources for maximum present and future benefit. Quantitative data and estimates on the various possibilities and relative position of gas compared to other forms of energy in overall energy balances have been prepared for study to provide background information for formulating policy decisions on the future of the gas industry and to identify specific areas for intergovernmental cooperation.

To establish a forum for the exchange of information and views regarding present and future prospects of natural gas, the Committee on Gas of the United Nations Economic Commission for Europe decided at its twenty-third session to organize a Symposium on the Gas Situation in the ECE Region Around the Year 1990. The Government of France offered to host this meeting which took place 2-5 October 1978 in Evian.

Representatives of 22 member countries* and 2 non-member countries participated together with intergovernmental organizations and both governmental and non-governmental international organizations.

R. Fort (France) and J. Danhelka (Czechoslovakia) were elected chairman and vice-chairman respectively.

The Director of the Energy Division of the Economic Commission for Europe opened the symposium by outlining the activities of the ECE and its principal subsidiary bodies. The organization of its secretariat was described along with priority areas

*Austria; Belgium; Byelorussian SSR. Czechoslovakia; Denmark; Finland; France; German Democratic Republic; Federal Republic of Germany; Hungary; Ireland; Italy; Netherlands; Norway; Spain; Sweden; Switzerland; Turkey; Union of Soviet Socialist Republics; United Kingdom; United States of America; and Yugoslavia.

of work. In addition, it was mentioned that the Economic Commission for Europe at its next session would be considering the establishment of a new subsidiary body to be called Senior Advisers to ECE Governments on Energy.

Participants were welcomed by G. Robert (France), Chairman of the Committee on Gas, who referred to work done earlier by the *Ad hoc* Meeting on Methods of Forecasting Gas Demand, now defunct but formerly a subsidiary body of the Committee on Gas, its functions having been incorporated into the terms of reference of the Group of Experts on Gas Statistics and Forecasting Problems. Following a brief historical review of the gas industry's efforts to match supply with demand, reference was made to various symposia held by the Committee on Gas in the area of forecasting gas demand and examining gas availabilities. Regarding availability, the natural gas resources in four regions were summarized. These regions were: northern Europe, where the North Sea contains almost half the proven reserves in western Europe; the east, specifically the USSR and Iran; the south, which includes Egypt, the Cameroons, Tunisia, Nigeria, Algeria and Libya; and the west, principally the USA and Canada.

Participants were also welcomed by O. Amilakvari (France) on behalf of the Government of France.

Three themes were selected for examination at the symposium: Production of natural gas and other gases; Consumption; Imports and exports.

To augment considerations of these topics, reviews of gas reserves and of gas resources were added along with an overall economic survey of the ECE region.

While the year 1990 was taken as a basic time horizon for reports, authors could, if they so wished, consider the situation up to the year 2000. General reports presented by the rapporteurs on each of the three themes led off discussions.

Conclusions and Recommendations

The following is a summary of discussions that took place on each topic, the conclusions reached and the recommendations made during the course of the symposium.

PRODUCTION OF NATURAL GAS AND OTHER GASES

Discussions under this topic covered several aspects: expected natural gas (NG) production schedules; the envisaged production of synthetic natural gas (SNG) from either liquid and/or solid fuels; hydrogen (both through electrolysis of water and thermochemical disassociation of water); and primary liquefied petroleum gas (LPG), so far as it has a bearing on natural gas markets.

The proven and expected gas reserves of the world are large; more than half of them are situated in the area covered by the Economic Commission for Europe. In the exploration of reserves, significant logistics problems are raised: costs are considerable, difficulties of access abound, whether deep sea, polar cold or, at the very least, extreme underground depths.

Of proven gas reserves of the world, 40 per cent consist of associated gases in fields of liquid oil and gas combined. A part of the gas thus produced, far from markets and lacking on-the-spot utilization possibilities, is flared. Not only is it not reinjected back into the field of extracted gas — often an expensive operation although it might have a salutary effect on the final recovery of more petroleum — but it is wasted and even harmful.

Natural gas production in 48 states of the United States is declining. The new natural gas legislation which appears imminent for passage by Congress may in time contribute to adequate new sources of gas supplies to reverse the recent trend in production. The discovery of hydrocarbons in the Baltimore Canyon on the eastern continental shelf indicates that there is a possible potential for the discovery there of significant quantities of oil and gas. Special economic incentives and improved technology are needed for the development of unconventional gas sources, such as tight sands, geopressured methane, coal bed, etc.

In 1977, 346×10^9 m^3 of natural gas was produced in the USSR, twice as much as in 1967. Natural gas production in the next decade is expected to double. The gas industry is one of the fastest-growing branches of the economy. The chief factor influencing this high growth is extensive resources. Gas resources, of industrial categories, in the USSR in 1976 were equal to 25.8 trillion m^3. Of this, the total European area resources accounted for 4.2 trillion m^3 (16.3 per cent); Siberia and the Far East produced 18.2 trillion m^3 (70.5 per cent), and Central Asia and Kazakhstan 3.4 trillion m^3 (13.2 per cent). In the future the level of around 30 trillion m^3 will probably be reached and substantial new discoveries are likely.

Considerable natural gas reserves are also contained in huge coal deposits. The problem remains, however, of how to make this gas readily available. At present, the greater part of natural gas resources are concentrated in large deposits, mainly in Siberia and the Far East. Development of these resources is connected with geological, technical and economic problems. Permafrost, extremes of temperatures and pressures, long distances of transportation, and the lack of infrastructure in newly developing areas are concerns that have to be overcome. On the positive side is the intensive development of the gas industry in the USSR which creates favourable conditions for export possibilities.

The known reserves of western Europe represent only 6 per cent of worldwide reserves and hope for any substantial discovery of important new fields seems to be limited to unconventional areas or places of difficult access. Associated gas represents only 5 per cent of known reserves.

Debate followed the assertion that the current production policy of some Governments was based on underestimates of reserves and appeared, consequently, to be unreasonably timid in that such policies do not sufficiently stimulate demand and tend to be unduly directed towards research and financing structures which would be needed to develop substitutes for natural gas. Joint action of member states of the European Economic Community could lead, as in the United States' new legislation, to stimulation of demand which leads to the search for further sources and new discoveries. Representatives of the three main producing countries in Western Europe pointed out that there was no evidence that official statements understate reserves; in fact, official estimates in at least two of the three countries took into account undiscovered gas on the basis of an analysis of only its probability.

Certain statistics, given in the paper "Constraints on the development of Western Europe's natural gas producing potential in the 1980s" and attributed to official forecasts, were considered erroneous: such criticism ignored the advisability of using gas as far as possible in those applications which suited its special qualities. Representatives of the Organization for Economic Cooperation and Development (OECD) pointed out that no responsible Government could conduct its resource management policy on the basis of hope which might prove to be in vain, adding that there was no advantage in short-term, reduced dependency on imports at the expense of greater dependence afterwards.

Liquefied petroleum gases could be further developed but their utilization must be facilitated by various incentives.

Unconventional sources also require economic incentives. Their exploitation alone involves additional costs and development of new technology. While these types of resources are not very well known in Europe, the gasification of coal at great depths, particularly, is being considered. Its exploration would be very difficult in densely developed areas. Doubts were expressed whether underground coal gasification could be developed commercially in this century.

As the countries of the ECE region (Europe and North America) depend largely on international trade for their import supplies, it was recommended that the Committee on Gas try to obtain statistical information on natural gas reserves and export plans of non-member exporting countries.

It would also be desirable that the Committee on Gas study the factors favourable to the development of LPG utilization and, along with other competent principal subsidiary bodies of the Economic Commission for Europe, study measures which would possibly facilitate the research and development of unconventional sources.

CONSUMPTION

Consumption was taken to include: expected gas consumption and possible variations of demand in the user sectors (industry, households, commercial, communal, etc.); the effect of measures to save energy in terms of both total gas consumption and gas consumption by sectors; and the place of gas in the total energy consumed.

The representatives of the Organization for Economic Cooperation and Development (OECD) informed the symposium that a new publication was forthcoming on the world energy outlook for the year 2000. Regarding forecasts of the 1985 gas consumption in the United States, there was an important difference between the last forecast made by the OECD — 570 million tons of equivalent petroleum (Mtep) — and that made by the Department of Energy, assuming the measures envisaged in the National Energy Plan would be put into effect — 18.2 trillion cubic feet (Tcf) or 470 Mtep.

It was granted that interruptible gas sales of limited volumes to thermal power plants could be a means of adapting gas deliveries to availabilities.

Development of heat pumps run on gas would lead to appreciable energy savings, as much as 50 per cent. A saving of 10-20 per cent might be obtained in existing central heating installations by employing specially-designed condensation boilers.

A number of factors fostering a balance between gas demand and supply around 1990 in the ECE area were considered. The commitment of the United States under the national energy programme to phase out national gas price control by 1985 and thus stimulate domestic production was one measure. Continuation in the USSR of efforts to develop gas resources in spite of the difficulties of exploitation in adverse climatic conditions and the necessity for long-distance transport was another. The search for energy and gas saving in France, the USSR and other countries of the ECE region was a significant factor. Balances between long-term supply and demand must, however, be the subject of vigilant examination necessitating the updating of forecasts for the ECE region.

IMPORTS AND EXPORTS

Possible import and export options, projects for importing gas from countries outside the ECE area and large-scale projects of common interest including their technical and economic feasibility were considered under this topic. The consequences of the technical evolution, particularly innovations and changes of scale, were discussed.

Reductions in costs due to technical changes are possible up to 1990 as far as LNG transport lines are concerned. Such reductions might come from liquefaction plants, for example. However, a countervailing tendency may be that of increased environmental and social constraints.

An increase in the scale of natural gas transport projects is likely to bring about a significant reduction in transport costs in the near future, although a correlative increase in risks might undermine gains.

Consortiums of purchasers, which would facilitate economies of scale and at the same time reduce the incidence of risk for each, could be formed. Circumstances have already led to various flexible forms of cooperation. These seem likely to develop further. They would enable each partner to make whatever individual decisions are necessary while dealing jointly with essentially common questions. Should such groupings generally remain a necessity for European gas companies, it is believed their forms will probably vary.

Interesting projections have been given on the possible evolution of the international trade of LPG. It is quite probable that world availabilities will greatly increase distribution to potential consumers (Japan, western Europe, the United States). However, prices and market are not unanimously agreed upon.

The future role of gas imports, particularly that of LNG in the United States' energy economy, was discussed. The status of the Pacific-Indonesia LNG Project was clarified by the United States' delegation; the Economic Regulatory Administration of the Department of Energy recently approved a revised FOB price escalation formula. The project awaited resolution of a number of issues including final selection of a site for an LNG terminal. United States gas import policy, specifically LNG imports, was defined by taking into account its possible impact on the evolution of the international market for this product and by considering these imports in the context of an overall United States energy policy. The discussion which followed clearly underlined the complexity of the problems with respect to FOB and end-use pricing.

GENERAL CONCLUSIONS

Delegates concluded that the non-conventional gases are likely to have a negligible influence in terms of gas availability around the year 1990. LPG will continue to be used as much as availability allows but, according to information presented, LPG production will not affect the equilibrium between requirements and availabilities of NG. Some LPG produced will be used by the chemical industry and some might be used as motor fuels.

Around 1990, the ECE region's NG availabilities will depend, on the one hand, on discoveries arising from exploration and, on the other, on the exploitation rate adapted to each field, taking into account technical and ecological difficulties. Debate showed that controversy arises over the speed of exploitation of fields. The expression "ultimate resources" perforce entails reference to a valorization limit and a margin for error.

Several factors may influence future gas production, it was observed. Studies on future hydrocarbon production possibilities are of a purely speculative character. Decisions of economic circles could influence and perhaps delay future gas availabilities.

The influence of campaigns to promote the saving of energy was considered difficult to measure in terms of their effect on the level of future consumption. Forecasts of gas requirements will differ widely depending on whether one considers only "specific requirements" or also includes "avoidable requirements" such as those of thermal power plants.

In future, countries reporting their gas requirements should distinguish "interruptible contract" requirements and "firm contracts" for gas requirements that must be satisfied with priority.

Technical progress might slightly reduce costs expected in the gas industry, but environmental considerations will undoubtedly increase costs.

Regarding future import/export schemes, several points emerged in the course of discussions. It is expected that negotiations will involve several considerations and, for example, may include diversification, industrial counterparts and the political climate between countries concerned. Such considerations make it impossible to foresee or evaluate future NG export/import deals. It is likely that equilibrium between requirements and resources will in the short term be realized by adapting the engagements to the availabilities, depending on changes or hazards.

In the medium term, the conclusion of new supply contracts may change the place of gas in the overall balance of energy. The significance of these contracts will increase according to the differential valorization attributed to the extra availabilities obtained.

RECOMMENDATIONS

It was suggested that special attention should be given to the following:

That the Committee on Gas consider holding another, similar symposium in three to four years' time in connection with large-scale import/export projects.

That non-member countries with NG export potential be invited to report annually on their NG reserves for consideration and publication by the Group of Experts on Natural Gas Resources, of the ECE Committee on Gas.

That, if a seminar on investments in the gas industry and financing possibilities is held, consideration be given to the financial aspects of any large-scale projects that are envisaged.

Regarding supplemental gas resources, it is expected that SNG from solid fuel gasification will gain importance in the late 1990s, and therefore it is recommended that close cooperation should continue with the ECE Coal Committee concerning coal gasification.

In connection with other non-conventional energy resources, e.g. geopressured gas, etc., the Committee on Gas might include in its programme of work subjects that would encourage the exchange of information and views on research and development in these fields.

Regarding future use of hydrogen, the Committee on Gas might explore the possibilities of cooperation with the ECE Committee on Electric Power.

In addition to energy-saving campaigns and their impact on future gas consumption, emphasis could be given to expanding the role of storage rather than increasing the use of interruptibles to match contracted supplies to seasonal demands of premium markets, particularly domestic and commercial markets. Furthermore, studies might be undertaken which aim at encouraging the use of gas and solar energy combinations for both space heating and hot water heating.

As for primary LPG availabilities, collection of associated natural gas in oil operations might yield large quantities of LPG in excess of that which existing LPG markets could absorb. With a view to finding uses for this primary fuel and to deal with any questions that may arise, the Committee on Gas might set up an *ad hoc* meeting. The European LPG Association should be consulted regarding the terms of reference to avoid duplication of work.

List of Contributors

M. A. Ban, Directeur Général, Köolaj-és Gazipari Tröszt Vezérigazgato, Szt. Istvánr krt. 11, H-1055 Budapest V, Hungary

E. M. Beale, SCICON, Sanderson House, 49-57 Berners Street, London W1P 4AQ, United Kingdom

R. Beggs, Gas Gathering Pipelines (North Sea) Ltd., 140 Tottenham Court Road, London W1P OAS, United Kingdom

M. M. Belguedj, Directeur, SONATRACH, Exportations Gaz, Division Commercialisation, 46 Boulevard Mohamed V, Alger, Algérie

A. Bogers, Energy Scientist, Centrale Organisatie TNO, Postbos 342, Apeldoorn, Netherlands

G. Bonfiglioli, SNAM S.p.A., Casella postale 3757, 20100 Milan, Italy

G. Bonvecchiato, SNAM S.p.A., Casella postale 3757, 20100 Milan, Italy

A. Bouleuc, Ingénieur, Chef de Service à Gaz de France, 23, rue Philibert-Delorme, 75840 Paris Cedex 17, France

N. Cojan, Conseiller Technique au Gaz de France, 23, rue Philibert-Delorme, 75840 Paris Cedex 17, France

J. Chapelle, Président de CEDIGAZ, 4, Avenue du Bois-Préau, Boîte postale 18, 92505 Rueil-Malmaison, France

P. H. Chaubernard, Compagnie française du méthane, 22, rue Boileau, 75016 Paris, France

P. Cruchon, Adjoint au Chef du Service Technique et Economique du Gaz, Ministère de l'Industrie, du Commerce et de l'Artisanat, 3-5, rue Barbet de Jouy, 75700 Paris, France

L. d'Andrea, Industrial Specialist, Chief, Natural Gas Division, Office of Oil and Gas, Department of Energy, 12th and Penn. Avenue, NW, Washington, DC 20461, USA

J. Danhelka, Directeur technique, Ceské plynárenské podniky, Revolucní, 2, 111 51 Prague 1, Czechoslovakia

P. Desprairies, Président du Conseil d'Administration de l'Institut Français du Pétrole, 4, Avenue de Bois-Préau, Boîte postale 18, 92502 Rueil-Malmaison, France

N. Fedorov, Ministry of the Gas Industry of USSR, ul. Stroitelei 8, Moscow 117939, USSR

G. Fournier, Directeur du Bureau des Economies d'Energie (Société Nationale ELF-AQUITAINE), 7, rue Nélaton, 75015 Paris, France

M. Garcia-Conde Ceñal, European LPG Association, 4, Avenue Hoche, 75008 Paris, France

F. Gendrin, Chef de la Division Etudes Générales à la Direction des Services Economiques et Commerciaux (Gaz de France), 23, rue Philibert-Delorme, 75840 Paris Cedex 17, France

F. Gläser, Vorstandsmitglied, Direktor der RHENAG, Bayenthalgürtel, 9, 5 Köln, 51 Federal Republic of Germany

J.-P. Hauet, Ingénieur en Chef des Mines, Délégation Générale à l'Energie, 35, rue St. Dominique, Paris, 7ème, France

Chakib Khelil, Sonatrach, Division des Hydrocarbures, Chemin du Réservoir, Hydra-Alger, Algérie

P. Leblanc, Chef de la Division Utilisation Industrielle à la Compagnie Française du Méthane, 15, rue de Lübeck, 75116 Paris, France

P. D. Lessieur, TOTAL CFP, 5, rue Michel Ange, Paris, 16ème, France

B. Machini, Utilities and Off-Sites Department Manager, SNAMPROGETTI S.p.A., Milan, Italy

M. Mainguy, Chargé de Mission à la Société Nationale ELF-AQUITAINE, 7, rue Nelaton, 75015 Paris, France

P. Millard, Chef de la Division Industrie à la Direction des Services Economiques et Commerciaux (Gaz de France), 23, rue Philibert-Delorme, 75840 Paris Cedex 17, France

C. Mulazzi, SNAM S.p.A., Casella postale 3757, 20100 Milan, Italy

P. Odell, Professor of Economic Geography, EGI, Erasmus University, Postbos 1738, Rotterdam, Netherlands

N. Potié, Ingénieur aux Services Commerciaux de la Société Nationale des Gaz du Sud-Ouest, 49, Avenue du Pau, 64010 Pau Université, France

S. A. Price, Economic and Statistics Division, Department of Energy, Thames House South, Millbank, London SW1P 4QJ, United Kingdom

J. Quakernaat, Energy Scientist, Centrale Organisatie TNO, Postbos 342, Apeldoorn, Netherlands

J. Ribesse, Ingénieur en chef, Distrigaz, S.A., Avenue des Arts 31, 1040 Brussels, Belgium

W. Schott, Bundesanstalt für Geowissenschaften und Rohstoffe, Alfred Bentz Haus, Postfach 51 01 53, 3000 Hannover 51, Federal Republic of Germany

J. Sewell, Secretary General, European LPG Association, 4, Avenue Hoche, 75008 Paris, France

Production of Natural Gas and Other Gases

A. Production of Natural Gas and Other Gases

M. Mainguy (General Rapporteur)

France

Natural gas production depends on the physical availability of gas (reserves), the geographical distribution of the reserves, logistic possibilities for exploitation and transmission to consumption areas, and the economic and political environment which helps to determine the rhythm of production of known reserves and the exploration work necessary for their renewal. Various resources which probably exist but which we do not know how to produce are called "unconventional".

"Other gases" can be produced from abundant raw materials, such as coal, and even from renewable sources, such as biomass or waste, or from very common sources such as water, the raw material of hydrogen. The essential condition for their development is the existence of economically acceptable processing techniques.

The General Report will examine these three aspects:

1. Natural gas resources which exist within the ECE region or can be imported.
2. Rhythm of production of these resources and of replacement of reserves.
3. Potential new sources.

REVIEW OF REPORTS

The paper by J. Chapelle and P. Desprairies (France) entitled "Possible sources of natural gas for France and the ECE area in 1990" covers in detail the question of natural gas resources. It places the question of resources for the ECE in broader context, estimating the outer limits of the world's ultimate reserves at approximately 200 to 300 Tm^3, or three to four times the proven reserves. From this the authors forecast the future trends in world production, which could culminate, in the year 2000, at a level also corresponding to three to four times the present level of production, and a possible pattern of world trade. The conditions governing existing resource availabilities depend partly on the use that may be made of associated gas in maintaining oil fields at pressure and also on the conditions governing economic transmission, since the cost of transmitting gas is high. In conclusion, the paper examines possible trends in the share represented by gas in energy supply in industrialized countries. In the United States, it will inevitably decline up to the year 2000; in the USSR and Japan it is increasing and will be virtually stationary from the 1980s; and in Western Europe it is likely to decline from 1985. Around 1990, it should be between 15 and 20 per cent in Western Europe and the United States, and between 25 and 28 per cent in the USSR. This paper relates mainly to resources and touches on rhythms of production; however, it extends beyond Topic A and will be taken up again in the discussion on Topics B and C.

The paper by P. Odell (Netherlands) concentrates on rhythms of production. Its title is "Constraints on the development of Western Europe's natural gas producing potential in the 1980s", and it describes Western Europe's resources, particularly those of the North Sea. Referring to former forecasts, the accuracy of which has been demonstrated by facts, it shows that official forecasts are always too low. Official forecasts of proven, probable and possible reserves in Western Europe as a whole for the beginning of the 1980s amount to 6380 Gm^3, whereas the author estimates them at 8300, or approximately 30 per cent more. The main criticism relates to production capacity, for which the OECD documents give a range of 230 to 260 Gm^3, for 1985, whereas the author considers that by 1985 it should be of the order of 400 Gm^3, i.e. one-twentieth of total reserves (proven, probable and possible) expected by him at the beginning of the 1980s, the ratio of reserves to production capacity being approximately 1 : 25 in the OECD forecasts. The author's forecasts lead him to support the projects for the transmission of North Sea gas by the two distribution systems currently under consideration, namely, a Norway-United Kingdom system supplying the United Kingdom, Belgium and France, and a purely Norwegian system supplying the Federal Republic of Germany, Sweden and, subordinately, Denmark and France.

The paper by L. d'Andrea (United States) "US gas supplies — an overview", outlines the present situation with respect to natural gas in the United States and the difficulties which the United States energy plan is attempting to tackle. It seems that, irrespective of the plan and its outcome, exploration work is petering out and reserves cannot fail to decline. The author discusses the possibilities of new sources. He brackets Alaskan gas with "unconventional" sources (tightsands and geopressured zones) because of the problems posed by its transmission. Coal gasification is a future source because of the abundance of coal in the United States, where it constitutes approximately 90 per cent of so-called "conventional" energy resources. Among possible sources of import (coming under Topic C), the author singles out overland imports and liquefied natural gas (LNG). The former will come from Canada and Mexico. Canadian policy was felt to be not in line with the considerable resources of that country; it is too restrictive. The McKenzie pipeline project might take over from the existing import contracts, which expire in 1992. Mexico is more promising and could furnish between 20 and 25 Gm^3 in 1985. Imports of LNG come mainly from Algeria and supply the East Coast. Economic and ecological problems have impeded development of these imports; the cost of imported gas in the form of LNG is still prohibitive. In conclusion, the paper emphasizes that sources are inexorably tending to decrease whereas demand is continuing to increase.

The paper from the American Gas Association (AGA), "The American point of view on the future of the gas industry", notes the increase in energy demand in the United States and the legislation and regulations that have hampered the development of American gas production. Maximum ultimate reserves, estimated at 1200 Tcf (trillion cubic feet), correspond to 60 years of present annual production; proven reserves, however, correspond to only 9 years of today's annual production and, even if regulations are relaxed, there will be no significant improvement before 1990. Until the end of the century, the major part of the gas used will be supplied by the 48 contiguous states: the contribution of Alaska, whose potential reserves are estimated at 200 Tcf (5.7 Tm^3), will increase; Canada's contribution will decrease but could be taken over by Mexico. The Association's view of the role of LNG imports is more optimistic than that of L. d'Andrea; it should help to prevent an excessive increase in the selling price of gas. These imports also have other advantages, including development of specialized ship construction, because of flag privilege, low pollution, diversification of sources (Algeria, Indonesia, Australia, Trinidad, Colombia, Iran, Nigeria, USSR and others) which help to reduce the effects of economic dependence. It is a matter for regret that, despite these advantages, imports remain blocked. It seems likely that for the world as a whole, resources yet to be discovered (200 to 225 Tm^3 as opposed to 60 to 70 proven) can meet needs for a further 50 years.

G. Bonfiglioli, G. Bonvecchiato and C. Mulazzi (Italy) examine possible new sources of energy under the title "Chances for alternative fuels in the gas industry". Production is declining, and the authors do not foresee that new "conventional" sources capable of halting the decline will intervene before 1985. At present, one-half of Italian consumption is imported and, assuming that consumption will increase by 4 to 5 per cent a year, imports will be of the order of 70 per cent by 1990. It may thus be asked whether it would not be preferable to resort to replacement sources. There is a very comprehensive review of energy sources, following which only those sources capable of replacing natural gas, wholly or partly, are taken into consideration. Their costs, for industrial and domestic uses, are compared under two sets of assumptions: one, that existing energy costs will increase only to keep pace with general inflation, and two, that they will increase more rapidly. In both cases, this comparison shows that most "unconventional" sources are at a distinct disadvantage and that they will not come into Italy's energy balance until the last decade of the century.

Also from Italy, a paper by B. Macchini, entitled "More energy from LNG", predicts spectacular development in trade in liquefied gas in the years to come: a tenfold increase between now and 1985. It is proposed to recover as much as possible of the energy contained in these enormous quantities of liquefied natural gas, using cold energy to make electricity. This process, which improves recovery of the energy content, can be regarded as equivalent to an increase of resource by an unconventional process.

At the time of writing only a brief summary was available of the paper prepared by the European LPG Association, "The place of liquefied petroleum gases in the energy situation around 1990". The role of LPG is marginal but may lead to substantial economies in other forms of energy and to a considerable reduction of pollution is one point made.

The subject of the paper by M. Ledent and J. Ribesse (Belgium) is "High depth underground coal gasification, a potential energy source for the future". Current known coal reserves do not take account of the gigantic reserves (1000 billion tons in the Ruhr, as against 10 billion tons at a depth of less than 1200 m) contained in deposits situated at depths inaccessible to human labour (1200 m) but exploitable only by gasification *in situ*. This has been attempted in the USSR, at shallow depths, for 40 years, and yields a very lean producer gas. At great depth, there is greater tightness and it is possible to work under great pressure and obtain a richer gas. The reactive gas would be either air, under high pressure (giving a lean gas: 4000 to 5000 kJ/m^3) or oxygen and steam (gas of 10 000 to 11 000 kJ/m^3) or hydrogen, giving a gas with a high calorific value (21 000 to 25 000 kJ/m^3). Total energy output is higher with air gasification (55 - 70 per cent). It is not more than 60 per cent for the two other methods, and may fall to 40 per cent. The costs could compete with those of imported natural gas (75 to 90 BF/GJ, or US$ 2.3 to 2.6/mc). An experimental programme is to be undertaken under an agreement between the Federal Republic of Germany and Belgium. It is to last 6 years and should produce results in 1984. By then it will be known whether it is in fact possible to put these technologies into effect.

The paper by M. Beale and R. Beggs (United Kingdom) on "Economic aspects of gas gathering in the North Sea" describes an economic study on the feasibility of a gathering system to drain gas from the United Kingdom part of the North Sea, the main aim being to limit the loss of associated gas. The study includes an estimate of reserves of eighty fields, a cost evaluation of the various components of the gathering system, a study of market possibilities, and a selection of the most promising systems. This study has used a series of specialized studies; it results in the establishment of an evaluation model of different plans for installing the gathering system, ensuring, *inter alia*, that each year there is sufficient capacity to transport all the available gas, and it calculates the corresponding cash flows. An optimization model makes it possible to select an installation plan that maximizes the net present value of the system. In cases where several possible connections can be made at a specific point, the model does not necessarily produce the best solution, but it does produce a good solution.

RESOURCES

Estimation of natural wealth reserves is a question so fraught with difficulties that miners are accustomed to say that it is only when exploitation is ended that a deposit becomes known. The only factor with a clear physical and geological meaning is the quantity on the site, and this by no means signifies that it is easily determined. On the other hand, the only factor that has a specific economic meaning is the quantity that it will surely — so far as one can be sure — be possible to extract from a deposit using known technologies and in prevailing economic circumstances: these are proven recoverable reserves which, because of technological progress and economic trends, are constantly being revised, mostly upwards but occasionally downwards. Besides proven reserves there are, at decreasing degrees of certainty, probable and possible reserves.

Defined as the ratio between recoverable reserves (the extent of which is subject to revision) and quantities on the site (the extent of which little is known) the rate of recovery suffers from this lack of precision on two counts. The Economic Commission for Europe is making a study on recovery rates. According to a note by the secretariat of the Committee on Gas, prepared by a Soviet expert, V. Pominov, the recovery rate for the majority of gas fields in the ECE area is between 60 and 80 per cent; it may be lower than 60 per cent, particularly in the case of small fields, but in a considerable number of cases it exceeds 80 per cent and may reach 95 per cent. The application of improved recovery techniques encourages the hope that the average rate will reach 80 per cent towards the middle of the 1990s. It will be noted that in the case of the Aquitaine Basin it is nearly 84 per cent.

The purpose of giving a figure for recoverable reserves is to take account only of that part of the reserves having an energy value, and for this it is necessary to know the composition of the gas. Unless care is taken in totalling crude gas reserves, reserves of fuel gas will be totalled up with non-energy by-products, some of which (such as sulphur) may be usable, while inerts such as CO^2 and nitrogen have no value. Existing reserves can be shown by their calorific (energy) content to the extent that the composition of the gases or their calorific value is known. In general, only the energy potential of domestic and industrial gas, which has already been partly purified, is known. At the World Energy Conference held at Istanbul in September 1977, several authors used the convenient equivalent of one Tcf for one EJ, which, translated into international units, corresponds to 35.5 MJ/NM^3 (8.4 Th/m^3) for a gas with a non-fuel content of approximately 20 per cent. Table 1 compares the data furnished by J. Chapelle and P. Desprairies, translated into energy content (Exajoules), with various data taken from the literature, including the report prepared by the CIA on the occasion of the formulation of the United States energy plan. This report is rather optimistic, although its revised version differs by only 5 per cent from that of the *Oil and Gas Journal* for the beginning of 1978; but the breakdown of reserves is very different: while optimistic for the Middle East and North America, it certainly underestimates USSR reserves. No evaluation can be given for China, the interest of which for the ECE zone is that it might supply Japan, thus liberating quantities from other countries.

TABLE 1

World Gas Reserves expressed in Exajoules*

	1	2	3	4	5
USSR — Eastern Europe	928	823	837	930	802
North America	274	310	289	268	275
Western Europe	142	176	176	138	138
Total ECE region	1344	1309	1302	1336	1215
Middle East	514	953	847	720	607
Africa	209	212	212	207	208
Far East	145	131	131	148	116
Latin America	95	114	134	108	95
(including Mexico)	(21)	(32)	(42)	(30)	-
Total other than ECE region	963	1410	1324	1183	1026
World Total	2307	2719	2626	2519	2241

Column 1 - Reserves at 1.1.77 according to the *Oil and Gas Journal*.
Column 2 - Reserves at 1.1.77 according to the CIA report.
Column 3 - Reserves at 1.4.77.
Column 4 - Reserves at 1.1.78 according to the *Oil and Gas Journal*.
Column 5 - Reserves at 1.1.77 according to Mr. Chapelle and Mr. Desprairies.

* 1 EJ = 1 Tcf or 28 317 Gm^3 for a gas at 35.3 MJ/m^3.

Gas *in situ* may contain liquid products (condensates). The presence of condensates poses technical problems with respect to exploitation and transport. When the liquid phase quantity increases, the question arises whether the field really is a wet gas field or whether it is a field of gas-rich light oil. We are then in the category of associated gas. The structure of associated gas is sufficiently special to raise the question whether it is reasonable to add together these two categories of natural gas; statistics differentiating between them are rare. At the UNITAR/IIASA Conference held at Luxemburg in July 1976, Robert L. Whiting emphasized that 40 per cent of the remaining world gas reserves consist of associated gas. Taking the figure 3400 Tcf (approximately 100 Tm^3), which is probably low, as the value of reserves yet to be discovered, he reckons that 45 per cent or slightly more than the existing proportion of the new discoveries will consist of associated gas. This seems rather unlikely, given the hopes for free gas entertained in the world. It may be felt, however, that the proportion of associated gas will not decline significantly. When it is considered that it may be desirable to reinject it (it will probably be only partially recovered) and that every attempt will be made to avoid flaring it, we perceive that only 60 per cent of the quantities mentioned as reserves are in fact completely under control. The overall figure does, however, conceal significant regional variations. The following figures, taken from various sources and giving the proportions of associated gas in the total reserves of various regions, can be cited:

United States	27 per cent (a)
Iran	56 per cent (a)
Iraq	65 per cent (a)
Saudi Arabia	78 per cent (a)
Abu Dhabi	*circa* 100 per cent (b)
Algeria	15 per cent (c)
Western Europe	16 per cent (d)
USSR and Eastern Europe	5 per cent (d)

(a) *Petroleum Economist*, July 1977.
(b) *Arab Oil and Gas*, 1 June 1978.
(c) *Ibid.*, 1 February 1978.
(d) Economic Commission for Europe statistics, 1978.

These figures may be compared with certain data concerning quantities flared. They are negligible in the industrialized countries but significant in crude oil-exporting countries. Table 2 gives some information concerning quantities of gas flared; the decline over time reflects efforts made to reduce this loss. The loss is relatively slight in comparison with the total quantity of energy produced. In Iran, for example, quantities flared in 1971 represented 6 per cent of total production (oil and gas) evaluated in energy content.

Few data on reinjection are available. Messrs. Chapelle and Desprairies consider that 80 Gm^3 were reinjected and 180 Gm^3 flared in 1976, as compared with a marketed production of 1350 Gm^3; they refer to the Iranian programme under which the gas equivalent of one barrel of oil must be injected to permit recovery of an extra barrel. On the average, the quantity required is probably lower: 300 m^3 of gas for an extra m^3 of oil. Those countries whose reserves consist mainly of free gas are in control of their production and export programmes. When the proportions of associated gas are considerable, the benefits of reinjection (imperfect storage) must be weighed against those of export, once domestic demand has been met.

This brings us to the main factor of uncertainty which is the major concern of the prospector, namely the physical presence of the resource, the technical possibility of discovering it, the critical gathering of items of information necessary for the evaluation of deposits discovered and, even more, of the possible size of deposits yet to be discovered. The different categories of reserves — proven, probable and possible — are not everywhere given the same definition. The Economic Commission for Europe's Committee on Gas is preparing a trilingual glossary which should cover the bulk of technical terms relating to the gas industry, including definitions of the different categories of reserves. It is to be feared that however great the attempt to clarify matters, not all the terminological ambiguities will be removed.

Ultimate reserves represent the idea that can be formed of the total quantity of hydrocarbons — in this case, of gas — existing in the world and capable of being found and recovered with the techniques and economy expected to be available within the time schedule under consideration. The apocalyptic term "ultimate reserves" should therefore not be taken literally.

A seminar on the geology and geophysics of prospecting for natural gas, organized by the Federal Republic of Germany under the auspices of the ECE Committee on Gas, is to be held in principle in the first half of 1981. It is only a few years since these technologies began to distinguish themselves from the general methods of

TABLE 2

Proportion of Gases Flared Off in Various Countries
(as a percentage of total gas production)

	Iran	Arabia[d]	Abu Dhabi[e]	Qatar[f]	Algeria[g]
1964	90[a]				
1968	82[a]				
1970	60.5[a]				
1971	59[a]		90		
1972			90		
1973			91		
1974	55[b]		91		
1975	51*[b]		91		
1976	54**[c]		91	72	
1977		66.7	80	64	48

Sources: (a) Symposium on petroleum resources in the ECAFE (ESCAP) area, Canberra, 1969.
(b) de Golyer McNaughton annual statistics.
(c) *Iran Oil Journal* 1975 (autumn) and 1976 (autumn).
(d) *Le Petrole et le Gaz Arabes*,16 March 1978.
(e) *Le Petrole et le Gaz Arabes*,1 June 1978.
(f) *Middle East Economic Survey*, 15 May 1978.
(g) *Le Petrole et le Gaz Arabes*,1 May 1978.

* 1975: 9 months.
**1976: 3 months (last quarter).

Breakdown of uses in Iran (last quarter 1976)

Flared	54.2 per cent
In situ consumption	7.2 per cent
Domestic and industrial consumption	20.3 per cent
Exports to the USSR	18.3 per cent

hydrocarbon prospection. The study of seismic responses is gradually being refined and it is now possible to derive from them certain data concerning the lithology of reflecting strata and the content of porous seams. Certain reflections, which at first sight seem strange, because they are horizontal and cut across geological levels, may in particular circumstances indicate contacts between fluids, mainly gas and liquid. These are detailed exploration techniques: they are valid for each trap taken individually, and their application calls for a minimum of geological knowledge of the region concerned. It is difficult to forecast whether, in the case of a basin as a whole or of a clearly defined part of a basin, mostly oil fields can be expected and whether they will be rich in gas. Knowledge of source rocks and of the maturation of organic matter is, however, improving. Taking account of temperature gradients, it is possible to assess whether areas are particularly favourable to gas formation or to oil formation, and it is now possible to distinguish between the primary gases which have been formed in the gaseous state and those obtained from the natural cracking of liquid hydrocarbons. Most of these techniques are, however, based on information obtained by drilling and although they permit fairly accurate determination of the chances of finding gas rather than oil, there is no really effective method of determining whether a given basin or sub-basin is rich in hydrocarbons, let alone of evaluating the quantities it is capable of yielding. An attempt must be made to do this, however, in order to obtain an idea of "ultimate reserves". By operating on sufficiently large units it is possible to obtain global results acceptable as a measure of relative size, but the greater the attempt to go into detail the greater will be the risk of error.

RHYTHMS OF PRODUCTION AND REPLACEMENT OF RESERVES

Oilfields have an optimum rate of production, generally known as MER (maximum efficient rate). Production at too high an annual rate, particularly at a rate higher than MER, involves the risk of jeopardizing the final recovery of the field. The production of associated gas in an oilfield will be limited by this technical consideration.

There appears to be no equivalent notion in respect of dry gas fields, especially deposits in a sandy reservoir with a very active acquifer. Some writers, particularly Soviet ones, even think that it might actually be advantageous to exploit such fields at a very fast rate, adapting, as it were, to the speed of the rise of the acquifer which tends to divide the field into compartments. Technical factors, by and large, will not play as important a role as in the exploitation of oilfields.

Other factors will have to be taken into consideration. The high level of investments for development and transport may slow up the start of production. At the commercial level, users want regular supplies over a long period. The outcome of this is that sale contracts tend, to some extent, to be based on the hope of renewing reserves.

This kind of reasoning may admittedly lead to a timid policy, based to an undue degree on the notion of conservation, which is tending at the present time to assume proportions inhibiting for ideas and action. It is sometimes argued that limited and non-renewable substances should be preserved for future generations. But just as Freud replied to the lady who asked him how she should bring up her child: "However you like, Madam; in any case you will do it badly", the world we shall leave to future generations will in any case be misjudged. If, on the other hand, we close down fields to provide warmth and heat for our children, perhaps they will appreciate this sacrifice so much that they will feel they should imitate it and leave the gas in the ground indefinitely.

However, if we need not aim at conserving the riches in the ground for the twenty-first century, we do need to avoid adopting rates of exploitation which lead to excessive consumption and create conditions for scarcity over the short term or medium term. Nature may not have set an upper limit to the rhythm of production for gasfields, but it has placed constraints on their renewal which must be taken into consideration.

L. d'Andrea and the AGA, on the one hand, and P. Odell, on the other, criticize production policies implemented in North America and Western Europe as being over-cautious. The papers by the former two are critical of the price regulations for gas delivery between one state and another in the United States (interstate gas). Its very low price is tending to promote excessive demand, to lead to low priority consumers being supplied with gas from a neighbouring state, and to curb exploratory initiative, whose successes are insufficiently rewarded. The most heavily penalized initiatives are the most expensive ones, namely, very deep drilling operations, which hold out the best promise for supplies of gas. The average depth of exploratory boring in the United States is still shallow (1800 m in 1975) and one might almost say that although it is one of the pioneer oil-producing countries and is often said to be exhausted, the surface has only begun to be scratched. It is a fact, however, that the quantities produced by exploratory boring are decreasing (see Fig. 1). Apart from the projects described as "unconventional", or short of a miracle, reserves can only dwindle.

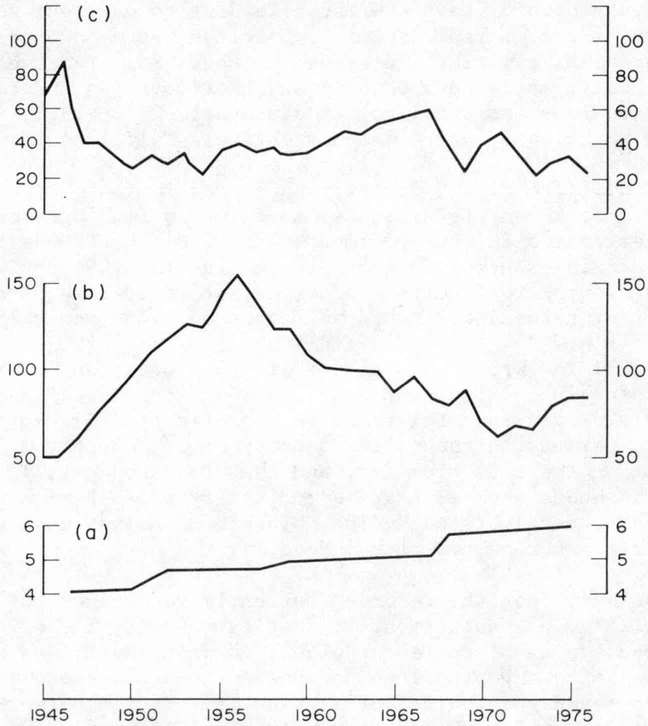

Fig. 1. (a) Average depth of exploratory drilling operations in the United States, 1945-1975. Unit: 1,000 feet. (b) Number of exploratory drilling operations *per annum* in the United States, 1945-1975. Unit: 1,000 drilling operations. (c) Amount of gas discovered per exploratory drilling operation in the United States, 1945-1975. Unit: 1 million m^3. (Source: *Twentieth Century Petroleum Statistics*, DeGolyer McNaughton.)

L. d'Andrea also deplores the Canadian policy of restricting exports. Following recent discoveries, that policy is also being disputed by the gas producers of Alberta. Supplying the Maritime Provinces of Canada does not appear to be a good exchange, since they form too restricted a market.

P. Odell criticizes the conservation policy of the European States, as shown in the forecasts contained in OECD documents. These estimate that in 1985 European gas production will range between 230/260 Gm^3, whereas the author thinks that it might reach 405 Gm^3 which would mean the exhaustion in 20 years of the proven, probable and possible reserves which he considers as likely by the early 1980s (Table 2). The divergence arises from the author's adoption of a slightly higher rhythm of production than the current one which OECD's figures indicate as being approximately 25 years, but mainly, as shown by various tables, from an optimistic assessment of the increase in reserves of the European countries. On the basis of an increase of 4 per cent per annum of production in Western Europe beginning with 1976, the replacement of reserves produced by 1982 (beginning of the 1980 decade) and their increase up to the level of 8300 Gm^3 from the accepted level of 6380 Gm^3, would necessitate the discovery of almost 3500 Gm^3 over the next few years.

During the period 1965-1975, the quantities discovered in Western Europe, including the North Sea — in other words the total of the quantities produced and the addition to reserves — were somewhat less than 3500 Gm^3. This quantity was discovered in 11 years, and the increase in reserves forecast by P. Odell would require an equivalent increase to be obtained in 6 years. Without leading to outright rejection of such hopes, the history of recent years seems to preclude basing a production policy on it. In Western Europe as a whole, excluding the North Sea, reserves are being maintained with difficulty; there have been no major discoveries and certain modest increases are due to reassessments, not to discoveries. In 1973, in Italy, great hopes were placed on Malossa, which has not entirely fulfilled its promise.

G. Bonfiglioli *et al.* (see Fig. 3 in their paper) estimate for 1985 a domestic production of 10 Gm^3 from known fields and 4 Gm^3 from new findings, or a total of 14 against 17.5 Gm^3 estimated in OECD documents and 20 Gm^3 by P. Odell. In France the production from Lacq is expected to begin to decline in 1983, and while there is still hope of discovering replacement resources, it would be a pleasant surprise if current production of natural gas could be increased by 50 per cent.

It may be argued that the known figures are official ones, and it is true that official figures are usually cautious and prudent. But it is hardly possible to keep major discoveries a secret, and the trend is significant. It can also be argued that even in north Germany, although the density of exploratory drilling is almost on the American scale, there is room for, and therefore hope of, large-scale discoveries. But such hopes are too fragile to justify a production policy over so short a term. Even the celebrated "golden blocks" of Norway may run dry, and the equivalents in western Europe are rare indeed.

It seems, nevertheless, that the reserves currently recognized, or those whose existence is highly likely, are sufficient to justify not only the existing pipelines but a more extended system of collector pipes. A paper by M. Beale and R. Beggs describes the "Gas gathering pipelines for the North Sea", where it is intended to gather the associated gas instead of flaring it. Such a network should probably be extended to the Continent, where there is a sizeable market.

Large, already-proven reserves — as well as very great hopes — are centred around the North Pole. Those reserves situated on the Continent give rise to no major exploitation problems. One of the major problems as far as the laying of pipelines in the permafrost regions is concerned is that of the ground being heated by the oil, and, conversely, of the oil becoming frozen. This problem does not arise for gas, and it is less harmful for the environment for a gas pipeline to cross Alaska than for an oil pipeline to do so. The gas reserves of Alaska as well as the land deposits in the Mackenzie Delta will be able to be exploited; none of the American papers submitted contains a precise forecast concerning the date on which gas from Alaska will be available.

The gas wealth of the Pechora and western Siberia can also be exploited. But other projects are bolder and are at the limits of the "conventional". Beyond the Mackenzie River delta (Beaufort Sea), exploitation will necessitate the erection of platforms able to resist the pressure of the pack ice. Gas from the Canadian Arctic would be able to compete with oil at \$20 (1978 value) a barrel, for minimum proven reserves of 600 to 800 Gm^3. After 15 years' work, however, only 400 Gm^3 have yet been found, distributed over seven fields, the largest of which contains 150 Gm^3. In Labrador a knowledge of protection against icebergs is needed, and thus far the only sure method has been to steer clear of them, which can be done during exploration but not at the exploitation stage.

On the other side of the Labrador Sea, the south-west coast of Greenland has thus far proved disappointing. Further to the north, no drilling has yet taken place in

Baffin Bay; the basin it covers might be promising, but the logistic difficulties are formidable: they are even more so on the east and north-east coasts of Greenland. By and large, the prospects there for the coming decade seem negligible.

In the Barents Sea — Norwegian and Soviet sectors alike — work will have to take place deep under water (at 200 to 500 metres). The problem is in fact already arising to the north of the Viking Basin, near 62°N latitude. It is likely, for economic as well as ecological reasons, that gas will be able to be exploited in deep water before oil.

Overall proven or possible reserves round the Pole would appear to be very considerable, but there is reason to fear that they may no longer be available by 1990.

Moving southwards, except as far as the Soviet Union is concerned, we leave the area of the Economic Commission for Europe and enter the domain of possible supply areas for imports: Mexico, the Maghreb, the Middle East. Mexico appears to offer the United States a nearer and more readily available source of supply than the Far North. The exploration of the new provinces and very deep strata (7000-8000 metres) is already under way and the prospects seem very good. A 122-cm (48 inches) pipeline extending 1200 km is planned, with an annual production of 20-25 Gm^3 of gas at 37.7 MJ/m^3.

In the Maghreb, the largest source of gas is Algeria, where 85 per cent of reserves are made up of free gas. A study in *Le Petrole et le Gaz Arabes* indicates that reserves amount to approximately 3600 Gm^3 and that domestic consumption should increase by 15 per cent per annum; and the production planned is calculated to exhaust 53 per cent of reserves, including 45 per cent earmarked for export, by the year 2000 or thereabouts. Large-scale exploration should be undertaken to renew these reserves. There are certainly zones which are under-explored, but the results published thus far point to a fairly cautious prognosis.

Black Africa does not seem to be very rich in gas; it will be noted, however, that there are ongoing projects whose object is the export of gas from Nigeria.

Existing resources in the Middle East are made up largely of associated gas, as we have seen earlier. However, deep drilling holds out great promise for free gas in Iran, in the Emirates and in part of Saudi Arabia. The management of these stocks may give rise to problems. It is likely that the Middle East will again become a source of hydrocarbons for the world and particularly for the ECE zone, with gas taking over from oil and Iran once more occupying a leading position after its oil production has fallen off. But it will have a serious competitor in the USSR, particularly in the zones where gasfields are already largely developed, as in Soviet Central Asia, or where development is taking place rapidly, as in western Siberia.

The total reserves of this vast basin are estimated as exceeding 20 Tm^3, two-thirds contained in four deposits (including Urengoy, which should produce 100 Gm^3 per annum from 1982 onwards). These four deposits alone contain half the total reserves of the USSR. The development of western Siberia (containing approximately 1 Tm^3 of proven and probable reserves) has been the subject of discussions between the USSR, the United States and Japan since 1972. Production could begin in 1985, supplying mainly the United States and Japan.

The USSR's total reserves are among the largest in the world, and production plans are bold. However, domestic consumption is expected to expand as oil reserves are exhausted or as increasing difficulties are encountered tapping basins where the acquifer is very active, as in Siberia. It is therefore difficult to forecast the extent to which the USSR will contribute to supplying the rest of Europe.

Finally, it will be noted that the Far East can contribute to provisioning the region, either by supplying Japan, thereby freeing quantities derived from other sources, or even directly, since there are projects for exports from Australia to the United States.

Table 3 reproduces data from the paper presented by the American Gas Association to the World Energy Conference at Istanbul in September 1977, indicating percentages and trends, not absolute values. Column 3 attempts to show the level of exploration of the country: the higher the ratio the greater the quantities to be discovered in relation to quantities already discovered. It will cause no surprise to see China in first place, since it has been little explored; South America comes second, but one geologist questions where almost 10 times more gas than already exists is to be found: perhaps partly in the unknown and inhospitable zones at the southern tip of the continent? North America still has more than 5 times more gas awaiting discovery than existing reserves; these new quantities may be found in Mexico, in deep strata in the United States and Canada, on the Atlantic Continental Shelf (although the results from the Great Bank of Newfoundland and from the first drilling in the Baltimore Canyon are disappointing) and, above all, in the Arctic. The renewal rate for Europe, including the USSR, is, however, lower than the world average; this figure is no doubt a pessimistic one.

TABLE 3

Proven and Ultimate Reserves, Variations in Production Potential (After the AGA paper submitted to the World Energy Conference, September 1977)

	(1) Proven reserves (% of total)	(2) Ultimate reserves (8% of total)	(3) Ratio of ultimate reserves to proven reserves	(4) %	(5) Annual variation 1985/2000 (%)
North America	13	18.5	6.3	3	- 0.9
South America, excluding OPEC	1.7	4.2	10.6	3	+ 4.5
Africa, excluding OPEC	0.5	0.5	5	3	+ 4
Western Europe	6.5	4.5	3.1	4.6	- 0.9
Middle East, excluding OPEC	0.6	0.4	3.4	4.6	+10.5
Eastern Asia, excluding OPEC	1.6	1.9	5.4	4.6	+ 4.5
Oceania	1.7	2.6	6.7	4.6	+11
China	0.9	3.8	19	2	+ 3.5
USSR and Eastern Europe	33.5	28.6	3.8	6	+ 4.2
OPEC	40	35	3.9	13	+10.1
World	100.0	100.0	4.5	4.6	+ 4.2

(1) Proven reserves - percentage of world total.
(2) Ultimate reserves (proven + probable + possible) - percentage of world total.
(3) Ratio ultimate reserves/proven reserves.
(4) Annual increase (in %) in production potential 1976-1985.
(5) Annual increase (in %) in production potential 1985-2000.

It is interesting to compare the data in column 3 with those in the succeeding columns which describe the foreseeable evolution in production potential. If we confine ourselves to the countries of the area, we see that the growth in production potential of the USSR is strong and sustained, and this is not surprising in view of the large number of giant deposits (almost three-quarters of the largest gasfields in the world are in the USSR). Western Europe and the United States, after seeing an increase in their potential, are likely to see it level off and then decline in the last decade of the present century. This explains the relative pessimism of the authors concerning the quantities to be discovered in Europe and the logistic and statutory obstacles in North America to the development of reserves, although these are believed to be high. Such views seem unrealistic: the European far north has prospects comparable to the American far north, and if considerable resources are found, and if the need is felt, the technical means of exploiting them will be forthcoming, and regulations will be relaxed. Outside the ECE area, it will be noted that there are large-scale development plans for two regions which could export to the member countries: South America and the Middle East. There is some scepticism regarding South America, but the development of the rich gas resources of the Middle East has hardly begun, and it is bound to continue strongly in the future.

NEW SOURCES AND OTHER GASES

"Unconventional" sources of gas have already been attended to earlier, and it was noted that the borderline was sometimes difficult to trace. Any attempt to improve ideas or techniques for exploration, production or use likely to lead to a direct increase in recoverable reserves or to a more efficient recovery of the energy contained may be described as "unconventional" so long as it has not produced positive results.

The techniques described in the paper "More energy from LNG" fall within this meaning of the term "unconventional". Prospecting operations in geological situations as yet not clearly defined may also be described as "unconventional". Such would be the case, for instance, of prospecting under the overthrust zones in the south-west of France, in the Apennines or along the edges of the Rocky Mountains, or in very deep strata with a complex structure as, for instance, in the south-east of France. There are considerable hopes for this area, but the seismic equipment for reconnoitring the structures is not available, and sometimes even the basic geological concept is still too fluid for the problem to be broached. Research projects, in strata of low permeability such as the Devonian shale in the United States for which the technical equipment for exploitation is lacking, are also described as unconventional. There may be similar problems in certain regions of Western Europe, but in many cases an intellectual guiding principle is still lacking, and this justifies these projects being classified as "unconventional". The exploration of the Franco-Belgian Devonian, the importance of which appears to have been proved by the drilling operation at Saint Ghislain in Belgium, is from this standpoint on the borderline of the "conventional". By and large, it is simply not possible to give any date or, *a fortiori*, any estimate of possible reserves for these various projects, which may equally well end in failure.

Geopressure zones are zones which have strata containing gas saturated in water, at great depths, and, consequently, under high pressure. These zones can be exploited by producing water and dissolved gas. It is the reverse problem from that of oil production by the expansion of dissolved gas, during which the liquid is recovered and the gas produced at the same time is reinjected or flared off. In exploiting geopressure zones, it is the liquid phase, the water, which is the by-product to be disposed of, after what is usable, namely the heat it contains, has been exploited. The problem is not altogether a new one. The Po delta produced a mixture of gas and water, as did certain off-shore basins of Japan. These shallow exploitations have been abandoned, since drawing off the water led to subsidence on land. The technical problem is different for geopressure zones. The deep coastal reservoirs of the Gulf of Mexico might well contain 2.5 to 10 m^3 of gas per m^3 of water, but complex problems of well completion, as well as facilities for methane extraction and brine disposal, must be solved. However, L. d'Andrea thinks that the United States might be able to produce between 15 and 30 Gm^3 by 1985 and between 45 and 75 Gm^3 by 1990 from all these unconventional sources taken together.

A crucially important source for Western Europe might be *in situ* coal gasification. Gas produced in a factory from coal still accounts for a by no means negligible percentage (17 per cent in 1976) of Western European production, but a large proportion of this is a by-product of coking plants. It still has a potential in the United States, to the extent that opencast mining is able to satisfy the requirements of protection of the environment.

Western Europe's coal resources are considerable, but a large proportion of these resources cannot be used directly, either because the coal is in thin seams with a complex geological structure, as in France, or because it is located at a great depth, and is inaccessible for the miner, as in the North Sea. It is generally recognized that the gasfields in the southern part of the North Sea originated in gas formed in deep-lying coal seams and deposited in the nearest overlying sandstone stratum. These deep and thick coal deposits may be processed by contemporary on-site gasification techniques. Modern techniques produce a gas with a very low calorific value; in the years ahead it may be possible to produce gases with a satisfactory calorific value, but there is little likelihood that these techniques will provide a significant yield before 1990. The experiment described in the paper by M. Ledent and J. Ribesse will provide an initial reply in 1984.

These are again resources of fossil energy of preserved sunlight. What is desirable is to be able to command sources of energy which are indefinitely renewable, and the most obvious of these is solar energy. The idea which then comes to mind is the use of photosynthesis, the source of all life; hence the idea of manufacturing gas from organic matter or from biological waste. This is quite clearly what occurred during the geological ages to create fossil energy. This fossil energy is very abundant and much more prevalent throughout the world than is generally believed, but if it is to be used rationally, it needs to be accumulated in suitable locations, landfills, whose minimum exploitable size varies according to geographic, logistic and economic conditions. The same considerations apply to renewable energy. It is very diffuse, and the major problem concerned with its use is that of local concentration. Solar energy can be concentrated in the form of wood, by cultivating quick-growing trees. Some writers speak of "energy plantation" and try to calculate the surface which should be allocated to this type of growth. In Sweden, it is considered that a net yield of 120 TJ/km^2 could be obtained; a surface equal to one-tenth the area of France would then be necessary to produce a supply of gas equal in energy content to that of Western Europe in 1976. A study by Clinton C. Kemp and George S. Szego ("The energy plantation", in *Energy Sources*, Vol. 2, No. 3) considers that an electric plant supplying 200 000 persons at the current rate of consumption might require between 450 and 1000 km^2, depending on the rate of insolation and the yield from the conversion of solar energy.

It may, however, be asked whether the use of photosynthesis for purely energy purposes is not an extremely ruthless process. Photosynthesis produces highly processed substances, and their conversion to the methane state to finish up as water and carbonic gas is perhaps wasteful. The raw material has more value than energy. This may change, but not around 1990 to 2000.

Biological waste and that from certain industrial activities can supply gas by fermentation. But there again, the problem is not to squander certain usable wastes.

No useful purpose would be served by producing gas from manure if at the same time fertilizers are manufactured with natural gas. Neither would it be worthwhile to convert into gas wood shavings capable of being agglomerated, and at the same time manufacturing plastic panels with natural gas. Except in the case of small isolated units, it is almost always possible to do better than merely produce energy, even with waste products.

It is difficult to forecast the real future of this type of energy. If satisfactory techniques for using and storing solar energy are worked out, the use of organic matter and biomass gas might be superfluous.

A few words should be said about hydrogen, which is seen as a means of storing and transporting energy. In his report on the conversion of primary energy for the World Energy Conference at Istanbul in September 1977 Dr. Osman Kemal Kadiroglu writes: "The energy from hydrogen is now viewed as a universal transporter of energy." As regards its use, this form of transport has the great advantage of not polluting. At the present time it is very costly; its development is seen as being linked to that of nuclear energy, and it is thought that it is not likely to have a significant role, at least in France, before the year 2000. It has been noted, however (in the paper by M. Ledent and J. Ribesse), that it can be used for *in situ* coal gasification.

CONCLUSION

A study made in the Economic Commission for Europe; "The overall economic perspective for the ECE region up to 1990", stresses the magnitude of the energy problems confronting the countries of the region. Since the demand for energy continues to increase and the energy sector is developing large-scale investment requirements to the detriment of other sectors, it might be advisable to increase efficiency in the use of energy and try to reduce demand.

Over and above the technical and economic issues related to the problem of energy, there is a broader question, which is not broached in the document mentioned above, and which, to adopt an expression often used nowadays, might be that of a social option, or more simply, that of redirecting technological as well as philosophical social trends. The rapid advances in communications and information may result in a growth rate in energy consumption lower than anticipated, or even non-existent or negative. If a trend in that direction should occur, it would certainly not yet be perceptible in the coming decade, and in any case such a discussion would exceed the scope of our discussion.

Topics having a direct impact on gas resources in the Economic Commission for Europe region during the next decade, or even the last decade of this century, and now being put forward for discussion relate to the volume of natural gas resources, the availability of those resources and price trends, the management of the stock they represent, and the use of "unconventional" resources to alleviate any pressure as a result of the exhaustion of conventional resources during the period in question.

The volume and availability of natural gas resources obviously dictate stock management policy. If the renewal rate for available reserves is insufficient, an over-audacious management of available stocks with a view to alleviating Europe's dependence on energy would bring about a shortage which would exacerbate that dependence. This was the reasoning behind the criticism of the interstate pricing regulations on gas in the United States. Over-cautious exploitation may, in some cases, and unlike what is happening in the case of oil, have technical drawbacks and bring about a rise in the aquifer. Economically, it is a useless accumulation: this is the reasoning behind the criticism of Canadian and European exploitation policies, while the Soviet exploitation policy is very dynamic.

Criticism of Canadian and European policies is meaningful only in so far as resources which are currently available or will be available by 1990 are really sizeable. Is this really the case as far as Canada is concerned? Is it really the case for Europe? Curiously, the answers overlap. Alberta seems rich, the Canadian Arctic, if not disappointing, at least still very uncertain. In Europe, the south is falling behind, while the North Sea and its extension to the north are rich in achievement and in promise. It is extremely difficult to say what the correct policy would be: does it still lie in the celebrated "golden mean" or is boldness not called for?

Lastly, it seems that the "unconventional" source most urgently in need of being brought into operation is that of *in situ* coal gasification. "Ultimate" world resources of fossil fuels, including all reserves belonging to this category, are ample. At the World Energy Conference in Istanbul they were assessed at 200 000 EJ, of which 80 per cent was accounted for by coal. The ECE region is fairly well endowed with very deep coal estimated at between 2000 and 3000 GT. These resources should not be left untapped, and although the future of natural gas is still bright, we should not put off learning how to make use of them until natural gas resources are on the verge of exhaustion.

Compléments au Rapport Général sur le Thème A

Maurice Mainguy

France

La discussion relative au Thème A a porté essentiellement sur les politiques de production. Le sujet est inépuisable, et la discussion n'a permis de toucher qu'à quelques uns de ses aspects.

Le cadre dans lequel se plaçait le Colloque, c'est-à-dire essentiellement l'horizon 1990, impose à la pensée d'utiles contraintes, qui permettent d'aboutir à des conclusions relativement claires. L'histoire du pétrole et du gaz est faite de périodes de crainte de manque alternant avec des périodes de surabondance. Sur longue période, la courbe des réserves connues et des ressources espérées (faussement dites réserves ultimes) est croissante; elle tend actuellement à plafonner, après une période faste. On peut penser que ce plafonnement est en grande partie artificiel, et qu'il reflète plus une attitude psychologique de peur, de pessimisme et de restriction qu'il ne représente le résultat d'observations non passionnelles. Si les nouvelles récentes concernant le Mexique sont confirmées, peut-être l'industrie pétrolière se trouve-t-elle à la veille d'un renouveau, peut-être même de nouveaux dangers provenant de la surabondance.

Si, par ailleurs il est exact, économiquement parlant, que la demande est antérieure à l'offre et la stimule, elle ne peut la porter au-delà des limites que lui imposent les contraintes auxquelles on se heurte. En créant une excessive demande par des prix trop bas - seul moyen de stimuler la demande - appliqués au gaz livré entre Etats (interstate gaz) la législation américaine antérieure aux mesures actuelles a conduit, au contraire, à une stérilisation de la recherche et à la raréfaction de l'offre.

Dans le cadre de l'Europe Occidentale où s'est limitée la discussion (et, faut-il le rappeler, à l'horizon 90) il ne semble pas possible de suivre le Professeur Odell lorsqu'il écrit: ". . . dans l'état actuel de l'histoire du développement du gaz naturel en Europe, les réserves sont une fonction de la demande, plutôt que l'inverse comme la plupart des communications semblent l'indiquer." Il est difficile d'imaginer, les choses étant ce qu'elles sont dans le domaine de la connaissance géologique et des moyens techniques et financiers, que l'on puisse découvrir dans les six ou sept années à venir les quantités nécessaires pour porter les réserves au début de la décennie 80, au niveau espéré considéré comme étant . . . "en partie fonction de motivations retrouvées et renforcées pour chercher du gaz et produire celui que l'on a trouvé."

Au début de l'histoire d'un bassin nouvellement découvert, les géologues se trouvent en quelque sorte en présence d'un monde neuf, plein de promesses et de risques. Les plus hardis font des prévisions largement optimistes, les timorés (ils sont rares) ne disent rien, les plus honnêtes donnent de très larges fourchettes d'estimation. Nul ne peut reprocher aux politiques de prendre en compte les prévisions moyennes plutôt que les plus élevées, et de prendre, pour élaborer leur action, une certaine marge de sécurité. Ces prévisions sont faites dans le cadre d'une prospection débutante, et l'on ne peut valablement partir de cette divergence entre prudence et optimisme pour, ultérieurement, fonder un choix systématique en faveur de ce dernier.

La connaissance progresse ensuite très vite, mais plus elle progresse, plus, en conséquence, la masse des informations s'accumule, plus il devient long et difficile d'assimiler ces informations et de les digérer, alors que se périme l'expérience acquise dans les débuts. L'accélération de la demande, et le renouveau des motivations à rechercher ne pourront pas accélérer le rythme de l'acquisition intellectuell des données. Un jour intervient une percée, un renouvellement des thèmes d'exploration, mais il est bien impossible d'en prévoir la date et bien léger de croire que l'on pourrait avancer celle-ci en agissant sur la demande en gaz naturel.

Parce que la discussion se plaçait dans le cadre de l'Europe Occidentale, et à l'horizon 1990, les prévisions faites à partir des réserves connues ou possibles, des données géologiques actuelles et des échecs enregistrés, ont une valeur qu'elles n'auraient pas à une échelle géographique plus vaste et à un horizon plus reculé. Les prévisions présentées par les délégations du Royaume-Uni, de Norvège et des Pays-Bas prennent en compte les possibilités raisonnables avec leur probabilité d'existence, et ne se contentent nullement d'extrapolations. Rien ne permet d'imaginer que les réserves marines des Pays-Bas pourraient atteindre mille milliards de mètres cubes; les géants y ont à peu près certainement tous été trouvés. Les découvertes à attendre de la Mer de Norvège (au Nord du 62ème parallèle) ne joueront pas un très grand rôle dans la décennie 80. Les gisements découverts ou espérés en mer profonde italienne ont été pris en compte par la délégation italienne. Enfin la nouvelle province du Golfe de Cadix est au stade initial de la découverte, et il serait prématuré d'en parler.

Etant d'un naturel optimiste que n'ont pas entamé trente cinq ans de carrière de géologue pétrolier, je pense que les prévisions actuelles se montreront inférieures à la réalité, mais elles sont seules à permettre de fonder une politique raisonnable. L'essentiel est que cette politique soit aussi souple et aussi adaptable que possible aux réalités changeantes, mais se placer aux limites du possible pourrait entraîner à la catastrophe.

On trouvera ci-après les reproductions de quelques figures présentées au cours du Colloque. La figure 1: Réserves et Productions en Europe Occidentale illustre la discussion ci-dessus. La carte fig. 2 localise les zones productrices et les zones d'espoir dans le monde. On a également noté quelques espoirs "non conventionnels": certaines zones à faible perméabilité notées F. P., ou de géopression, notées G. P.; enfin celles où l'on pense trouver du charbon susceptible d'être gazéifié in situ, notées G.C.

Les figures 3 et 4 illustrent les questions de prospection. La fig. 3 représente schématiquement les conditions de formation des gaz précoces, des gaz liés à l'huile et des gaz tardifs. Température et enfouissement sont fonction de la profondeur, mais différemment selon les régions. La répartition des gaz secs, des gaz humides, des gaz associés peut être modifiée par des migrations. Sur la fig. 4 sont représentées schématiquement les modifications apportées aux réflexions sismiques par la présence de gaz dans une roche réservoir, ce qui permet de supposer la présence de gaz dans une structure.

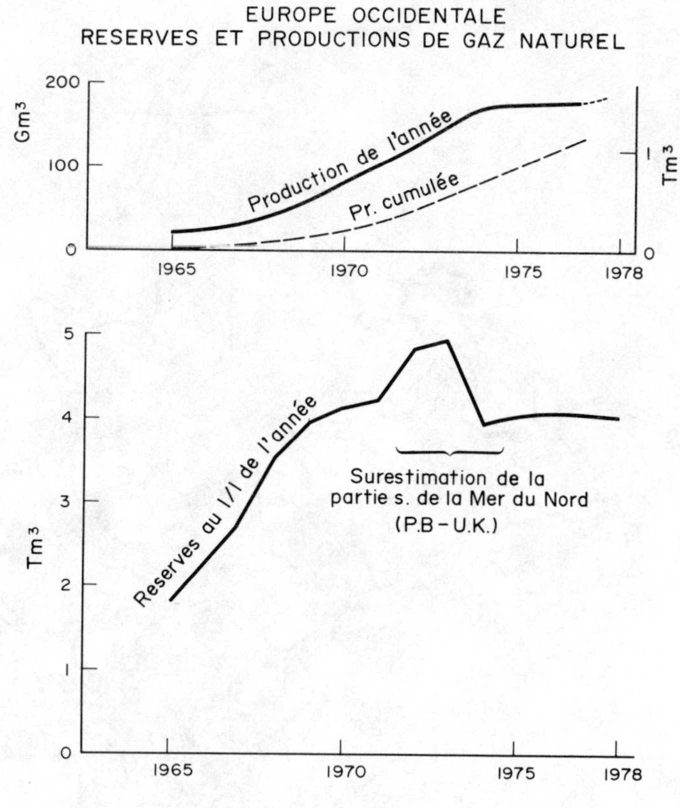

Fig. 1.

Un aspect fondamental de la conservation des richesses naturelles est la recherche de moyens économiquement valables permettant d'éviter de brûler à la torche le "gaz fatal" lié à la production du pétrole. Les torches ont disparu du paysage pétrolier américain parce qu'il existe des réseaux de distribution et un marché proche pour le gaz. Mais, malgré les efforts faits, elles sont loin d'avoir disparu du Moyen Orient et d'Afrique; elles existeraient sans doute en Mer du Nord si elles n'étaient pas interdites par les autorités britanniques et norvégiennes.

Si l'on veut éviter la destruction par brûlage, le meilleur emploi qui puisse être fait de ce gaz est l'emploi sur place, lorsqu'il existe une capacité d'absorption des quantités souvent considérables rendues ainsi disponibles. L'exportation reste une solution attrayante, surtout lorsqu'elle peut se faire par pipeline.

Reste enfin à renvoyer en terre le gaz involontairement produit. Cette opération a un double aspect: d'une part, c'est un acte de conservation, de stockage pour utilisation ou vente dans l'avenir, d'autre part, cela peut constituer un moyen d'améliorer la récupération du pétrole auquel le gaz est associé, la fonction de stockage restant vraisemblablement préservée. Il ne s'agit que d'un moyen parmi d'autres d'améliorer la récupération; il peut être efficace (Iran) mais ce n'est pas toujours le meilleur, et dans certains cas, comme en Arabie, la réinjection peut avoir un effet négatif sur la récupération finale du pétrole. On a pu noter que, aux Etats-Unis, les quantités réinjectées avaient en quelques années diminué de moitié, passant de 50 à 25 G (m^3) par an, tandis que des méthodes plus complexes et plus efficaces étaient mises en oeuvre.

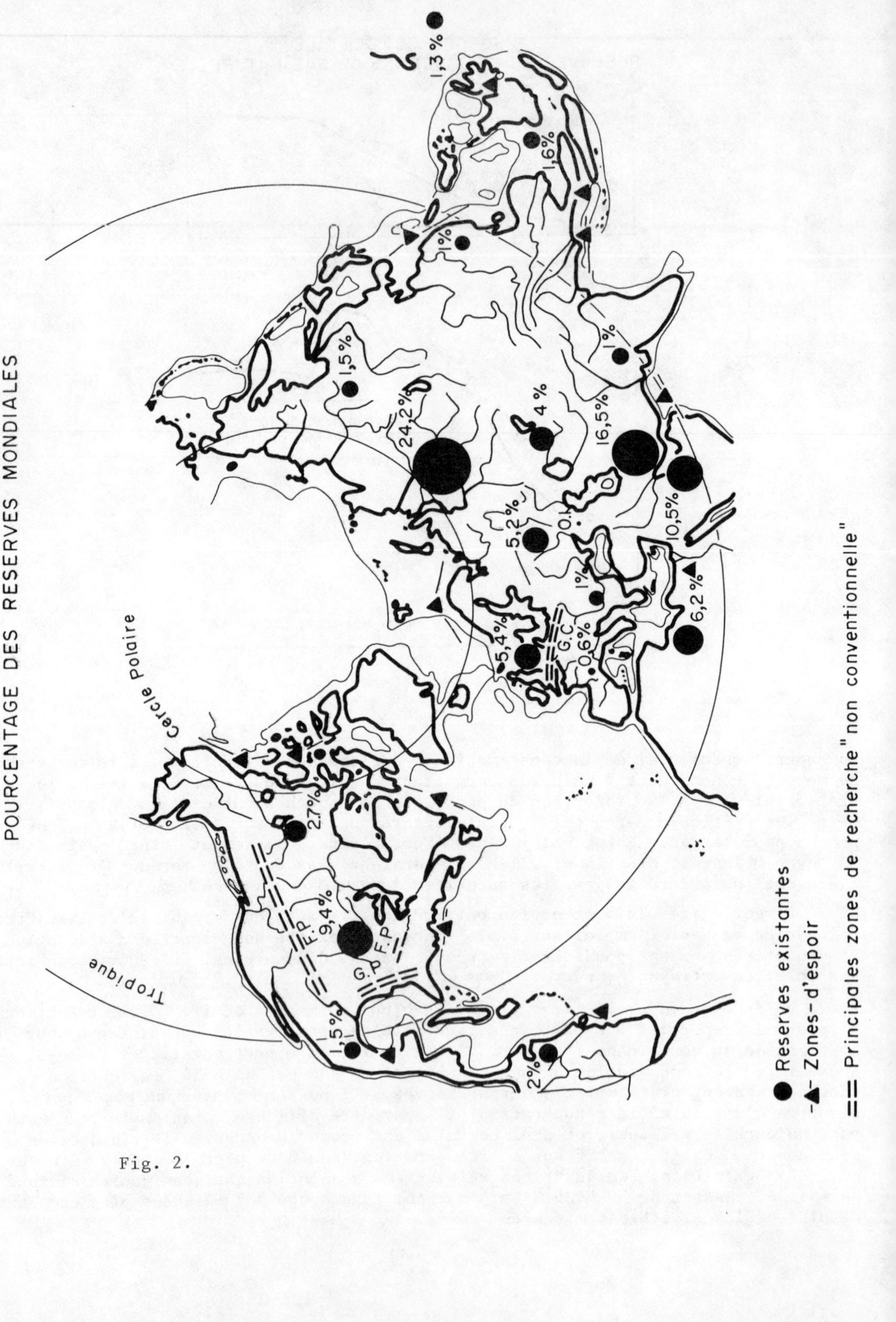
POURCENTAGE DES RESERVES MONDIALES
Cercle Polaire
Tropique
1,3 %
1,6%
1%
1,5%
24,2%
4 %
16,5%
1%
10,5%
5,2%
0,1
1%
6,2%
5,4%
G.C
0,6%
2,7%
9,4%
F.P
F.P
G.C
1,5%
2%
Reserves existantes
Zones-d'espoir
Principales zones de recherche "non conventionnelle"

Fig. 2.

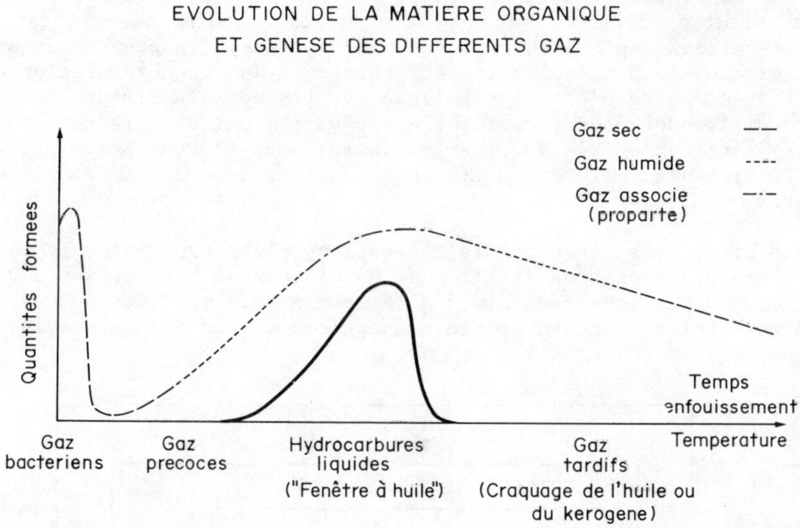

Fig. 3. Modifié d'après Rumeau et Sourisse, *Bulletin du Centre de Recherches de Pau* (Groupe ELF AQUITAINE), No.9, juin 1975.

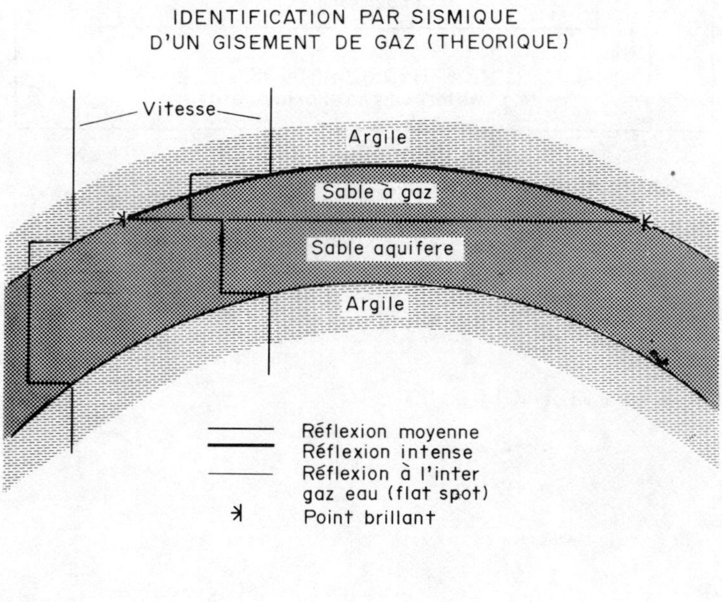

Fig. 4.

Les figures 5, 6 et 7 (présentées par l'AEGPL) ont trait aux gaz de pétrole liquéfiés, (GPL, ou LPG en anglais) bien définis sur la fig. 5. Ils se situent à la frontière entre les domaines du gaz et du pétrole, soit qu'ils proviennent du fonctionnement des raffineries, soit qu'ils constituent la fraction lourde des gaz humides, et soient produits de façon plus ou moins "fatale" avec le gaz ou avec le pétrole. Ils ont des emplois spécifiques très variées; ils peuvent se substituer au gaz naturel dans le domaine de la pétrochimie et dans la carburation automobile, où ils ont l'avantage de n'être pas polluants. Ils sont facilement mobilisables et peuvent être distribués sous forme très divisée; ils ont une grande flexibilité d'empl Par ailleurs, ainsi qu'on l'a vu, les gaz lourds sont l'agent principal de récupératio assistée dite miscible. Il ne s'agit pas en fait de produits de substitution de masse du gaz naturel.

Les figures 8 et 9, présentées par la délégation belge, concernant la gazéification in situ du charbon, et ses possibilités de développement à l'horizon 2000. La délégation de la République Fédérale d'Allemagne a émis quelques doutes sur la possibilité de développer cette source d'énergie d'ici la fin du siècle.

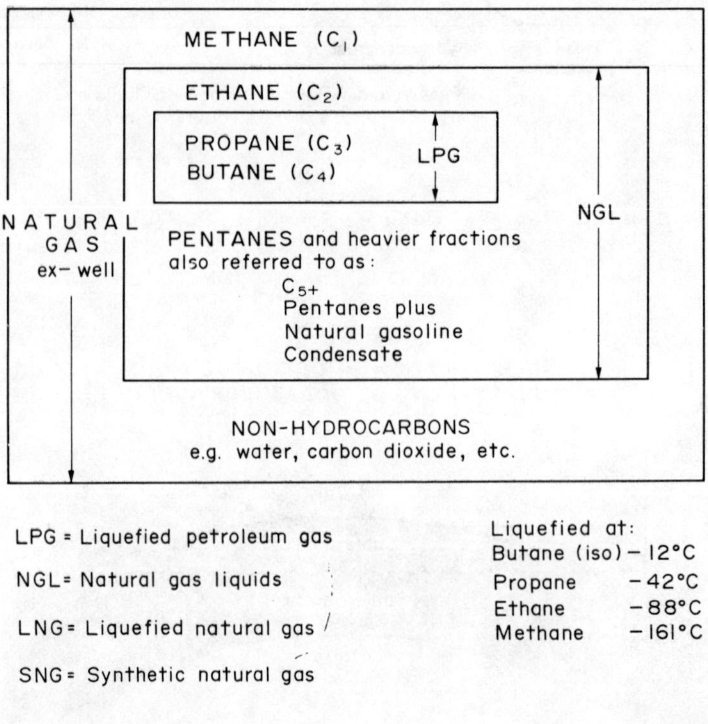

Fig. 5. Submitted by A.E.G.P.L

WORLD WIDE LPG AVAILABILITY 1978 THROUGH 1985

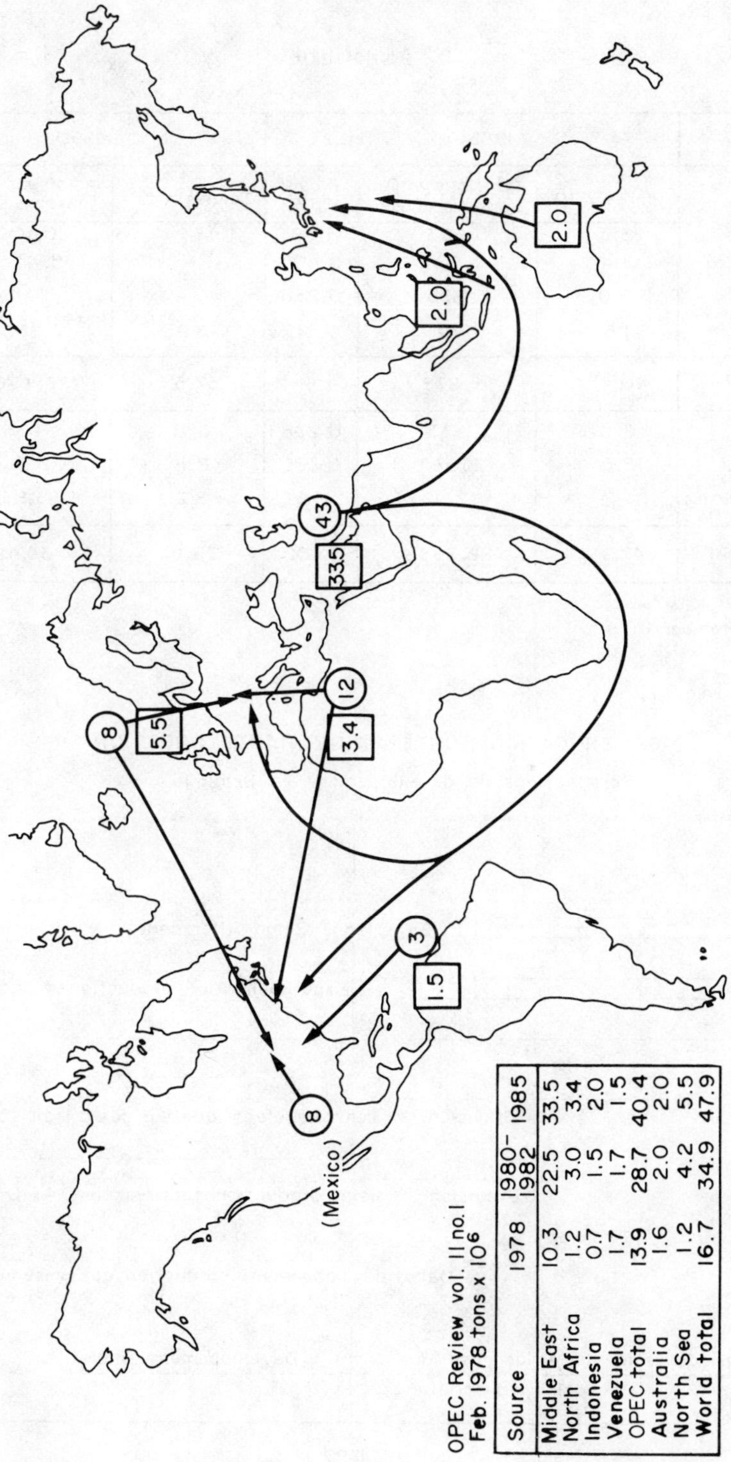

OPEC Review vol. II no. I
Feb. 1978 tons x 10^6

Source	1978	1980–1982	1985
Middle East	10.3	22.5	33.5
North Africa	1.2	3.0	3.4
Indonesia	0.7	1.5	2.0
Venezuela	1.7	1.7	1.5
OPEC total	13.9	28.7	40.4
Australia	1.6	2.0	2.0
North Sea	1.2	4.2	5.5
World total	16.7	34.9	47.9

Fig. 6.

LPG Availabilities

	1985			1990 (High line)		
	Tonnes x 10^6	m^3 of gas x 10^6 (gaseous phase)	T cals x 10^5	Tonnes x 10^6	m^3 of gas x 10^6 (gaseous phase)	T cals x 10^5
Middle East	33.5	15, 653	3.99	43	20, 100	5.119
North Africa	3.4	1,589	0.405	11.6	5, 420	1.381
Indonesia	2.0	935	0.238	0	0	0
Venezuela	1.5	701	0.317	2.9	1,355	0.345
OPEC	40.4	18, 878	4.81	57.5	26, 870	6.845
Australia	2.0	935	0.238	2.0	935	0.238
North Sea	5.5	2,570	0.655	8.1	3,784	0.964
Canada/Mexico	—	—	—	8.2	3,831	0.576
World total	47.9	22,383	5.702	75.8	35,420	9.024

28th September 1978

Fig. 7.

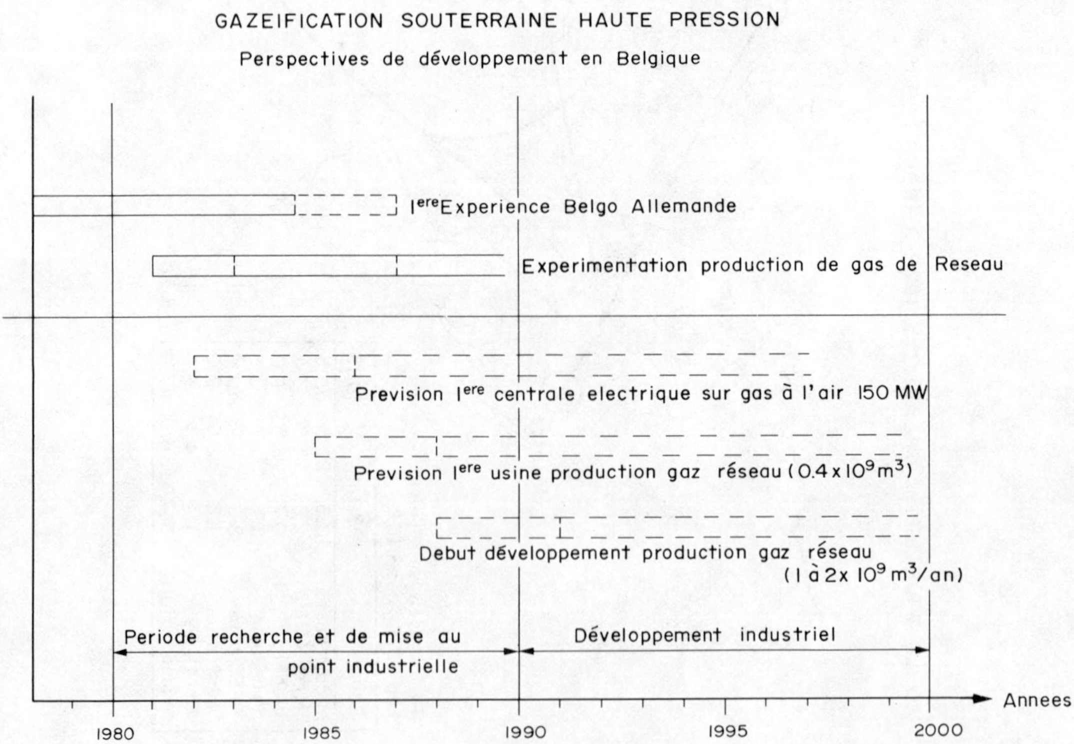

Fig. 8.

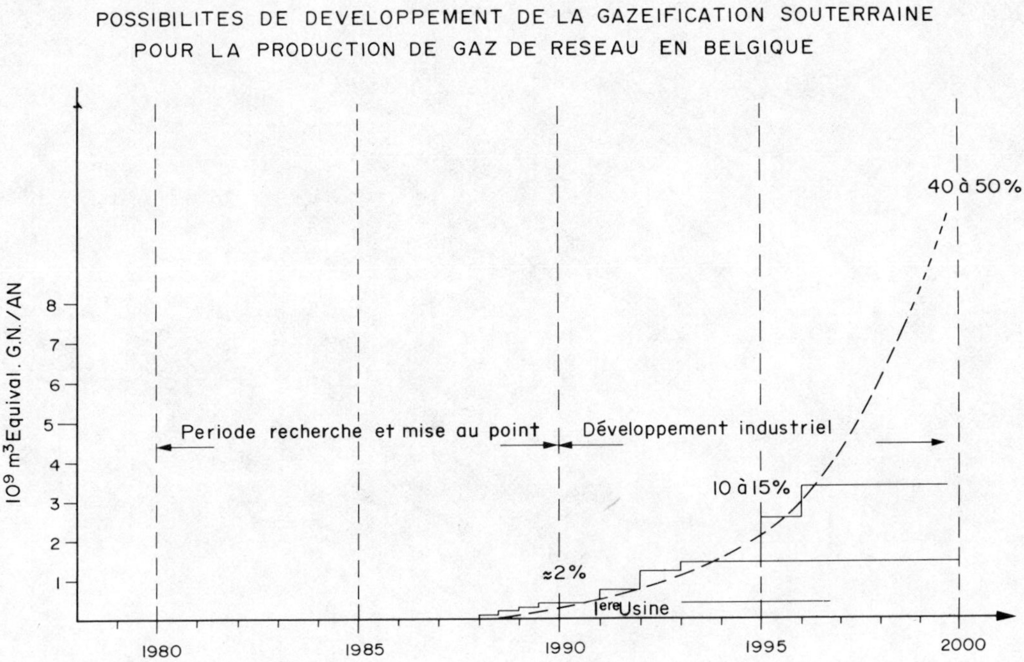

POSSIBILITES DE DEVELOPPEMENT DE LA GAZEIFICATION SOUTERRAINE
POUR LA PRODUCTION DE GAZ DE RESEAU EN BELGIQUE
10⁹ m³Equival. G.N./AN
8
7
6
5
4
3
2
1
Periode recherche et mise au point
Développement industriel
40 à 50%
10 à 15%
≈2%
Iere Usine
1980
1985
1990
1995
2000

Fig. 9.

Possible Sources of Natural Gas for France and the ECE Area in 1990

J. Chapelle and P. Desprairies

France

INTRODUCTION

The high cost of transporting gas has caused the world natural-gas market to develop later than that of oil. None the less, the extent of proven reserves 63 500 billion m^3 and ultimate reserves (over 200 000 billion m^3) makes it reasonable to expect an increase of 50% in the worldwide marketed production between now and 1990, bringing it to over 2000 billion m^3. The Economic Commission for Europe area, possessing about 55% of the proven and ultimate reserves, would then produce more than 1500 billion m^3 in 1990 and might well import more than 200 billion m^3 of natural gas to meet its full needs.

However, in bringing the world's gas reserves to the market one may encounter various technical and economic obstacles. Thus in 1976 only 10% of marketed gas production was exported, and liquefied natural gas (LNG) amounted to only 12.5% of the gas exported. But in the future the volume of gas available for export may be reduced because of local use, i.e. production of energy, raw material for petroleum chemicals, reinjection into oil-bearing strata. Progress in international trade will require enormous investments in the construction of gas pipelines under difficult conditions (arctic regions, subsea pipelines) or in liquefaction chains, the cost of which has gone up about 40% in the last 2 years.

Consequently the contribution of natural gas to the energy needs of the industrialized nations should continue to increase in Europe and in 1990 should reach nearly 30% in Eastern Europe and 18% in Western Europe. In North America, however, particularly in the United States, it should decrease to 20%.

In France the requirements in 1985 will be 37 billion m^3 according to the goals established by the Planning Council. In 1990 they may be expected to be from 42 to 44 billion m^3.

The discovery and production of natural gas are closely related to those of oil. However, although oil normally meets 44% of the world's energy requirements, natural gas supplies only rather less than 20%. On the international market, gas does not have the advantages of oil as concerns flexibility in transport and marketing.

For this reason the use of gas has been confined to a small number of countries in which large reserves are found in the vicinity of steady markets. The first example was the United States which in 1938 accounted for 90% of the world's marketed production of natural gas. Thus natural gas came to play a role of prime importance in the American energy balance; today it supplies 30% of the energy consumed in the United States, as against 44% supplied by oil.

From the 1950s onwards, the discovery of rich gasfields in Western Europe stimulated demand in this part of the world. Concurrently, progress in long-distance transport by pipeline and by LNG tankers opened the international market to the gas available in the USSR, North Africa, the Middle East and Indonesia.

The world-wide marketed production of natural gas is now at 20 times the 1938 level, whilst oil production has risen elevenfold since that year. What is more, the 1973 crisis hardly interrupted the upward trend of natural gas, as it did in the case of oil production; it only slowed the rise by the effect of economic recession and energy-conservation policies.

In spite of all this, the part played by natural gas in the world's energy requirements is not proportional to its reserves as compared to those of oil. Indeed natural gas accounts for less than one-third of the world production of hydrocarbons, whilst gas reserves amount to 40% of the world hydrocarbon reserves. Moreover, gas reserves are increasing faster than those of oil; in the 10 years from 1968 to 1977, the proven world reserves of natural gas have doubled. What is more, the deeper the strata reached by wildcat wells, the greater the likelihood of finding gas rather than oil. Finally, some gasfields have not been exploited for want of accessible markets, whereas all oil-producing countries do what they can to diminish the waste of associated gas flared off. It is thus reasonable to expect that all countries, with the probable exception of the United States, will see natural gas satisfying an increasing share of their energy needs.

Current world gas potentialities are reviewed on the basis of the location and volume of proven and ultimate natural gas reserves. The world-wide export movements are then outlined for 1990 in the light of expected consumption. Various economic and political obstacles which lie in the way of the exploitation of these resources are examined, particularly in the fields of production and transport. Finally the particular case of France is taken up.

PROVEN AND ULTIMATE RESERVES OF NATURAL GAS

Since our aim is to sketch the main movements of supplies in 1990, we present the forecast for world reserves rather than only those of the countries of the ECE area.

The reserves to be examined first are those that have been proven by wells already in production or on the point of being so. The data are backed up by production curves or results of industrial tests. However, it is not always possible to know whether a reserve has been proven to be recoverable or merely proven to exist in the formation. Although the recovery percentage is very high compared with that of oil, the difference between *in situ* and recoverable reserves may reach 20%. (The figure may be 70% non-recoverable in the case of oil.) However, there is a factor that affects estimations in the opposite direction, i.e. the total associated gas content of crude oil is not always taken into account in published gas reserve figures.

As of 1 January 1977 the proven world reserves of natural gas were estimated to be 63 500 billion m^3, which corresponds to about 39 years of gross production at the 1976 rate (1617 billion m^3). The market production in 1976 was 1357 billion m^3 if it is considered that the volume of gas flared or lost was 180 billion m^3 and the volume reinjected into formations was 80 billion m^3. It should nevertheless be noted that the relation between proven reserves and gross production in 1965 corresponded to only 28 years.

The ECE Area

The countries belonging to the Economic Commission for Europe (Western Europe, Eastern Europe, North America) represent 54.2% of the world's proven natural gas reserves (34 400 billion m^3; see Table 1).

TABLE 1

Location of Gas Reserves as of 1 January 1977

	Proven reserves*		Ultimate reserves**	
	10^{12} m^3	%	10^{12} m^3	%
USSR and Eastern Europe	22.7	35.7	70 to 80	35 to 29.6
North America	7.8	12.3	35 to 50	17.5 to 18.5
Western Europe	3.9	6.2	10 to 15	5 to 5.6
Total of ECE area	34.4	54.2	115 to 145	57.5 to 53.7
Middle East	17.2	27.1	40 to 60	20 to 22.2
Africa	5.9	9.3	15 to 20	7.5 to 7.4
Far East	3.3	5.2	15 to 25	7.5 to 9.3
Latin America	2.7	4.2	15 to 20	7.5 to 7.4
Total of other than ECE area	29.1	45.8	85 to 125	42.5 to 46.3
World total	63.5	100.0	200 to 270	100.0

*World Oil, 15 August 1977.
**Institut Francais du Pétrole, 1977, according to various estimates.

Thanks to the enormous gas potential of the USSR, *Eastern Europe* is of course the best provided in gas resources: 22 700 billion m^3 or 35.7% of world supplies. Almost the entire proven reserves of Eastern Europe are in the USSR: 22 100 billion m^3 or 34.8% of world reserves. Furthermore, at the Symposium of the Gas Committee in Belgrade in 1974, M. Faoustov reported that at that time the USSR possessed 20 000 billion m^3 of proven reserves.

The other Eastern European countries account for only some 600 billion m^3:

Romania	160 billion m^3
Poland	140 billion m^3
Hungary	120 billion m^3
East Germany	90 billion m^3
Yugoslavia	40 billion m^3
Czechoslovakia	16 billion m^3
Bulgaria	14 billion m^3
Albania	12 billion m^3

North America comes next with 7800 billion m^3, i.e. 12.3% of the world total. While the *United States* possesses most of these reserves, 6130 billion m^3 or 9.6% of the world total, it should be noted that the ratio of reserves to annual production is only about 10 and that the situation has deteriorated considerably in the last few years, since in January 1972 the United States's proven reserves were 7900 billion m^3 giving a reserves/production ratio of 13. Alaska alone now represents 15% of the proven reserves of the United States, 82% of these reserves being in the form of associated gas.

On the other hand, *Canada* shows a very slight positive trend, its proven reserves now being 1650 billion m^3 as against only 1614 billion m^3 last year.

Finally, the mere 3900 billion m^3 of Western Europe correspond to only 6.2% of the world's proven reserves, and the situation has deteriorated in the last 2 years. The *Netherlands* alone, with 1760 billion m^3, has about 45% of the reserves of Western Europe, which is no more than 2.7% of world reserves. The other main countries possessing gas reserves in Western Europe are:

United Kingdom	810 billion m^3
Norway	725 billion m^3
West Germany	210 billion m^3
Italy	185 billion m^3
France	135 billion m^3

Other than the ECE Area

The proven gas reserves of the countries not belonging to the ECE area amounted to 29 100 billion m^3 on 1 January 1977, i.e. 45.8% of world reserves. The *Middle East* leads with 17 200 billion m^3 or 27.1% of the world total. *Iran* has the greatest potential with 10 700 billion m^3, i.e. 16.8% of world reserves or nearly two-thirds of Middle East reserves. At present most of these reserves, about 65%, consist of associated gas, but an increase in the proportion of non-associated gas has been noted, not only in Iraq, but in Iran.

The other countries having large reserves are:

Saudi Arabia	2280 billion m^3
Qatar	1300 billion m^3
Kuwait	1060 billion m^3
Abu Dhabi	865 billion m^3
Iraq	760 billion m^3

In Saudi Arabia 22% of the proven reserves consist of non-associated gas, but it is known that other Gulf countries have announced large reserves of non-associated gas that are now in the process of estimation or confirmation.

Africa, with 5900 billion m^3, corresponds to only 9.3% of the world's proven natural gas reserves. Of this, *Algeria* has the largest share: 3250 billion m^3 or 5% of the world total. There follow:

Nigeria	1400 billion m^3
Libya	790 billion m^3

Egypt	62 billion m^3
Gabon	50 billion m^3

The *Far East* has only 3300 billion m^3 of proven reserves, which is rather less than Western Europe and constitutes 5.2% of the world total. But this estimate takes into account a reserve of only 600 billion m^3 for *China*, information being incomplete. The other principal countries having gas reserves are:

Australia	915 billion m^3
Indonesia	680 billion m^3
Brunei-Malaysia	530 billion m^3
Pakistan	310 billion m^3
New Zealand	170 billion m^3
India	100 billion m^3

Finally *Latin America*, with 2700 billion m^3, represents only 4.2% of the world's reserves. About half of this is in *Venezuela* (1200 billion m^3), followed by *Mexico* (580 billion m^3), *Argentina* (200 billion m^3), *Colombia* (180 billion m^3) and *Bolivia* (110 billion m^3).

Since 1 January 1976, world-wide proven reserves of natural gas have increased by about 2220 billion m^3, i.e. 3.6%, most of which increase is in the USSR (+2000 billion m^3) and Mexico (+240 billion m^3), while a decrease of 340 billion m^3 was recorded in the United States.

On 1 January 1978 the world reserves of natural gas were reported to have reached 71300 billion m^3, representing an increase of 7800 billion m^3 with respect to the previous year. This recent increase is essentially due to the USSR (+4000 billion m^3) and Iran (+3450 billion m^3).

It is to be noted, nevertheless, that a considerable portion, 18% (the figure is 22.5% for oil) of the reserves detailed geographically in the above paragraphs, are offshore fields (11 500 billion m^3 out of 63 500 billion m^3, Eastern European countries excluded). Of these 11 500 billion m^3, 43.4% are in the Middle East, 21.3% in Western Europe and 15.4% in the Far East. The marketed production of offshore natural gas has grown very quickly in recent years, having gone from 121.5 billion m^3 in 1970 to 170.4 billion m^3 in 1976, an increase of 40%.

The countries of the ECE area are thus better off in proven reserves of natural gas (over 54% of the world total) than in proven reserves of oil (only 24% of the world total).

What is more, probable and possible reserves may be added to the present proven reserves. Estimates of the resulting *ultimate reserves* vary by a factor of 2, from 160 to 320 billion m^3 (see Fig. 1). In the present survey we have used the narrower range of 200 to 270 billion m^3.

The ultimate gas reserves (Table 1) occur in approximately the same regions as the proven reserves. Thus the ECE area represents between 53.7 and 57.5% of the world's ultimate reserves, and North America is better off in ultimate reserves than in proven reserves. Nevertheless, with these relatively rich ultimate reserves, it would be a mistake to lose sight of the long time required to develop new fields and of the problems of transport, as will be seen later.

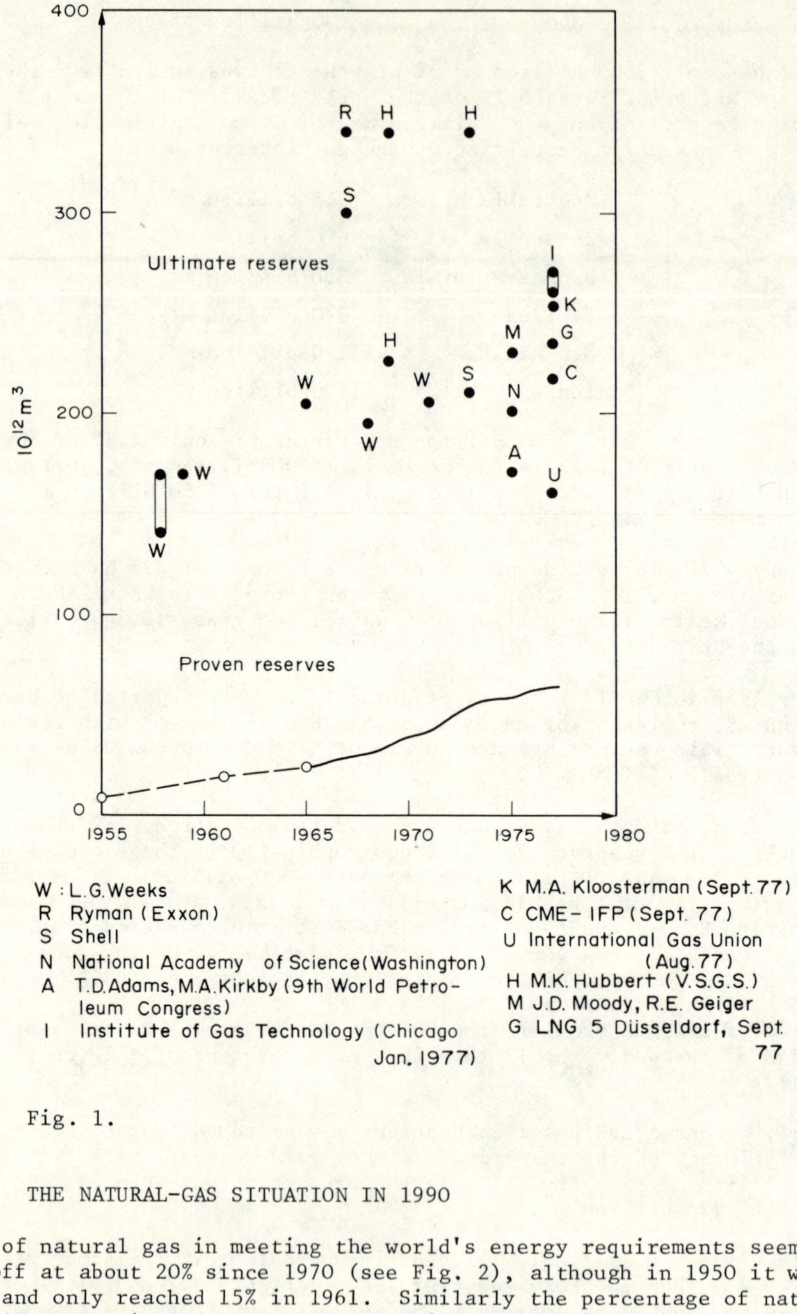

Fig. 1.

THE NATURAL-GAS SITUATION IN 1990

The share of natural gas in meeting the world's energy requirements seems to have levelled off at about 20% since 1970 (see Fig. 2), although in 1950 it was still only 10%, and only reached 15% in 1961. Similarly the percentage of natural gas in the world production of hydrocarbons, which grew rapidly after the Second World War, has levelled off at about 30% since 1965. The world production of natural gas, which already satisfies a large part of the energy requirements of the industrialized countries, will certainly grow rapidly from now until 1990. The production forecasts that were put forward at the 1977 World Energy Conference seem, however, rather optimistic. According to estimates made by the Institut Français

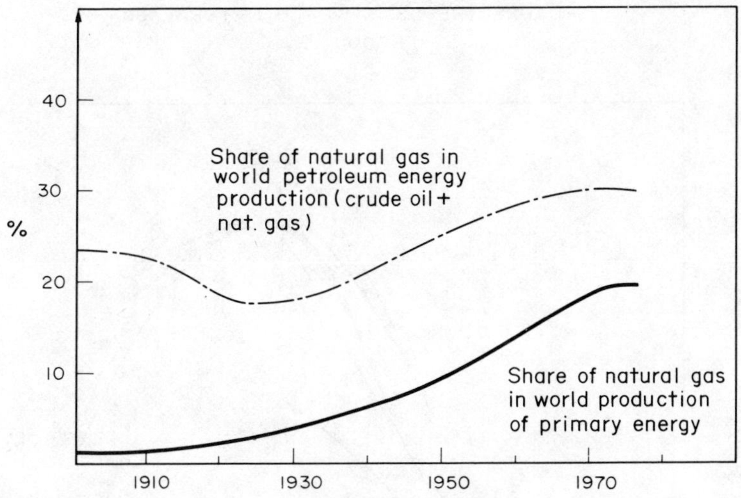

Fig. 2. L'Industrie du gaz dans la monde, I.F.P., 1977

du Pétrole, the world production of natural gas in 1990 should be between 2000 and 2500 billion m^3, rather than between 2500 and 3000 billion m^3 (see Fig. 3). The countries of the ECE area should produce about 75% of the world total in 1990 as opposed to 89% in 1976 (Table 2).

The expected production increase in countries outside the ECE area will be due partly to increased local use and partly to increased international trade. These exports, which in 1976 amounted to only 135 billion m^3 (17 of which were LNG), may be expected to rise to 305 to 370 billion m^3 (83 to 135 of which will be LNG) in 1985 and to 371 to 480 billion m^3 (130 to 175 as LNG) in 1990 (see Fig. 4).

Available natural gas in the ECE area countries in 1990 is expected to break down as shown in Table 3, imports from countries outside the area included. Production levels forecast for 1990 are consistent with the proven reserve situation of today, but the exploitation of these gas potentialities may well encounter a certain number of obstacles, particularly since natural gas constitutes a rather dispersed form of energy whose transport is difficult and costly.

THE CONDITIONS OF EXPLOITATION OF GAS RESERVES

The 1990 outlook for the total available gas in the ECE area takes into account average total imports amounting to 11 or 12% of requirements. But in Western Europe imports may well constitute a third of the consumption. Obstacles of two kinds are likely to restrain the exploitation of the world's reserves.

Technical Obstacles

The marketed production of natural gas today amounts to only 84% of gross production. Although this utilization ratio has improved appreciably, having been only 66% in 1950, in some regions far removed from consumer markets it is still much lower: 46% in Latin America, 27% in the Middle East and 20% in Africa.

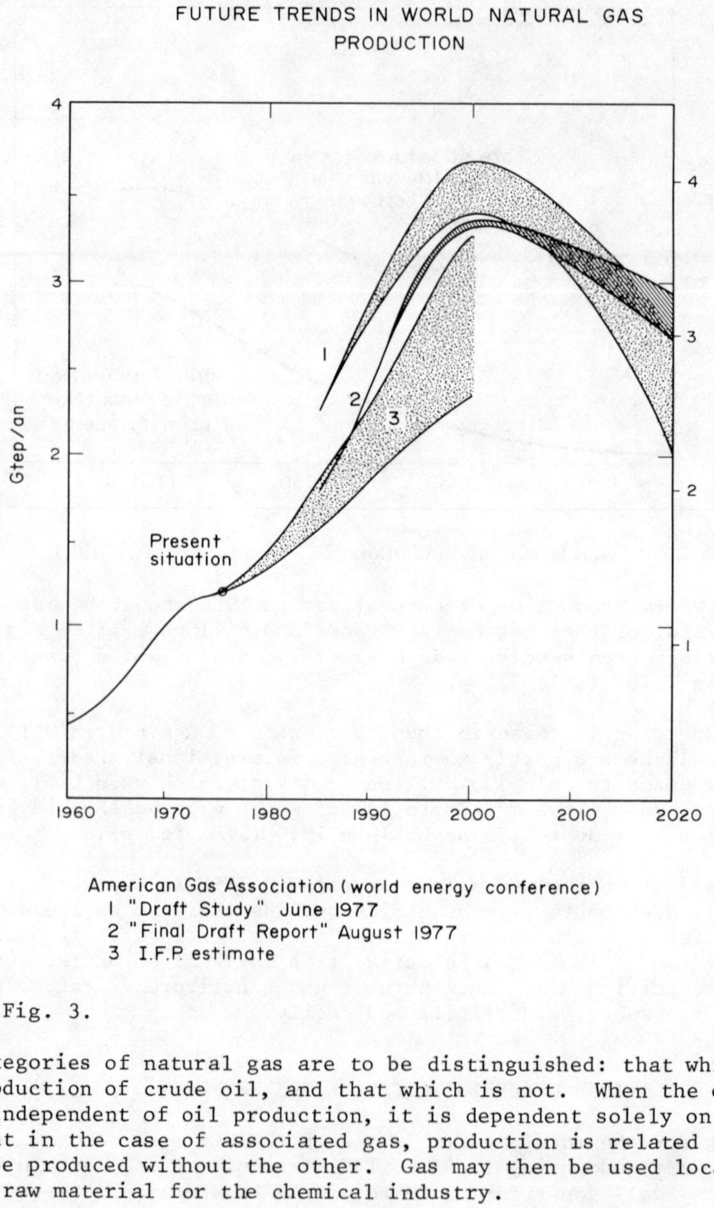

Fig. 3.

Two main categories of natural gas are to be distinguished: that which is associated with the production of crude oil, and that which is not. When the exploitation of a field is independent of oil production, it is dependent solely on the market for the gas. But in the case of associated gas, production is related to that of oil; one cannot be produced without the other. Gas may then be used locally as a source of power or raw material for the chemical industry.

If associated gas is not used locally, it may be reinjected into the formation (at the price of installing expensive compressor units), or exporting it may be considered, rather than letting it be flared off.

The relative excess of crude oil produced in the world early in 1978 led for the first time to a situation exactly opposite that of associated gas until now. The Emirate of Kuwait, which makes local use of the associated gas from its oilfields for generating electricity, for desalting sea water and for the manufacture of fertilizers, found it necessary to lower its price of crude oil in order to

TABLE 2

World Production of Natural Gas
Regional forecasts in billion m^3 per year

	1976	1980	1985	1990
North America	652	580- 660	580- 730	550- 800
Eastern Europe	375	490- 500	640- 680	800- 900
Western Europe	181	220- 230	210- 250	180- 250
Total ECE area	1208	1290-1390	1430-1660	1530-1950
Middle East	39	65- 75	100- 130	150- 200
Africa	12	25- 30	80- 100	130- 160
Far East	48	60- 70	80- 90	110- 130
Latin America	50	60- 70	80- 90	110- 130
Total outside ECE area	149	210- 245	340- 410	500- 620
World total	1357	1500-1635	1770-2070	2030-2570

maintain production at a level consistent with the country's need for associated gas. At this point it would be appropriate to speak of "associated oil" or "wasted oil" as one speaks of "wasted gas" or "flared gas".

The distance between gas supplies and consumption points is one of the chief reasons why the production of gas lags behind that of oil. Effectively, gas-transport problems are responsible for the fact that international trade in gas is still at a moderate level.

Although about 59% of the world's oil production enters into international trade (1705 MT out of a world production of 2926 MT in 1976), only 10% of the marketed production of gas was exported in 1976. The liquefaction chain, essential to the operational sea-going transport of gas, corresponds to only 12.5% of gas exports, or 1.25% of worldwide marketed production (see Fig. 5).

Even though the construction of liquefaction chains is expected to be considerable in the coming years, LNG is not likely to constitute more than 4 to 6% of gas production in 1985, i.e. 83 to 135 billion m^3 as against 17 billion m^3 in 1976. In 1990 the flow of LNG may attain 130 to 175 billion m^3, which would still represent only 6.4 to 6.8% of gas production at that date.

However, the volume of gas available for export may be limited by a factor other than local utilization, in the case of both associated and non-associated gas. This is the decision of a certain number of countries to reinject part of their gas resources into their oil formations. Thus in Iran, where about 25 billion m^3 is now being flared, a reinjection programme is aimed at a final goal of 130 billion m^3 per year. The intention is to maintain, by means of an initial phase of reinjection of 70 to 80 billion m^3 per year, a crude-oil production of 300 MT per year in 1983.

In some oil-bearing formations, an additional barrel of crude may be produced by injecting the equivalent of a barrel in the form of gas. The cost is of the order of 2 to 2.5 dollars, without counting the intrinsic value of the injected gas, the value of the oil being about $12.5.

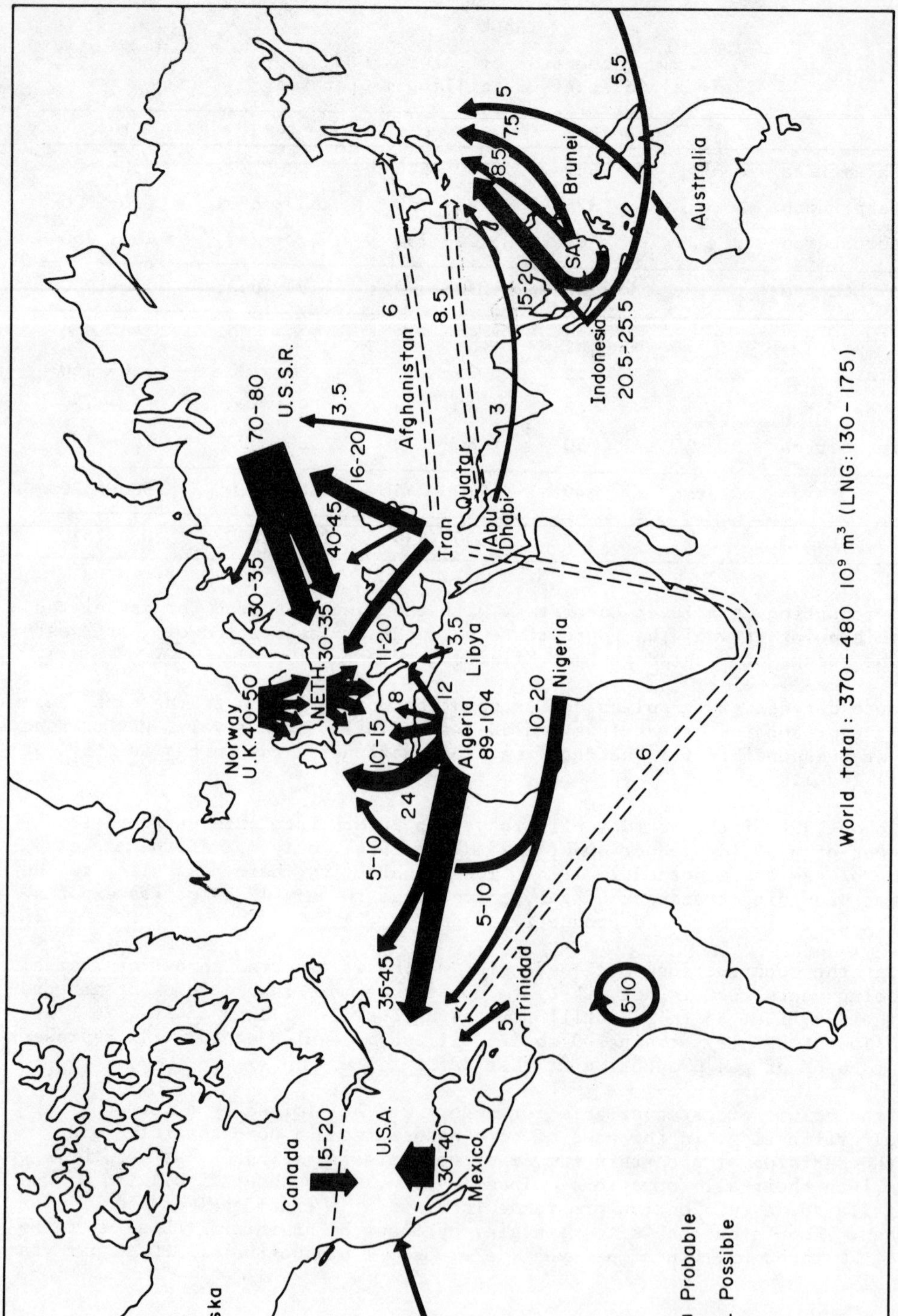

MAIN NATURAL GAS EXPORTS – 1990 (10^9 m^3)

Fig. 4

TABLE 3

1990 forecasts in billion m^3	Production	Imports	Total Available
North America	550- 800	80.5-121.5	630.5- 921.5
Eastern Europe	800- 900	18.5- 22.5	818.5- 922.5
Western Europe	180- 250	103.5-127.5	283.5- 377.5
Total ECE area	1530-1950	202.5-271.5	1732.5-2221.5

TRENDS IN THE INTERNATIONAL GAS TRADE

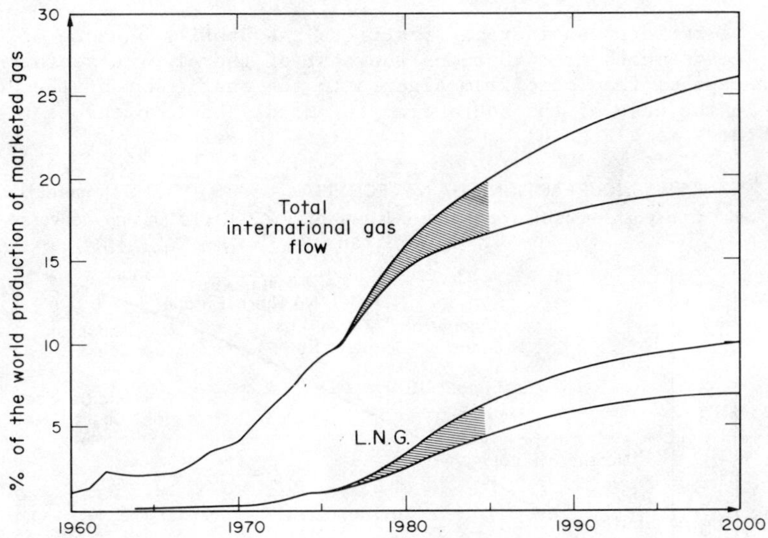

Fig. 5.

Thanks to the reinjection technique, the profit from gas may be raised to 3 to 7 times that obtainable using a liquefaction chain. In addition there remains the volume of injected gas recoverable at the end of the operation to a degree that cannot yet be easily estimated.

No general rule can be established as to the choice between transport to the consumer market and reinjection to aid and prolong oil production. But the latter does imply a limit to the volume of gas available for export.

Economic Obstacles

If the international gas trade is to be increased by a factor of 2.7 to 3.5 between 1976 and 1990, the increased transport facilities will require the construction of thousands of kilometres of gas pipelines, in the main leading to North America, Europe and Japan, and new liquefaction chains with a total transport capacity of 235 to 345 billion m^3 annually.

The construction of these new pipelines means overcoming not only political problems but serious technical difficulties, as in the case of pipes from Siberia to Western

Europe, from Alaska, the Beaufort Sea or the Canadian Arctic islands to North America markets, and seabed pipelines to collect North Sea gas or to transport Algerian gas to southern Europe. All this construction will require enormous investments, because, for a given transport of energy, a gas pipeline costs twice as much to build as an oil pipeline.

As to the installation of new liquefaction chains, not only is the construction time long, but so is the time taken to obtain preliminary authorization, the technology being still recent.

Such time lags, added to continuing rises in costs, may well render some Middle East projects economically unfeasible.

During the past 2 years, transport costs on the Middle East-United States route have risen by 43%, whereas liquefaction costs have increased by 40%.

The distance to markets considerably restricts gas profits because of the increase in transport costs which are 4.3 times the cost of the Algeria-southern Europe route in the case of the route from Algeria to the east coast of the United States, and 7 times in the case of the route from the Middle East to the east coast of the United States (see Fig. 6).

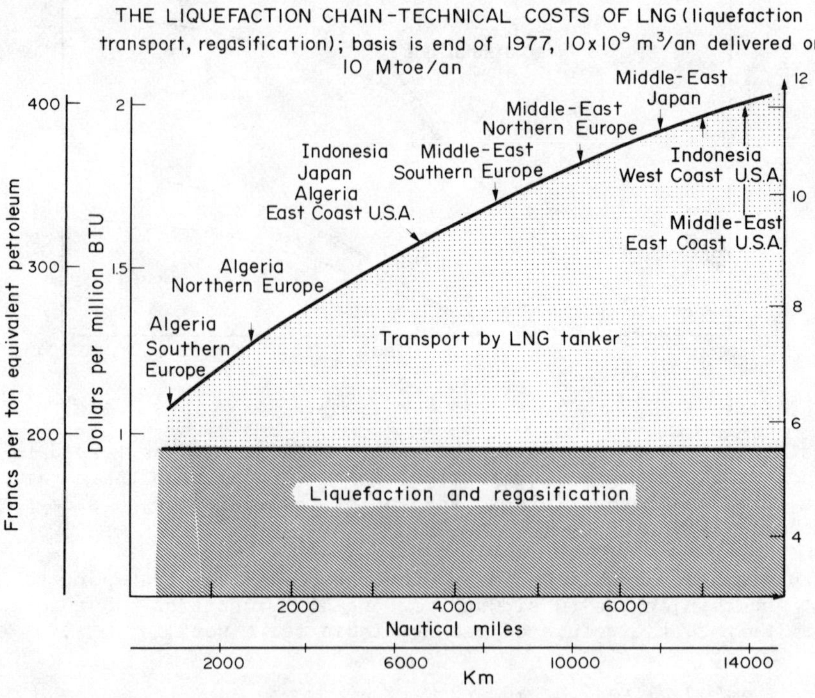

Fig. 6.

The duration of international gas contracts, which may be 15 to 25 years, is evidence of the lack of flexibility in large-scale gas transport. Contracts involve investments such that installations must exceed a certain critical size and generally that the project must be internationalized, as concerns both the financial structure and the actual construction.

OUTLOOK FOR 1990 IN SEVERAL COUNTRIES

The development of the contribution of natural gas to the energy supply of the industrial countries is shown in Fig. 7.

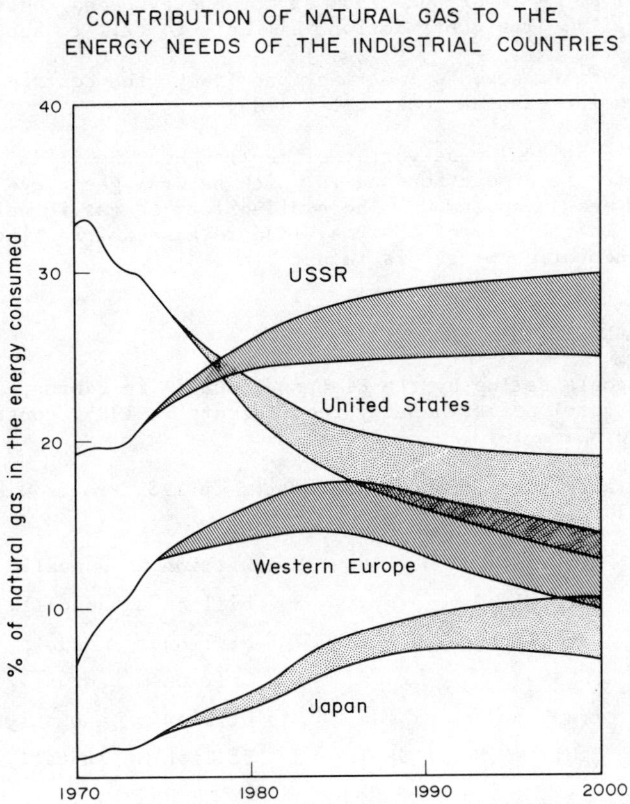

Fig. 7.

North America. The growth of production from Arctic fields will probably not make up for falling production in the southern United States until 1985 or 1990. A production of 450 billion m^3 in the United States and 150 billion m^3 in Canada can be expected for 1990. Consequently the contribution of natural gas to the energy supply of the United States should fall off gradually, to reach 20% in 1990, allowing for pipeline imports from Canada and Mexico and LNG imports from Algeria, Nigeria, Iran and Indonesia.

In Canada the contribution of natural gas should be able to level off at about 20% from 1976 to 1990 thanks to production from the arctic islands and the Beaufort Sea.

Eastern Europe. The production of natural gas in the USSR should continue to grow at the rate of 5 or 6% per year, so that in 1990 the USSR will be one of the world's biggest producers with about 40% of the world total. Concurrently with this evolution, gas should come to play a more important role in the country's energy balance, leading to the opposite tendency of that of the United States, with 30% in 1990 in place of the present value of 22%.

Western Europe. Unless rich new discoveries are made, the production of gas in Western Europe is expected to decline from 1985 onwards, which means that the contribution of gas to the energy needs of this area will reach a maximum of about 18% from 1985 to 1990, then fall off to about 15% which is the present level. The decline of gas should be much more marked in some countries such as the Netherlands where it is now of prime importance (55% of the energy needs) because of the quantity produced locally. Its contribution is likely to fall to about 30% in 1990.

On the other hand, in France, West Germany and Italy, the contribution of gas should continue to increase at least until 1985.

Japan is a very exceptional case worthy of mention because it is a country having practically no domestic production and in which natural gas plays a very minor role, 2.5% of the energy consumed. The availability of gas in neighbouring countries, exportable in the form of LNG, may lead to gas contributing 7% in 1990. Here again the penetration of gas is rather slow.

NATURAL GAS IN FRANCE

According to the goals set up by the Planning Council in February 1975, gas should meet about 16% of total primary energy requirements in 1985, contrasted with 10.8% in 1976 and only 8.5% in 1973.

According to contracts signed by Gaz de France, in 1985 France will have the following resources available:

Algeria	9.1	billion m^3 annually
Netherlands	8	billion m^3 annually
Lacq (France)	7	billion m^3 annually
USSR	4	billion m^3 annually
Iran	3.5	billion m^3 annually
Ekofisk (North Sea)	3	billion m^3 annually

in all 34.6 billion m^3 contract as against requirements of 37 billion m^3 per annum.

In 1990 the French requirements may reach 42 to 44 billion m^3. But production at Lacq will be declining, as will be imports from the Netherlands. The lack may be made up by gas available from the Norwegian sector of the North Sea and by LNG imports from Nigeria, for example. Mention must also be made of the Segamo project (Sonatrach 50%, Enagas 25% and GDF 25%) for a seabed gas pipeline from Algeria to transport 15 billion m^3 per year in the first instance, of which 10 billion m^3 will be for Gaz de France.

Even though French natural-gas resources are larger than oil resources, in 1976 France was only in fourteenth place among the world's producers; none the less she occupies one of the top places in gas technology as concerns gas liquefaction, construction of LNG tankers, installation of gas pipelines (particularly undersea pipes) and underground storage.

THE FUTURE OF NATURAL GAS IN THE ECE AREA

Between now and the end of the century, the position of younger sister now occupied by natural gas with respect to oil is hardly likely to undergo any change on a world scale. Often more expensive to produce, always much more expensive to transport, requiring heavy investments for liquefaction and delivery by pipeline or tankers, natural gas, on the basis of the same price as oil for a given heating value, gives the producing country a profit margin of only about one-fifth to one-half that of oil, varying according to the distance to be covered to the consumer market. After the oil price rises of 1973/74, the equivalence between prices of oil and gas was destroyed, and the situation of gas relative to oil is today still rather worse. Such are the reasons behind the modest forecasts for the development of international trade in natural gas. It appears that in 1990 world trade in gas will still be 5 to 7 times less than that in oil.

Focusing, on the other hand, on the extensive ECE area, it is seen that its reserves and impressive resources, notably those of the USSR which constitute about 55% of the world's natural gas, give an initial advantage to this choice raw material. In addition, this region is highly industrialized. A close network of gas pipelines already exists and is waiting to be used. The demands of the anti-pollution effort are in favour of the cleanest of all fuels, particularly for household use. Looking at the future beyond 1990, it would be surprising if the tendency toward a strong demand and a relatively high price of natural gas did not become established in the ECE area. At a time when excessive consumption of oil points to a critical supply problem before the end of the century, the almost untouched reserves of natural gas are ready and waiting to take their place. It is hardly contestable that for many years to come the gas resources to the ECE area will remain one of its vital energy sources, more so even than today. And when the need comes, the great reserves lying close to its southern border — those of North Africa, of the Persian Gulf, probably those of Mexico — and the vast promises of the arctic, the Mackenzie Delta and the mouth of the Ob give reason to believe that these great riches of precious fossil energy will all be gradually and methodically exploited over a long period, perhaps in a more rational way than have been our oil resources.

BIBLIOGRAPHY

1. *The Gas Industry in the World*, November 1977, published by Technip.
2. *Quelques éléments de base concernant la situation du gaz naturel dans le monde en 1976*, April 1977, Cedigaz.
3. *The Future for the World Natural Gas Supply — A report prepared for the 10th World Energy Conference on Natural Gas Resources — 1985 to 2020*, American Gas Association, August 1977.
4. Periodical Surveys by Cedigaz:
   The Development of the Soviet Gas Potential (January 1977).
   Natural Gas in France (November 1975 and July 1976).
   Dissolved Associated Gas in the Near East (January 1976).
   Iran (January 1975).
   The Netherlands (October 1974).
   Algeria (June 1974).
   Indonesia, Malaysia, Brunei (February 1974).
   Great Britain (November 1973).
   Australia (September 1973).
   Italy (May 1973).
   Nigeria (February 1973).
   West Germany (January 1973).
   Venezuela (November 1972).

More Energy from LNG

B. Macchini

Italy

INTRODUCTION

In industrialized countries the energy-consumption patterns show a trend, currently and in the forecasts for the next 10 years, towards increasing demand for natural gas as a primary energy source. This trend is consistent with the diversification of energy supply as to both source and type of energy. For example, the EEC policies, aimed at reducing dependence on crude oil from 57% down to 49-41%, also foresee an increase in natural-gas requirements, which would be supplied partly from domestic sources and partly by imported LNG. It is expected that in 1985 natural-gas supplies will meet 18-23% of EEC total energy consumption. The present amount is only 15%. Similar criteria are used in the energy-supply programmes of countries with large energy requirements and insufficient domestic availability, such as Belgium, Japan, Italy, etc.

In the light of these trends there is a lively interest in the LNG market, as well as a wealth of programmes for medium- and long-term production.

LNG MARKET

In 1975 LNG world production was equivalent to 12.5 million tons/year, corresponding to about 15 000 million Nm^3/year of natural gas (1700 MM SCFD).

According to specialized sources, it is expected that in 1980 LNG production will grow to 40 million tons/year and will reach a figure of about 130 million tons/year by 1985.

This is equivalent to supplying the world market extra quantities of natural gas amounting to 50 000 million Nm^3/year (5600 MM SCFD) and 170 000 million Nm^3/year (19 000 MM SCFD) in 1980 and 1985 respectively. Consequently, regasification terminal capacities are increasing.

Global capacities are growing from the present 1000-4000 million Nm^3/year (110-450 MM SCFD) to values in the order of 8000-10 000 million Nm^3/year (900-1100 MM SCFD).

The increased installed capacity requires regasification modules of increased larger individual capacity. The present traditional modules have a regasification capacity of 1000-2000 million Nm^3/year (110-220 MM SCFD).

SP together with Brown Boveri have now designed modules having capacities ranging from 2000 to 4000 million Nm^3/y (220-440 MM SCFD) at 8000 operating hours/year. These are based on the principle of energy recovery with simultaneous production of electric power and regasified LNG.

MORE ENERGY FROM LNG

The cost of buying LNG on the international market today is strictly linked to the cost of crude and consequently of fuel oil.

The recent instability of the energy market has also involved the LNG market. In the past few years the purchase price of LNG has risen from 0.49 $/million Btu to the present 1.5-2 $/million Btu.

Consequently the enormous price increase of this primary energy source makes it necessary to recover the cold energy contained in it, by means of technological operations with a much higher thermodynamic efficiency than that obtained in traditional plants.

Among the different possibilities of utilization of the "cold energy" contained in LNG, SP and BBC have chosen to recover it by producing electric power at lower cost and higher yield than is possible in large thermal power stations.

Moreover, the electric-power production allows the complete utilization of the cold energy available at the terminal, as well as an easy means to transport it to utilization areas.

TYPES OF REGASIFICATION UNITS

The types of regasification units are subdivided into two main categories:

processes which degrade energy,
processes which recover energy.

Commonly Used Processes

The most commonly used processes which result in energy degradation are:

vaporization by sea water,
vaporization by submerged combustion.

We describe hereunder the main characteristics of each process and their most significant advantages and disadvantages.

Vaporization by sea water

<u>Advantages</u>

Operating continuity and safety,
low investment if site is favourable.

<u>Disadvantages</u>

completely degrades the LNG cold energy,
high investments for sea water discharge and intake systems,
considerable consumption of pumping energy,

consumption of fuel for heating the sea water in the periods when its temperature falls below 6-10°C (42-50°F),
thermal pollution of waterways,
chemical pollution of waterways.

Vaporization by submerged combustion

Advantages

low installation costs

Disadvantages

completely degrades LNG cold energy,
high operating consumptions: 1.7-2% of gas produced is used as fuel.

Among the energy-recovery processes there are the ones licensed by SP/BBC which regasify LNG with simultaneous production of electric power.

SP/BBC Processes

Figure 1 illustrates the Snamprogetti/Brown Boveri process no. 1.

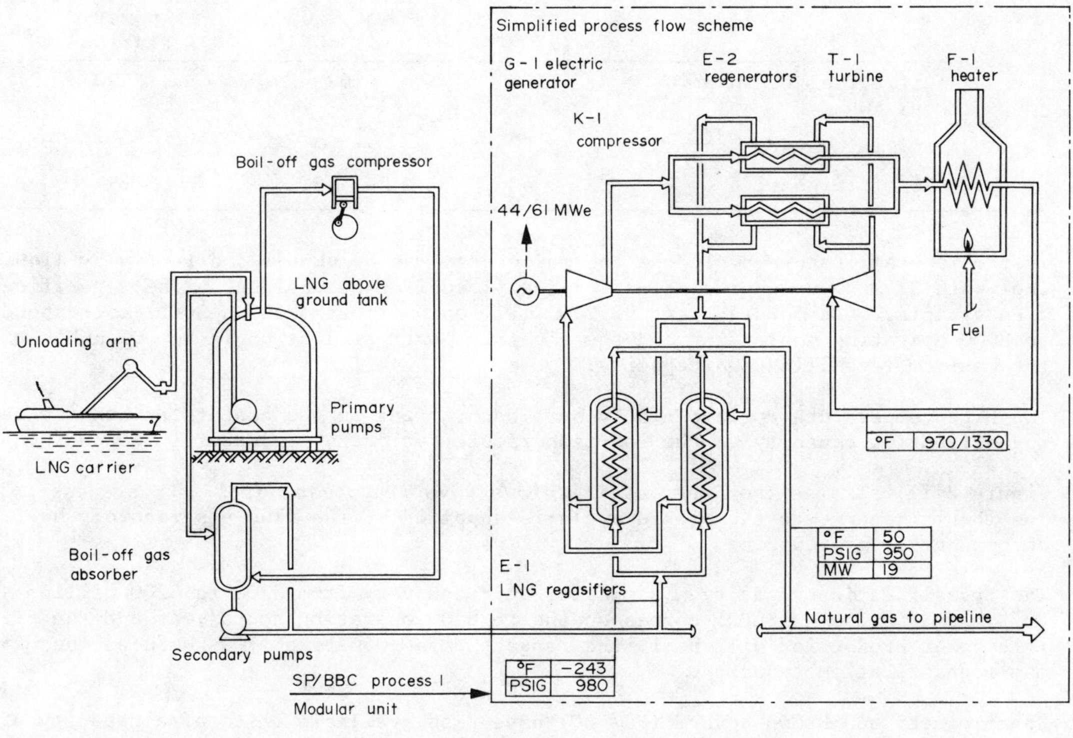

Fig. 1. Electric energy from LNG regasification SP/BBC process 1.

The principal advantage of the regasification of LNG using nitrogen in a closed cycle is the complete recovery of the thermal energy available. Energy recovery at low temperatures allows compression at cryogenic temperatures, resulting in a significant reduction in the compression power required and thus improving the efficiency of the thermodynamic cycle. The nitrogen, after compression, passes through a heat-recovery unit then through a natural-gas-fired heater which it leaves at 720°C (1330°F).

The gas then enters an expansion turbine, which is directly coupled to a compressor and an alternator. The alternator generates power required by the process and also for export.

The efficiency of the cycle is 53% and the specific fuel consumption is 1620 kcal per kWh produced. A module using the SP/BBC thermodynamic cycle has been designed with a regasification capacity of 2000 million Nm^3/y (220 MM SCFD) with a corresponding electrical power generation of 61 000 kWh/h. A plant scheme is based on standard modules in order to provide flexible operation of regasification terminals.

The module size is also determined by the present availability of proven equipment technology. The diagram of process no. 1 is the same for cycle temperatures of 720°C and 520°C (1330°F and 970°F) but the cycle characteristics will vary according to Table 1.

TABLE 1

	Cycle temp. at 520°C (970°F)	Cycle temp. at 720°C (1330°F)	Traditional power station
Specific fuel consumption, kcal/kWh	1950	1620	2200
Saving in energy, kcal/hWh	250	580	—
Cycle efficiency, %	44	53	39

In the low-temperature cycle the thermal energy can be supplied directly by liquid fuel with 3% max. sulphur content, by fossil fuel or natural gas. The regasification capacity of a modular unit is 2000 million Nm^3/year (220 MM SCFD) corresponding to 8000 operating hours/year and the electric power production is 44 000 kWh/h at the maximum regasification capacity.

The high-temperature cycle can only burn natural gas in the heater; it has the same regasification capacity as the low-temperature cycle and produces 61 000 kWh/h.

Figure 2 illustrates the Snamprogetti/Brown Boveri process no. 2. In process no. 2 the thermal energy to the closed cycle is supplied by the flue gas recovery heat of an open-cycle gas turbine.

The regasification capacity of a modular unit can vary from 2200 to 3200 million Nm^3/year (240-350 MM SCFD) corresponding to 8000 operating hours/year and the electric power production will be in the range from 56 000 to 86 700 kWh/h at the maximum regasification capacity.

Snamprogetti and Brown Boveri (SP/BBC) have also available units of a capacity of 2800 or 4000 million Nm^3/y (305 or 440 MM SCFD) with a production of 69 000 or 120 000 kWh/h of electricity when operating at maximum capacity. The main characteristics of processes no. 2 are shown in Table 2 compared with a traditional power station.

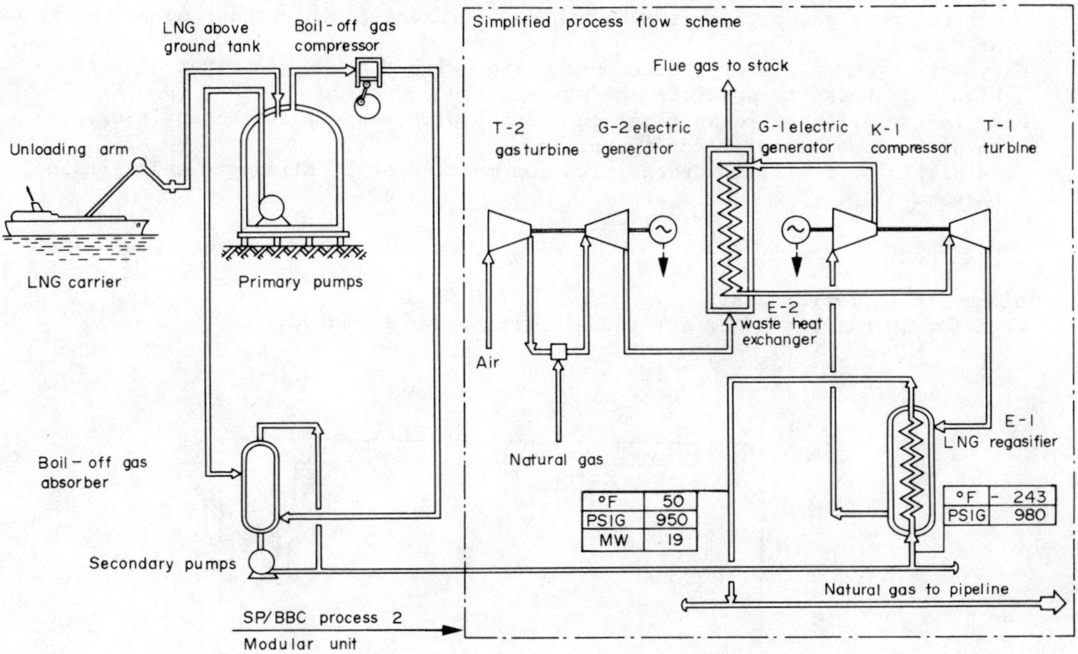

Fig. 2. Electric energy from LNG regasification SP/BBC process 2.

TABLE 2

	SP/BBC - processes 2				Traditional power station
Regasification capacity, million Nm^3/y	2200	2800	3200	4000	-
Specific fuel consumption, kcal/kWh	1755	1830	1790	1685	2200
Saving in energy, kcal/kWh	445	370	410	515	-
Cycle efficiency, %	49	47	48	51	39
Electricity generation, kWh/h	56 000	69 000	86 700	120 000	-

The most significant advantages and disadvantages for this new type of units are the following:

Advantages

total recovery of LNG cold energy,
high operating safety and continuity,
no environmental pollution if fuel gas is burned,
possibility of total recovery of the heat to be dissipated in closed cycle and consequently "total energy" operation,

adaptability of the cycle in Fig. 1 to any type of fossil fuel for thermal power stations,
electric power production at much higher efficiencies than those obtained in power stations,
easy variation of electric power production with the LNG vaporization rate. Figure 3 shows the possible operation range for cycle at 520°C (970°F),
easy installation of regasification modules with respect to the environmental conditions and coastal configuration,
possibility of installing regasification modules on floating and self-sinking barges.

Disadvantages

higher installation cost,
need for an external user for the electric power produced.

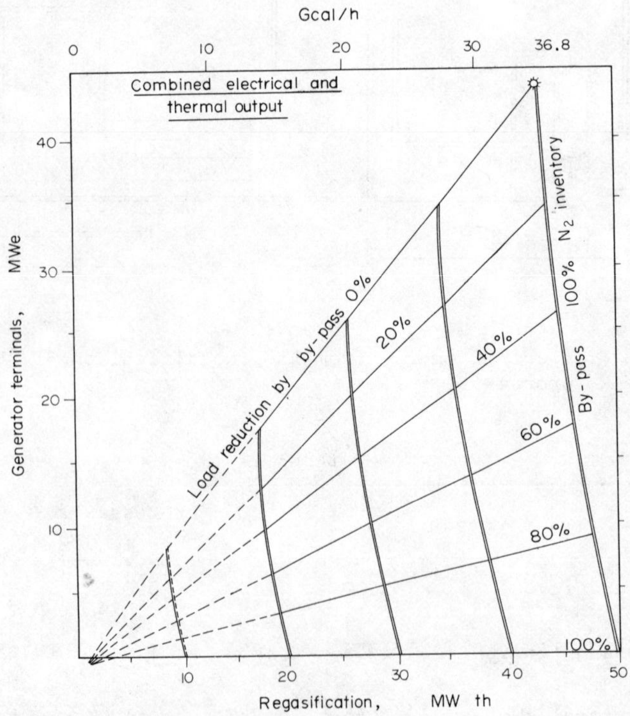

Fig. 3.

PROCESS ECONOMICS

The economics of the processes has been studied in detail and we report hereunder the considerations made for the high-temperature process 1. The economic study has been carried out on the following bases:

part of the capacity of a traditional LNG regasification terminal is replaced with an SP/BBC high-temperature 2000 million Nm^3/y (220 MM SCFD);

a comparison has been made with the traditional systems of sea water and submerged combustion;

the additional investment has been obtained by deducting from the investment for one SP/BBC unit the investment for a medium-capacity traditional plant minus the investment for a medium capacity traditional plant after reduction of its capacity of 2000 million Nm^3/y (220 MM SCFD).

The cost of LNG storage and general facilities have not been taken into consideration, as these are the same for both traditional and SP/BBC process plants. The results of the study are shown in Figs. 4, 5, 6 and 7 in which the various indicators of profitability (annual savings, pay-out time, DCF rate of return) are expressed as functions of the selling price of electricity.

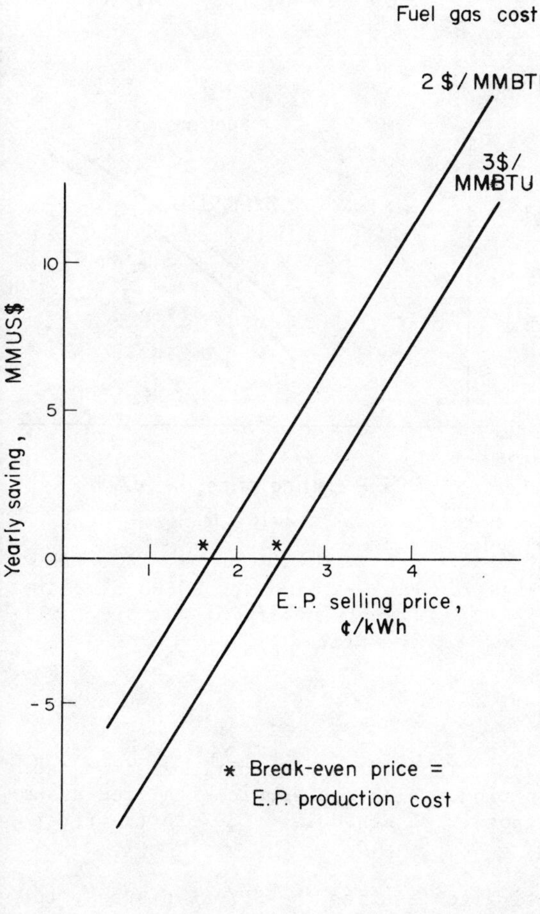

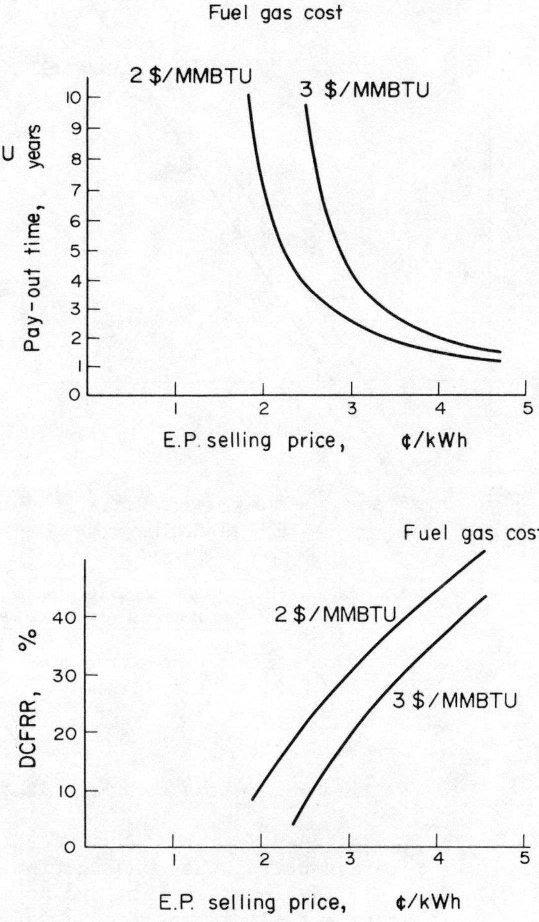

LNG REGASIFICATION SP/BBC PROCESS 1 AT 720°C CAP. 2 x 10^9 Nm^3/y

Economics vs. Sea water

Fig. 4. LNG regasification yearly saving vs. E.P. selling price.

Fig. 5. Pay-out time and DCFRR of additional vs. E.P. selling price.

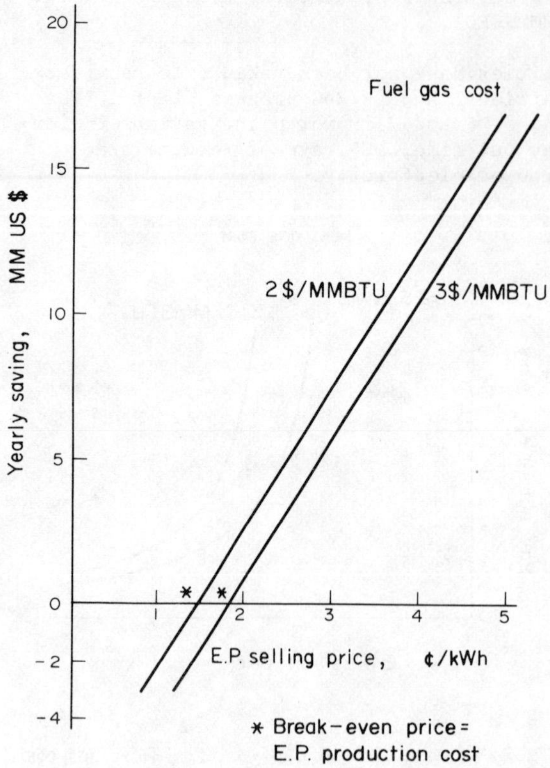

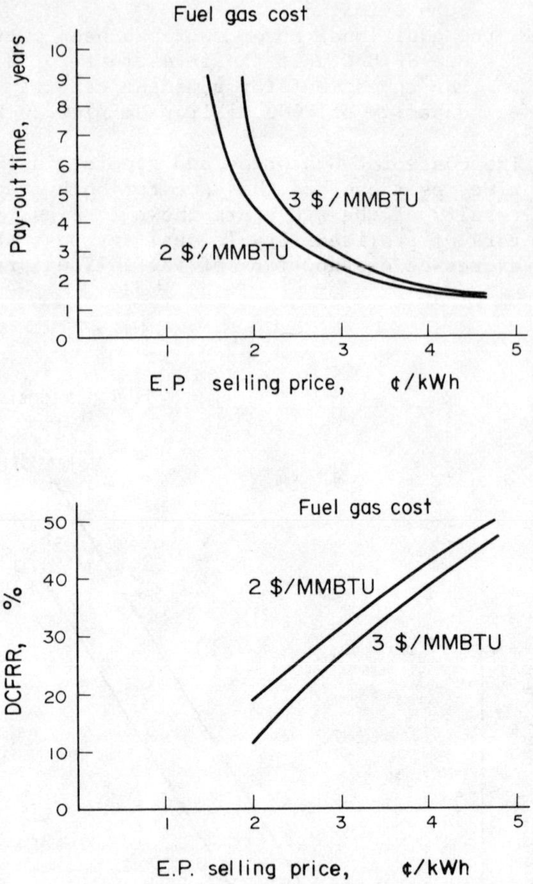

LNG REGASIFICATION SP/BBC PROCESS 1 AT 720°C CAP. $2x10^9$ Nm^3/y

Economics vs. Submerged combustion

Fig. 6. LNG regasification yearly saving vs. E.P. selling price.

Fig. 7. Pay-out time and DCFRR of additional investment vs. E.P. selling price.

Data Used in the Economic Study

The main figures used in the study are shown in Table 3. The additional investment includes interest during the construction period, which has been capitalized assuming 100% loan at 8% interest rate and an expenditure schedule of 30% in the first and third years and 40% in the second year.

In the calculation of the additional operating costs using the SP/BBC process, the following assumptions have been made:

Labour, 12 men at an average yearly cost of 20 000 \$/man.
Utilities, estimated on the basis of 8000 operating hours per year.

TABLE 3

	Versus	Sea water	and Submerged combustion
Additional investment for SP/BBC process	MM US $	21	25.8
Additional operating cost for SP/BBC process:			
Fuel gas cost at 2 $/MM Btu	MM US $	9	7
Fuel gas cost at 3 $/MM Btu	MM US $	12	9
Available electric power	MM kWh/y	492	485

Depreciation, straight line during a period of 20 years.
Maintenance 2% of capital cost.
Overhead and interests on working capital have not been considered as the SP/BBC unit does not involve any increase in these costs.
Financial charges, a 100% loan at 8% interest rate over a period of 20 years.

The electricity available for sale in using SP/BBC unit has been assumed equal to the total E.P. produced, less the amount consumed within the unit, plus the power saved in respect of a traditional unit. With this assumption, the credit, allowed for pumping power saved, is less than the actual saving when the selling price of electricity is lower than the market price.

Figures 4 and 6 show the yearly savings (in million US $) that can be obtained using the SP/BBC process instead of sea water and submerged combustion systems respectively. To arrive at the yearly saving, the additional operating costs for the SP/BBC unit, including the fuel gas costs, have been subtracted from the revenue obtained from sales of electricity.

Savings are shown against electricity selling price and for two fuel gas costs. Break-even price corresponds to the production cost of electricity.

Figures 5 and 7 show the variation of the DCFRR (discounted cash flow rate of return) and the pay-out time, beginning from the startup of the plant, as a function of the selling price of electricity. To obtain the DCFRR and the pay-out time of the additional investment, the capital cost has been compared with the cash flow (revenues minus cash outflow) obtained using the SP/BBC unit.

In particular, if the high-temperature SP/BBC unit replaces part of the capacity of sea water process, assuming the electricity selling price equal to 3 ¢/kWh and a fuel gas cost of 2 $/MM Btu the following results are obtained:

Yearly saving	MM US $	5.5
Pay-out time	years	2.5
DCFRR	%	30

The DCFRR values may be compared with the interest on capital assumed (8%). With regard the submerged combustion system, the profitability figures are even more attractive.

The same calculation for process no. 2 of 3200 million Nm^3/y (350 MM SCFD) regasification capacity has given the following result:

Yearly saving	MM US $	8.2
Pay-out time	years	1.8
DCFRR	%	39

With regard to the submerged combustion system, the profitability figures are more attractive. The results of the study are shown in Figs. 8, 9, 10 and 11.

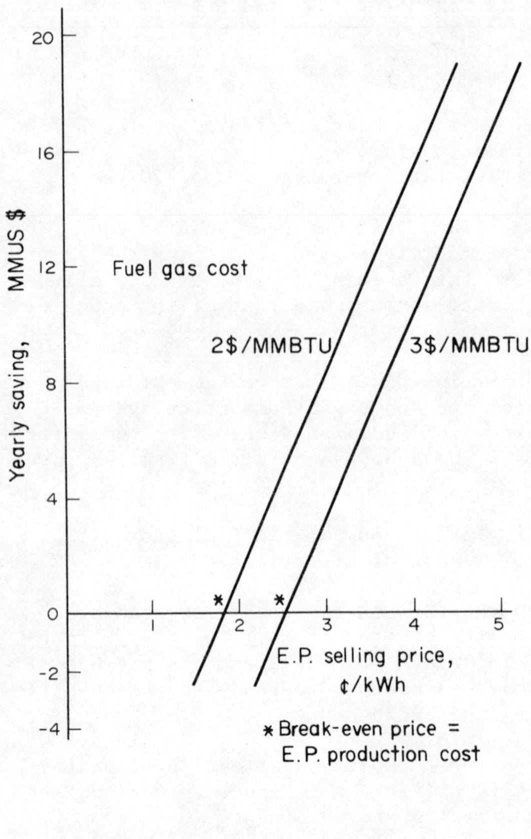

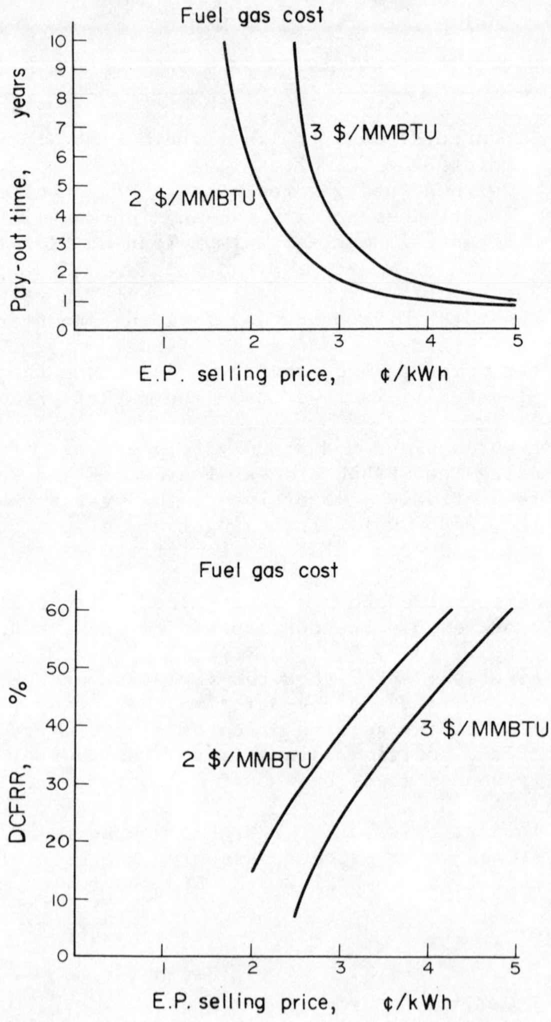

LNG REGASIFICATION SP/BBC PROCESS 2 CAP. 3.2 BILLION Nm^3/y

Economics vs. Sea water

Fig. 8. LNG regasification yearly saving vs. E.P. selling price.

Fig. 9. Pay-out time and DCFRR of additional investment vs. E.P. selling price.

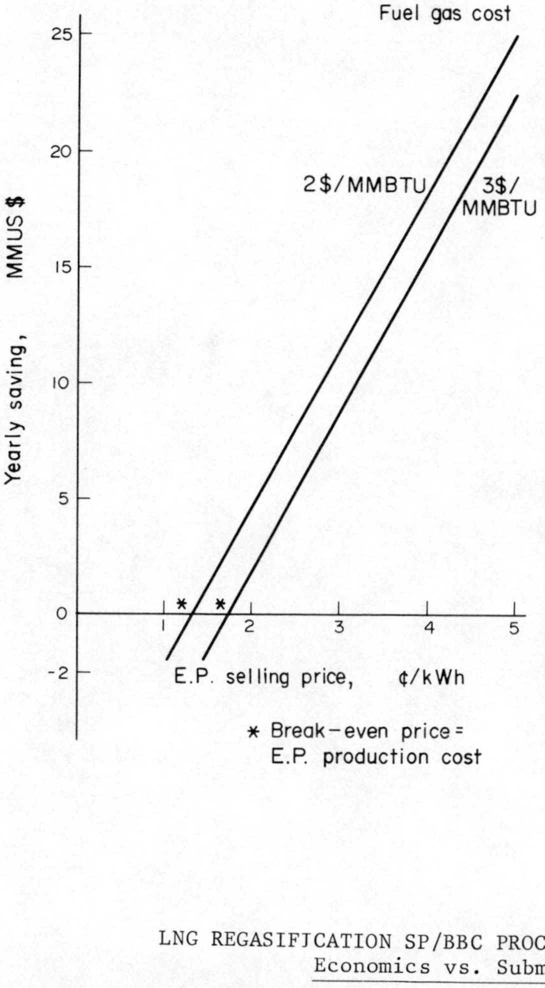

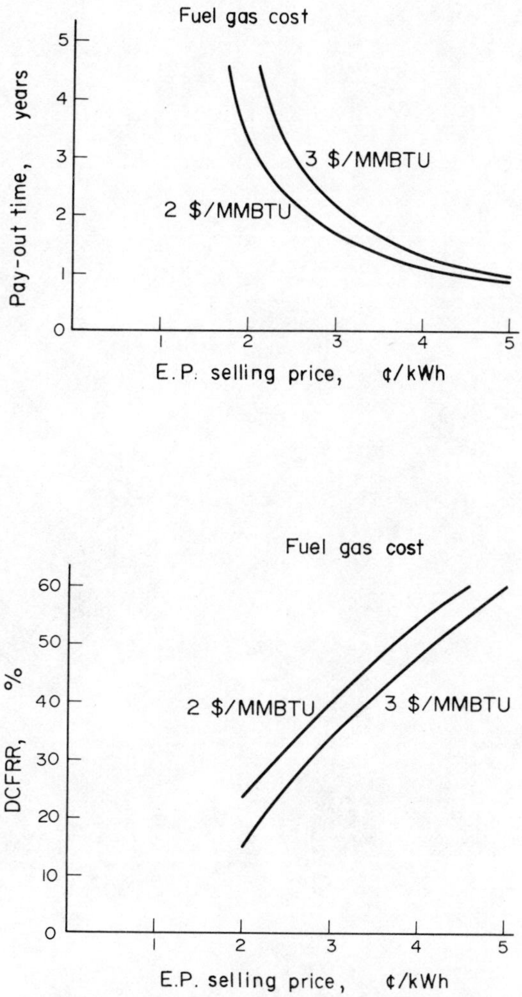

LNG REGASIFICATION SP/BBC PROCESS 2 CAP. 3.2 BILLION Nm^3/y

Economics vs. Submerged combustion

Fig. 10. LNG regasification yearly saving vs. E.P. selling price.

Fig. 11. Pay-out time and DCFRR of additional investment vs. E.P. selling price.

Chances for Alternative Fuels in the Gas Industry

G. Bonfiglioli, G. Bonvecchiato and C. Mulazzi

Italy

INTRODUCTION

Based on the Italian natural gas picture, this paper purposes to present an outlook on possible new contributions to be added to the existing availability in order to meet market demand up to 1990. Available gas includes both natural gas produced by domestic fields already known and imports either operative or under way. This paper considers contributions from new domestic findings and possible new imports and also investigates the possibility of having recourse to SNG and other alternative sources. Such a recourse would be on the basis of an economic comparison between the various sources whether at present costs or, on the assumption that, as regards energy, the rate of price escalation will be higher than the rate of general inflation.

The vista of 1990 as a reference point has remarkable significance for the European gas industry, since it stands on the borderline between the long-term planning and the broader perspectives.

As a matter of fact, it is relatively easy to forecast with a certain degree of approximation the availability of the known European gas fields though, sometimes, they give rise to significant problems of a technical-economical-political-financial nature (North Sea or deep structures). One can in the same way forecast the availability of imports, both operative or under way, whereas a considerable degree of uncertainty remains as far as other availability is concerned.

This paper attempts to answer the following questions: What will be the contribution from new local natural gasfields that will concern the second half of the 1980s? What will be the contribution of additional imports from extra European countries? What will be the impact of other forms of energy, gaseous or otherwise?

While this report is focused on the Italian case, it may show some analogy to the case of other European countries.

THE ITALIAN CASE

Natural gas has attained a penetration of 16% in the Italian energy market, thanks to the contribution of imports that have added to the national production. On the whole, more than 26 billion cubic metres of natural gas were consumed in Italy in 1977, 49% of which was imported.

The following imports are operative at present:

from Libya (LNG)	3 billion m^3/year,
from the USSR	7 billion m^3/year, under full rate conditions (5.5 in 1977),
from Holland	6 billion m^3/year.

In addition, the preliminary works have already started for importing 12 billion m^3/y from Algeria through the Transmediterranean gas pipeline, that will reach depths of about 600 metres, by applying the technique set up by the ENI Group Companies during the last years (Fig. 1).

THE CROSSING OF SICILY CHANNEL
TO IMPORT ALGERIAN GAS TO ITALY

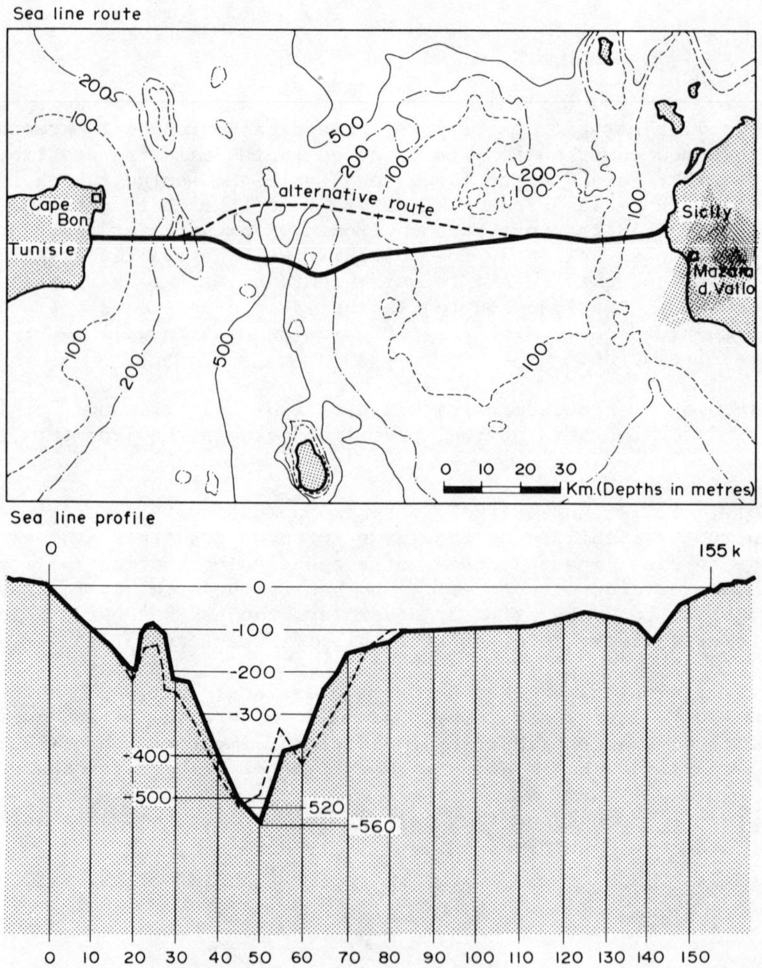

Fig. 1.

The volumes available through imports covered by contracts will be increasing up to 28 billion m^3/year; this volume will be attained in 1984. On the contrary, the domestic production from known natural gasfields will be decreasing as it can be noticed from Fig. 2 which illustrates the acquired availability up to 1990.

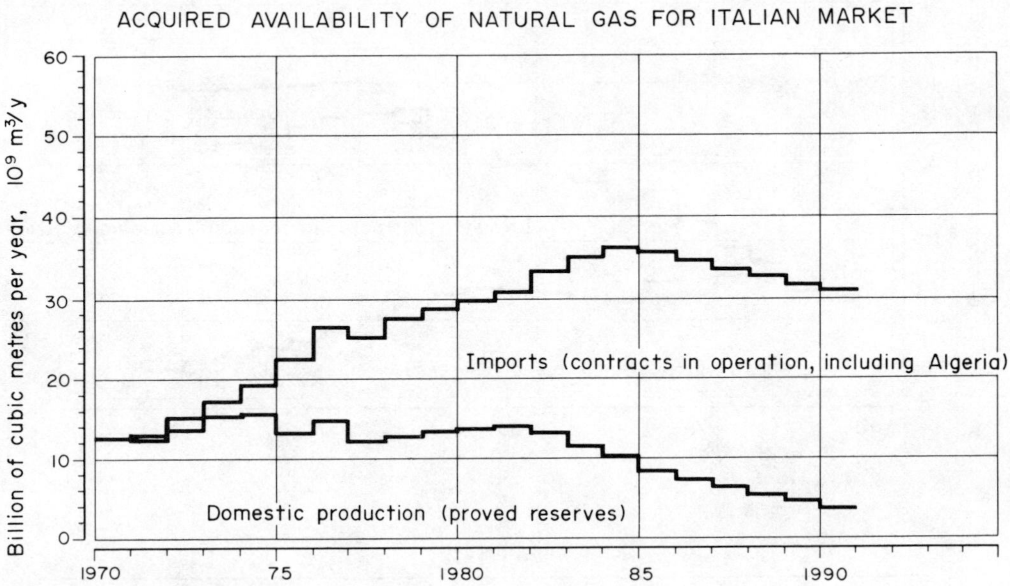

Fig. 2.

On the other hand, the overall demand of energy (140 million toe in 1977) will develop as a function of the gross national product (GNP), once the present economical crisis is overcome. The coefficient of elasticity between energy consumption and GNP will depend upon the actions that will be taken in the field of the conservation of the energy. Such actions are, in their turn, a function of the price of energy in real terms.

The Energy National Plan envisages an average rate of development for energy consumptions within 4 + 5%/year, up to 1985. Even if more cautionary evaluations ranging between 3 + 4%/year are assumed, it seems quite reasonable to forecast a significant demand for other natural gas, in addition to the amount assured through the known domestic sources and from the imports already acquired by contracts. Relying on a possible contribution from new domestic findings in increasingly deeper sea and in new on-shore strata, it can be assumed that the overall domestic production will remain at the present level even in the future. In this case, the overall availability shown in Fig. 3 and in Table 1 will be reached.

The natural gas supply system for the Italian market in the middle 1980s is shown in Fig. 4.

On the other hand, it is believed that gas can be in a position to suitably cover, up to 1990, a share of 16 + 19% of the national requirement of energy.

In order to cover the deficiency between demand and availability of natural gas (NG), recourse could be had to new imports or, if convenient, to alternative sources.

It appears that it will be possible to import from North Africa and the USSR and later on, from the Middle East and from West Africa.

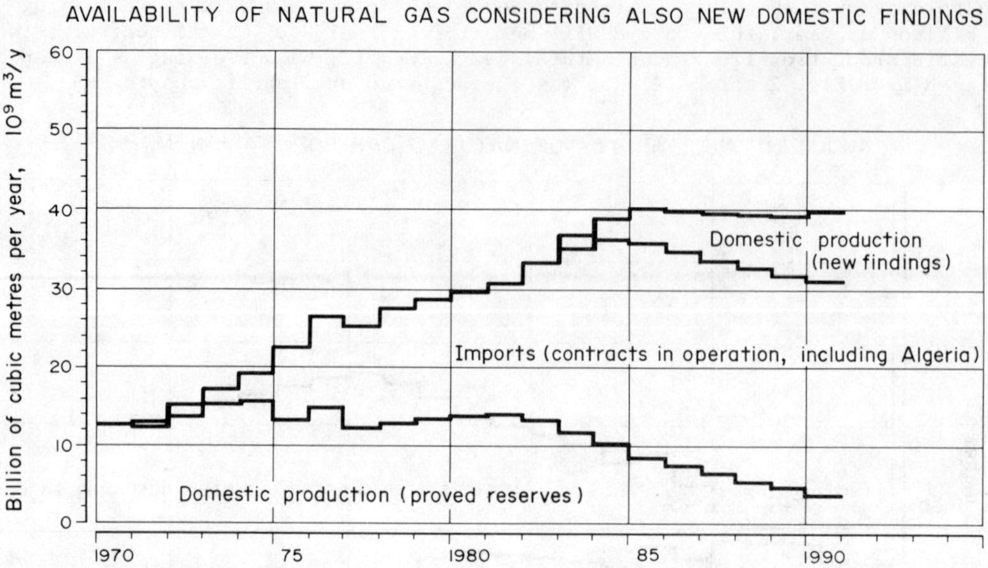

Fig. 3.

TABLE 1

Forecast of availability (billion m^3/y)	1977	1980	1985	1990
National production	13.0	14	8	4
Imports: Libya	2.6	3	3	3
USSR	5.5	7	7	7
Holland	4.9	6	6	6
Algeria	–	–	12	12
Total of certain availability	26.0	30	36	32
New findings	–	–	4	8
Grand total	26.0	30	40	40

The aforegoing resulted from an investigation of the known availability of the areas from which the supply of natural gas for Italy is possible. Due account was taken, in the investigation, of the significance of long-distance transport costs for natural gas, either transported via gas pipeline or as LNG. The demand of both producing and importing countries was considered as well.

THE NEW SOURCES OF ENERGY

The possibility for new sources of energy having a footing in the energy market depends upon the cost, the availability, in due time, of adequate quantities and — as regards the types of energy of interest to the gas industry — upon their capability to replace NG. The elements featuring their capability to replace natural gas are shown in Table 2. The elements relating to cost can be estimated through

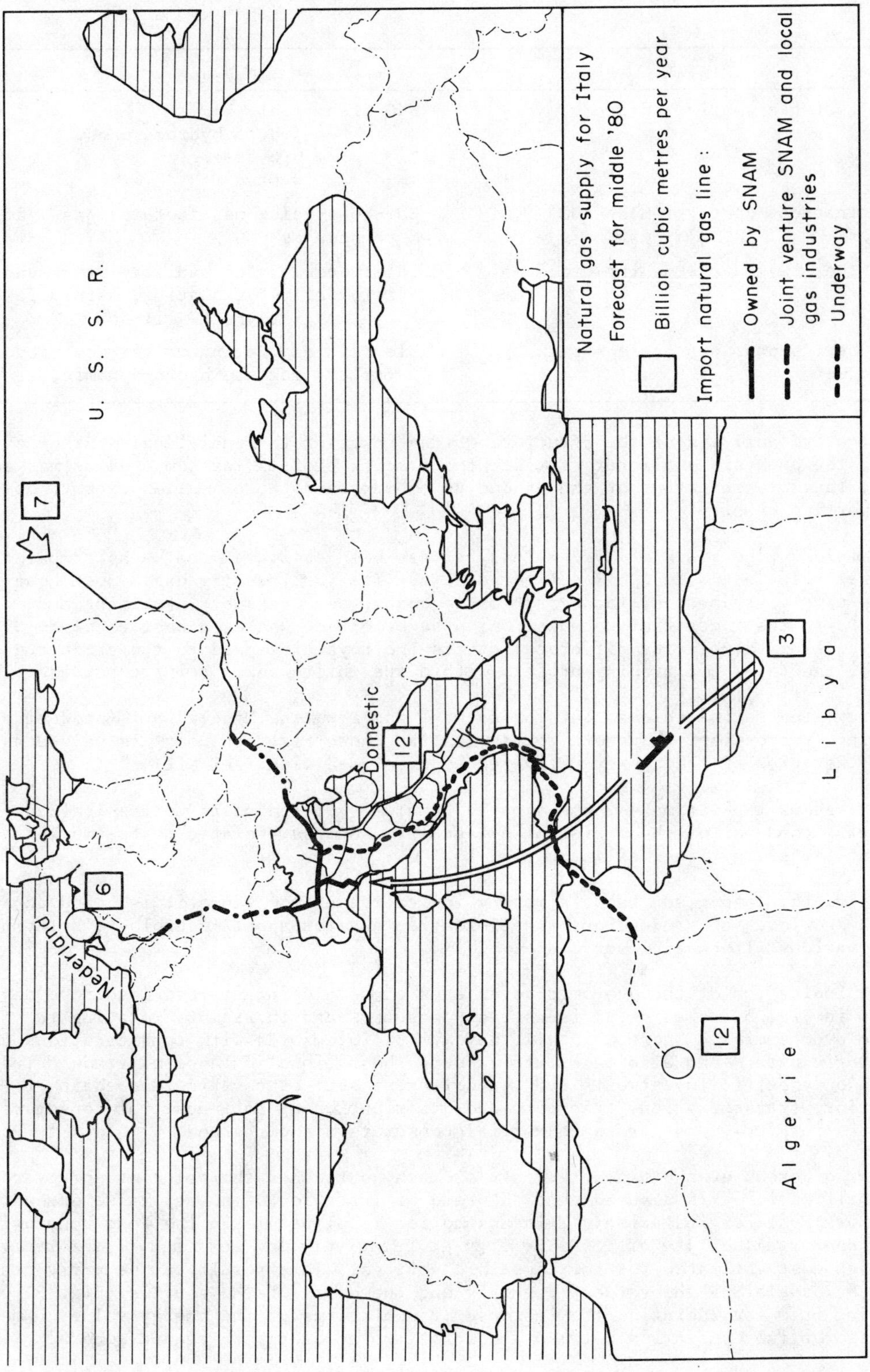
U. S. S. R.
7
Nederland
6
Domestic
12
Libya
3
Algerie
12
Natural gas supply for Italy
Forecast for middle '80
Billion cubic metres per year
Import natural gas line :
Owned by SNAM
Joint venture SNAM and local gas industries
Underway

Fig. 4.

TABLE 2

Class	Energy
Energy that can wholly replace NG	SNG from: coal liquid hydrocarbons, bio-masses, wastes
Energy that can partly replace NG	Low-heat-value manufactured gas, same sources as SNG
Energy that may have some influence on NG markets	Solar energy, for sanitary water and eventually for heating, mainly for systems integrated with NG
Energy having poor connections with NG market	Electric energy, no matter how produced (solar, magneto-hydro-dynamic, etc.)

an indicative analysis of the structure of the costs of the individual sources of energy; the analysis shall not consider the trading profits, as compared to market values, that are in favour of the producing country and, to a smaller extent, of the importing country.

For example, on the basis of 1977 values, it has been estimated that – as compared to market value – the fuel oil delivered to the Italian industry has a cost equal to 30% (namely production, transport to the coast, sea transport, refining and transport to the industrial consumer which have, on the whole, a cost equal to 30% of the selling value); the difference covers the royalties paid to the producing country, the taxes and trading profits within the entire chain being considered.

Similarly, the estimated cost for the natural gas from the Middle East (production, gathering, processing, transport to the Italian industrial consumer) is valued as 85% of the value of the energy delivered (equivalence with fuel oil).

Figure 5 shows the estimates, based on 1977 values, of the costs to the Italian industrial consumer of the different alternative sources, related to the market value of the energy (fuel oil) equalled to 100.

Obviously, the comparison is of a merely indicative nature since it is based on present knowledge of production, transformation and transport technology, relating to the various alternative sources.

A first indication of the economic advantages of developing different sources may be seen in Fig. 5, relating to industrial consumers and in Fig. 6, relating to household consumers. These data will have to be integrated with considerations of a strategic nature and also take into account the national balance of trade. Indications on specific investments for the different sources covering the chain of production, transformation, transport and distribution up to industrial consumers are given in Fig. 7, as far as industrial consumers are concerned.

The year of break-even between the cost of each source and the value of energy can be identified if it is assumed that the rate of inflation of energy value exceeds the general rate of inflation. This assumption is plausible in the long run, owing to the poor availability of fossil energy at relatively low cost and to the remarkably high cost estimated for inexhaustible sources. This result can be arrived at either by escalating the values of energy and the costs of the sources (Fig. 8) or by taking into consideration the increase in real terms of the energy value, namely the differential escalation between the inflation of energy value and general

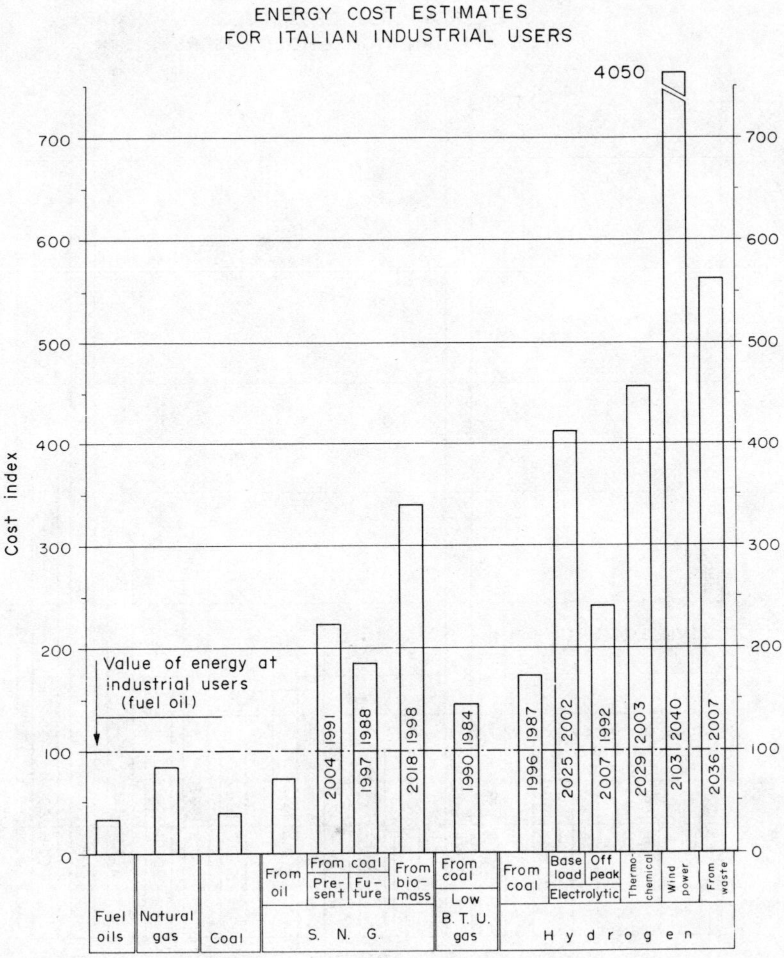

Fig. 5.

inflation (therefore of the investment goods and of the production, transformation, transport, distribution chain) as shown in Fig. 9.

RESULTS

In each column of Figs. 5 and 6 two values are shown: they indicate the years of break-even between energy value and cost of the individual source in relation to a differential escalation (energy-general inflation) of 3%/year (value shown at the bottom) and of 6%/year (value shown on the top), corresponding respectively to a mean-low and to a high hypothesis.

The examination of the years of such break-even between costs and prices shows that

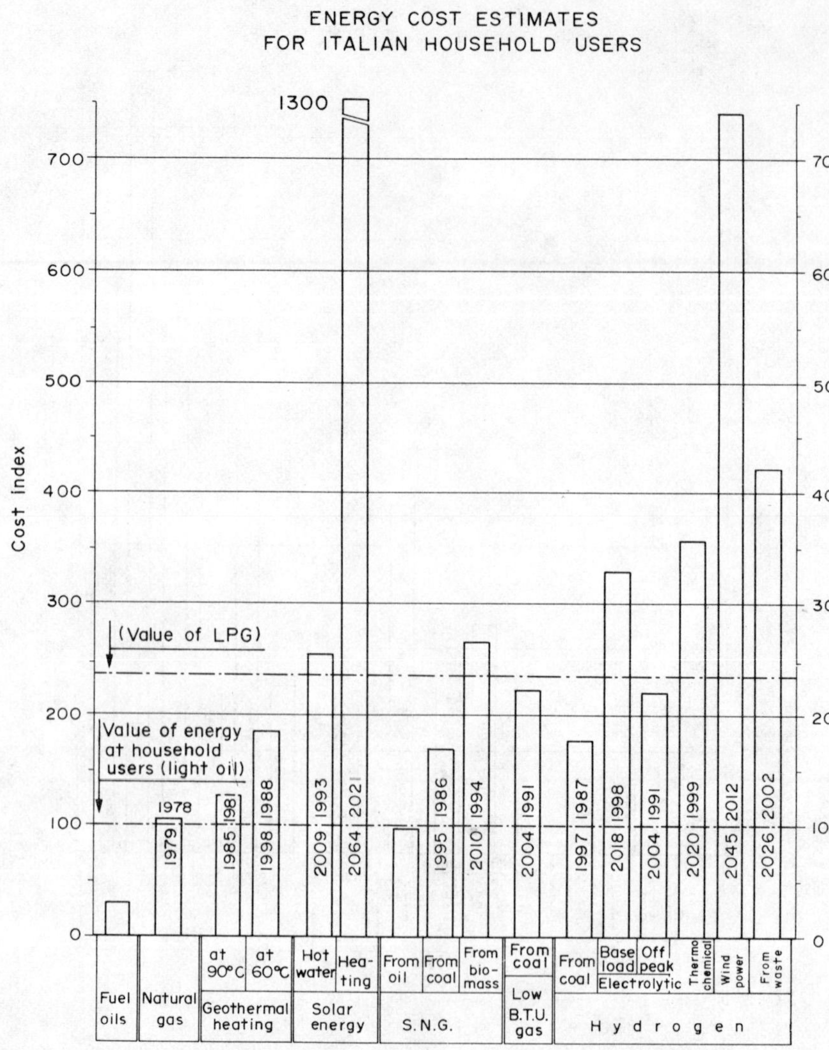

Fig. 6.

for the years around 1990 the Italian gas industry should not feel the effects of the presence of new sources of energy. Such sources are deemed limited within the national market of energy, unless the differential escalation between the industrial costs and the prices of energy should reach very high values for many consecutive years.

In this case, however, the time required to set up the prototype and to build production plants operating on an industrial scale would be a factor.

It can be forecast that the first among the new sources to have a footing in the market directly concerned with gas industry should be SNG from coal, imported from countries where there is a large availability of strip mines. Presumably, the development will be hindered by difficulties of an ecological nature in installing the gasification plants along the coasts.

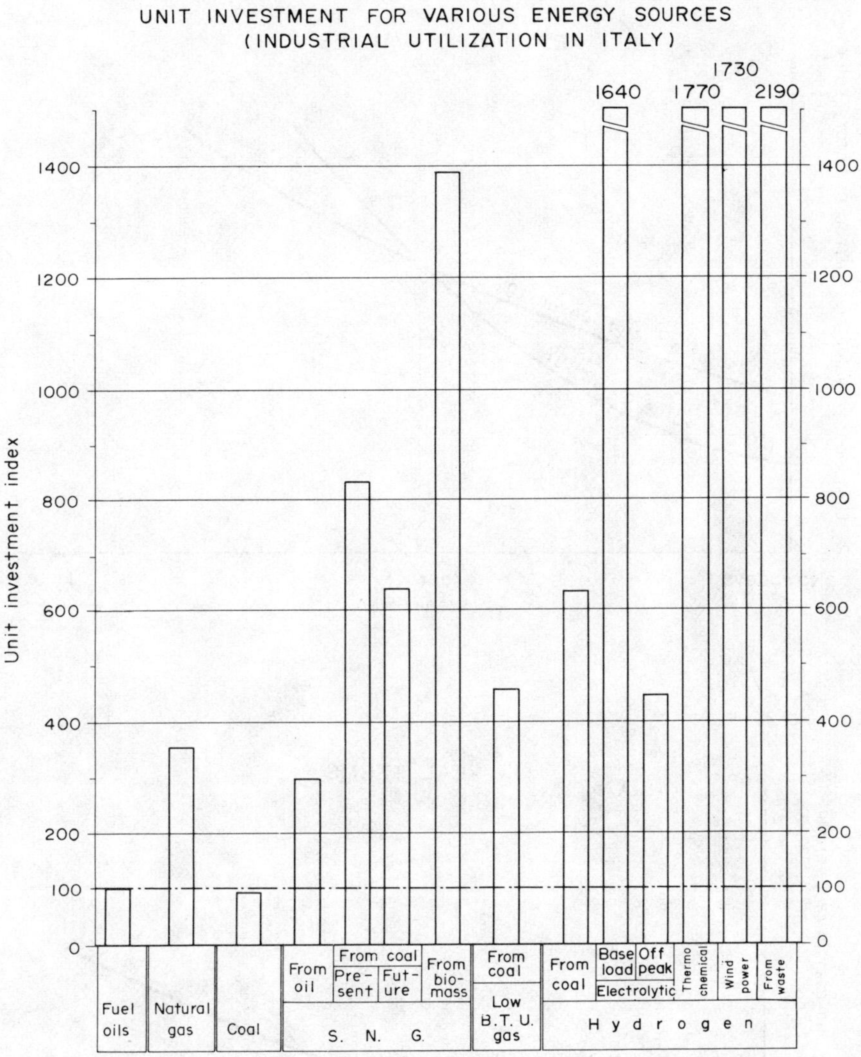

Fig. 7.

Later on, hydrogen could play its role in the first decade of 2000, hindered in its turn by its non-interchangeability with the natural gas distributed through the existing networks.

CONCLUSIONS

With a view to the aforegoing and in connection with the forecast of gas availability in Italy around 1990, the following considerations can be made:

Further contributions may be made by natural gas imported from North Africa or the USSR (Fig. 10) in addition to that which is assured by domestic production, presumably at a level equal to the present level (as a result of the new findings that

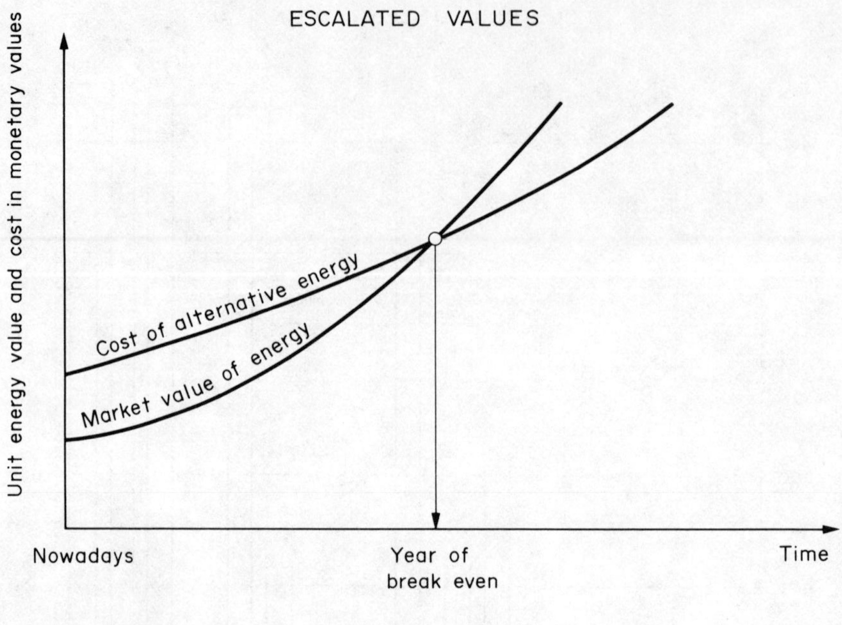
ESCALATED VALUES
Unit energy value and cost in monetary values
Cost of alternative energy
Market value of energy
Nowadays
Year of
break even
Time

Fig. 8.

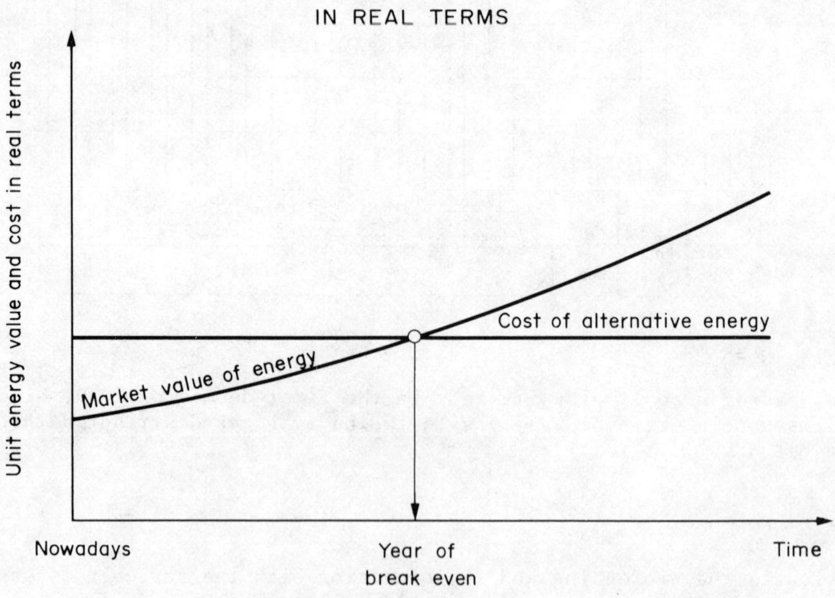
IN REAL TERMS
Unit energy value and cost in real terms
Cost of alternative energy
Market value of energy
Nowadays
Year of
break even
Time

Fig. 9.

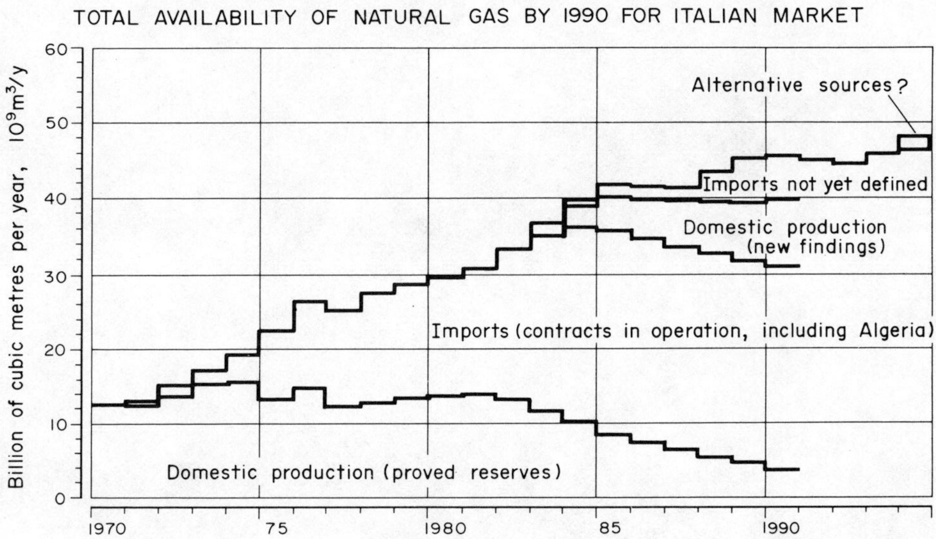

Fig. 10.

will make up for the depletion of the known fields) as well as the operating imports (from Libya, Holland and the USSR) and to those covered by the contract, from Algeria.

Within the time frame it is unlikely that a significant contribution could be added by alternative sources such as coal SNG or bio-mass gas which could develop later. This development may, however, take place when it is no longer possible to integrate availability through natural gas imported from more distant sources such as Middle East and eventually from West Africa. In the long term these gases could also replace the natural gas imported from Holland, at the expiry of the contract.

The Use of Gas in the Netherlands after the Year 2000

A. J. Bogers and J. Quakernaat

The Netherlands

INTRODUCTION

By the extensive use of natural gas, being the most convenient and least polluting of the fossil fuels, the Netherlands have a unique energy supply system. More than half of its energy comes from natural gas and the distribution grid reaches into nearly every village.

In a few decades the stock of indigenous gas will have decreased considerably. Also the availability of oil, the Netherlands' second energy source, will be limited by that time. Therefore the question arises of what the Netherlands should do about the situation of gradually losing its two sources of energy.

The role of the so-called alternative energy sources can only be expected to be moderate by that time; solar energy, wind energy and possibly the underground gasification of coal may have some significance.

Large-scale use of nuclear energy in a hybrid system based on the energy carriers hydrogen and electricity is not considered to be very likely, at least not during the first years of the new century. It would be very difficult, both technically and financially. The whole idea, moreover, lacks public acceptance.

Substantial contribution to the energy supply may come from the importation of coal and natural gas. Coal especially is given much thought lately. This energy raw material can be used in two, principally different, ways: directly by combustion or after conversion to coal gas. Possible developments in this whole field have been surveyed in a study on the future role of gas and electricity in the Netherlands. The study[1] was co-ordinated by two non-profit research organizations, the Organization for Scientific Research TNO and ESC/ECN, and carried out in cooperation with about forty contributors from industries, utility companies, gas companies, ministries and others. This paper will present some of the main results of the study and discuss some consequences for the Netherlands' household energy.

DEVELOPMENTS IN THE ENERGY-SUPPLY SYSTEM

Diversification

The energy system in the Netherlands will highly diversify in the coming years (see Fig. 1), both with respect to the kind of energy carriers and to their applications.

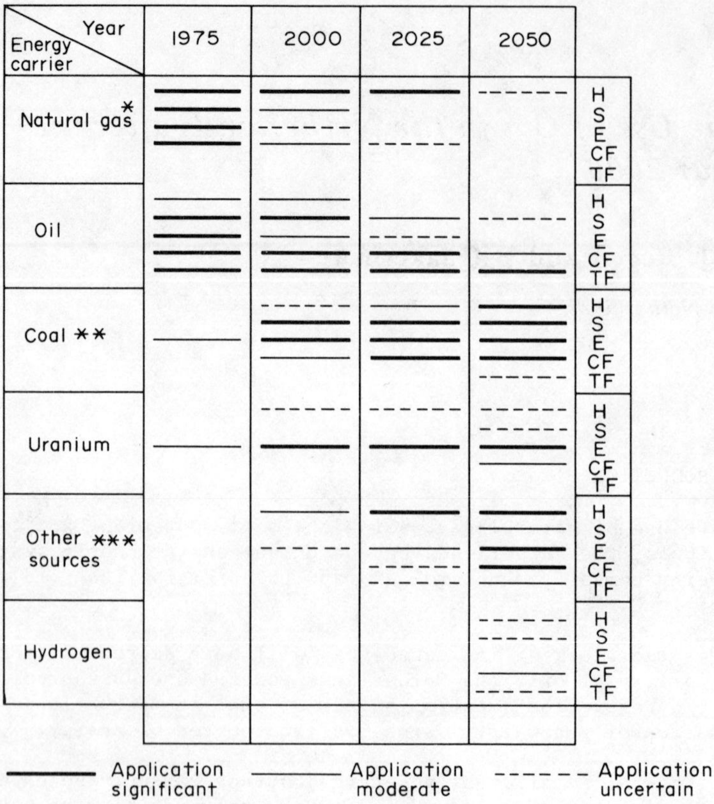

Fig. 1. How the energy carriers will change their roles over time in the Netherlands.

* = indigenous natural gas + imports of natural gas

** = coal + coalgas

*** = such as solar and wind energy and underground coal gasification.

H = heating

S = steam

E = electricity

CF = chemical feedstock

TF = transportation fuel.

Mainly, this is a consequence of the policy to restrict the use of natural gas to those applications where, compared to other energy carriers, natural gas is the most desired. This is done to stretch the period of availability of the indigenous gas and thereby to secure a supply of gas to those sectors to which gas is considered to be essential.

The expectations of new major gas strikes in the Netherlands or on the Netherlands part of the Continental Shelf have been considerably lowered since the second half of the sixties. It is believed it will be possible to obtain contracts to import LNG or pipeline gas, albeit prices will have risen because of the distance between the Netherlands and the regions of origin of the gas.

In the future, power stations and larger industries now using natural gas to fire their boilers will look for cheaper fuels on their own initiative. On a somewhat longer term this also applies to the chemical industry for its raw materials.

A straight shift from gas to oil cannot be a solution, not only because of the increase of dependence on oil-producing countries, but also because of the limited world oil reserves and other constraints that are expected to squeeze the oil supply in the coming decades. For these reasons, in addition to nuclear energy, more use will have to be made of coal.

The pattern of diversification, expected to develop in the Netherlands, not only in the near but also over the longer term, is given, of course in a qualitative sense, in Fig. 1. The significance of each energy carrier for purposes like heating, steam production, electricity generation, chemical feedstock and transportation fuel is indicated together with the way in which this will change over time.

The Demand for Energy

The development of energy demand in the Netherlands is given in two scenarios based on the relation between economic development and energy consumption. The two scenarios represent extreme situations; one represents ongoing economic growth, the other zero economic growth. Moderate economic growth is supposed to be the most likely development.

In both scenarios it is taken as a starting-point that there will be an increased set of measures in order to conserve energy. The measures will not only have to include better insulation of homes and other buildings, but also improvement of the efficiency of all kinds of energy-involved processes.

The growth scenario implies an annual growth of GNP that amounts to 3.5 or 4.0% in the period 1975-1980, then decreases to 2.7% in the year 2000 and subsequently stabilizes at that level. In the zero-growth scenario a significantly different ratio is assumed between material production and leisure time. In this assumption, between 1980 and 1990 a situation will be reached in which increased labour productivity will not result in more material income such as more goods and services, but in more leisure time. The difference in energy consumption in this scenario is lower not because of better efficiencies but simply because of less production.

The figures from both scenarios are given in Table 1.

The Supply of Energy

A survey of the supply possibilities for the next 40 years was based on well-known studies like those made by the Workshop on Alternative Energy Systems (WAES) and

TABLE 1

The Demand for Energy in the Netherlands for the Years 2000 and 2015 in Millions of Tons of Oil Equivalent

	Year				
	1975	2000		2015	
Sectors		Ongoing growth	Zero growth	Ongoing growth	Zero[a] growth
Small consumers	21	35	30	42	30
Industry	19	52	37	86	40
Energy conversion[b]	11	21	17	28	17
Transportation	7	14	11	20	11
Total domestic consumption	58	122	95	176	98

[a]Though there is zero economic growth, due to shifts in production there still is some growth in energy consumption.
[b]Losses during conversion in power stations, refineries, etc.

by the 10th World Energy Conference (WEC). This among other things led to a ceiling of 53 Mtoe of oil to be imported from 1990. In the zero growth scenario even less increase of oil import is foreseen, as in this case the price of scarce oil will be more of a factor.

With regard to the gas situation, both indigenous and imported, existing estimates for both stock and sales were used.[2] The figures for the year 2015 are new estimates. Concerning the gas imports a strong buying policy is presumed.

Regarding coal it is assumed that a lot of effort will be made to have the necessary infrastructure ready on time and to have long-term contracts secured.

The overall availability of coal may pose a problem, because estimates of world trade volume are lower than expected previously.

Furthermore, it is assumed that nuclear energy's position with the public will improve, so that in the 1980s an extensive investment programme can be initiated.

The "other energy" sources will be more developed in case of ongoing growth, because more funds will be available. No attention is paid to wind energy as its contribution, even the maximum estimate, remains only marginal. Funds for research being available, the longer term (2015) may see some new energy technologies like underground gasification of coal, nuclear fusion, solar energy from space, etc.

The figures concerning the energy supply are given in Table 2. It should be mentioned that the estimates for both oil and coal in the growth scenario in the year 2015 represent a very tight situation and may be very difficult to be realized, a choice being very hard to make right now.

A Working Scenario

Both the ongoing-growth and the zero-growth scenario, as previously indicated represent extreme cases. In order to be able to compare a number of suggestions for

TABLE 2

The Supply of Energy in the Netherlands for the Years 2000 and 2015 in Millions of Tons of Oil Equivalent

Energy sources	1975	2000 Ongoing growth	2000 Zero growth	2015 Ongoing growth	2015 Zero growth
Oil	23	53	49	53	45
Indigenous gas	32	27	25	27	25
Imported gas		9	5	19	3
Coal	2	18	10	37	14
Uranium	1	11	6	20	9
Other sources		4		20	2
Total domestic consumption	58	122	95	176	98

diversification there was a need for a working scenario. The various participants to the study obtained consensus on a set of figures for total domestic consumption that could be labelled "most reasonable". These figures tend to moderate growth.

Table 3 shows the interaction between supply and demand in the Netherlands for the first three decades of the next century. These figures have been rounded off in multiples of 5 Mtoe as they are only meant to be trend indicators.

TABLE 3

The Interaction between Energy Demand and Supply in the Netherlands for the Years 2000-2020 Represented as a Working Scenario. Figures rounded off in multiples of 5 millions of tons of oil equivalent

Energy sources	Residential and commercial	Power and light	Industrial process heat	Chemical feedstock	Transportation	Σ
Coal		10	5	5		20
Uranium		10				10
Natural gas	25	0	5	5		35
Oil		5	20	15	10	50
Sub-total	25	25	30	25	10	115
Other sources			5			5
Total			120			120

THE GAS RESERVES IN THE NETHERLANDS AT THE TURN OF THE CENTURY

It is the policy of the Nederlandse Gasunie to give ample supply of gas to the public system, premium use in industry and special energy-saving applications during the next 25 years.[2] According to the estimates of the Gasunie the share of natural gas in total domestic supply will be about 30% at the turn of the century, being about 55% in 1977. For the period 1978-2002, according to the Gasunie, a total of 1783 billion m^3 of gas will be needed including 650 billion m^3 for export. The expected reserve is 2573 billion m^3 leaving a reserve of 790 billion m^3 after the year 2000. This estimate takes into account more energy conservation by small consumers, e.g. by insulation of homes, less sales to large consumers, re-evaluation of existing reserves and the acquisition of 80 billion m^3 of high calorific gas from Algeria. The figures used are not certain, of course, since the expectation is partly based on the results of exploration. The search for and the development of gas fields in the Netherlands remains of undiminished importance.

THE ALLOTMENT OF GAS AFTER THE YEAR 2000

As already mentioned, there is a well-established and operational infrastructure for gas in the Netherlands. This, in addition to the general applicability and convenience, is a strong argument in favour of maintaining a gas component in the energy system.

From an environmental point of view gas is a special fuel that can always be burned in urban areas, because there is hardly any harmful emission. Finally, it should be stated that gas, in contrast to electricity, can be stored in large quantities both above and underground and transmission losses are low.

The composition of the gas can vary within a range. This will allow, as time goes on, for other gases among them synthesis gas, to be introduced in addition to the currently used Groningen gas. As explained elsewhere even hydrogen can take over this part.[3]

Table 3 shows that in the years 2000-2020 the Netherlands will need about 35 Mtoe (about 40 billion m^3) of gas for the sectors residential and commercial, industrial process heat and the chemical industry. Use of the reserve of gas at the beginning of the next century, being approximately 800 billion m^3, in these sectors would mean, arithmetically at least, that there is sufficient supply for the next 20 years. This is a very attractive suggestion, even more so since mixing the gas with gas from coal may help stretch the period of supply considerably. The sooner the coal gas is introduced the more time will be gained to take appropriate measures to prepare for the hydrogen era.

For whatever reason, if it turns out to be impossible to reserve the 800 billion m^3 of gas entirely for the above objective, one would have to consider securing, at least, for long periods of time, the seasonally variable part of supply which is mainly room heating. This would be necessary because of the optimal possibilities that are offered by the indigenous stock to accommodate the seasonal peak. For tens of years this will be the cheapest way of providing heat to homes, at a minimum of pollution.

Restricting the use of natural gas to home-heating purposes implies that the seasonal peak — an amount of about 5 billion m^3 — will have to be produced properly. Production rate might become a problem. At this moment the capacity of the Groningen field is such that it can still produce at a high rate. However, if extraction continues, the capacity will decrease to such an extent that finally not even by the drilling of more wells, the installation of compressors or modified gas treatment will the Groningen field be able to produce at the required rate. Some other form of underground storage to accommodate the winter situation will become necessary.

Consumption is low during summer. The minima are especially attained during the holiday period and also during weekends and nights. The importation of LNG or pipeline gas will not allow for much variation in shipment, coal gas can only be produced efficiently on a continuous basis, and so there will be quite a difference between supply and demand during this period of the year. Underground storage will then be needed to accommodate these fluctuations. The necessity of such storage will increase if the consumption is more restricted to markets with a strong seasonal variation like room heating, or a strong variation in use per day such as commercial. The necessity will also increase with the increasing operation of gas fields in the North Sea that require a high regularity of production. The solution for storage to produce a seasonal peak of 5 billion m^3 of gas could be found by creating an underground storage room, possibly in the form of a gas field or aquifer with the size of about 25 billion m^3. During summer this storage can be supplemented by imported gas, North Sea gas, or even Groningen gas.

The Groningen field is, though it has excellent properties as a reservoir, not suited for use as underground storage because of the huge quantity of cushion gas that would have to be left in this reservoir. For the underground storage a gas field with an original content of 25 billion m^3 would be most appropriate; the quantity of cushion gas would only have to be 10-15 billion m^3. There will be no point in above-ground storage because of the really huge winter quantities involved.

CONSERVATION MEASURES

It is certain that a great many measures are to be taken in order to keep as much gas as possible available after the year 2000.

Reduced Consumption

As mentioned earlier, energy conservation to the greatest extent possible should be striven for. In addition to thorough insulation of homes, use will also have to be made of co-generation both in distance heating and in larger industries. From a number of model calculations it was shown that direct combustion of coal can play an important part in this respect. Finally, the gasified heat pump looks very promising as applied to room heating.

The Direct Combustion of Coal

Because of the restricted use of gas in power stations, apart from oil, coal will again be used, as nuclear energy cannot be seen to be able to replace gas in the near future, for whatever reason. In the case of coal the abatement of air pollution will possibly be carried out by gasifying coal at the site of the power plant by an integrated process. Largely based on the scale at which these processes will be carried out the environmental problems are very likely to be solved.

Industry can burn coal in two ways. Firstly, in slagging furnaces for large-scale production of steam (about 200 tons of steam per hour) in combination with stack gas cleaning. Secondly, in fluidized bed combustion units for smaller-scale production of steam (about 15-50 tons of steam per hour). Both methods still lack a final solution with respect to the removal of both volatile and solid wastes. The slagging technique as yet still faces the NO_x- problem. Fluidized bed combustion produces a solid waste, of which it is highly uncertain whether it may be dumped or not. Taking into account the extent to which these problems will have to be faced, it is necessary to mention them and to state that a lot of research still has to be done to solve them.

The Gasification of Coal

By gasifying coal it becomes possible to supplement or replace gas in the existing grid. Experience with this technique is badly needed in the Netherlands, as there are still a great many uncertainties with respect to the regulation of the emission to air, water and soil in relation to the type of gasification process. The Netherlands will have to look for a type of process that can be operated with various kinds of coal with a minimum of pollution.

Coal gasification conflicts with the policy of energy conservation. The efficiency of coal gasification for the public gas-supply system is such that it leads to an increase of primary energy consumption. The relatively high-cost and low-energy efficiency urge one to look for integration possibilities with the production of chemical feedstock and with power generation.

The Importation of Natural Gas

It should be again stressed that the importation of gas is, of course, a very appropriate way of conserving the domestic gas reserve. Together with other European gas companies the Gasunie tries to direct supplementing quantities of Norwegian gas to northwestern Europe. Furthermore, there are contacts with producers in more remote areas: the Soviet Union, Iran (overland pipeline) and Nigeria and the Middle East (LNG by ship).

BUYING NATURAL GAS AS AN ALTERNATIVE TO BUYING COAL

Regarding the large-scale introduction of coal into the energy system of the Netherlands there are a great number of uncertainties. The problem of the availability of coal on a sellers' market has already been mentioned.

By the use of coal in power stations, air pollution will increase by the emission of sulphur and nitrogen oxides, certain hydrocarbons and trace elements among which are heavy metals. Air pollution will be less when the coal is gasified but this may be counteracted by the fact that some of the gasification processes involve water pollution. So on a regional scale there might be the problem of the protection of the surface and the ground water and this not only against the direct draining of the process water but also the draining of leak water of large quantities of stored waste from power stations and gasification plants. The size of the area needed for controlled waste disposal might be a limiting factor for the use of coal. From a national point of view there is no lack of room for the storage of coal but on certain locations the conditions of the soil may be prohibitive to the storage of large quantities. It appears that the infrastructure in the Netherlands needs only minor adjustments to facilitate the logistics of many tens of tons of coal annually. Problems in this respect might arise for industrial locations in densely populated areas.

From the foregoing, it is obvious that the option coal has its advantages and disadvantages and that it is not simply a choice between direct combustion and gasification. Imported natural gas might be an appropriate alternative. Tables 2 and 3 show that the Netherlands would have to change to buying large quantities of gas abroad. Global availability of gas would allow this; the production of most of the areas is not expected to peak before 2020. There is a lack of harbour facilities, liquefaction installations, ships, pipelines, transit stations and the like, however. If the Netherlands decides to enter the international gas market as a buyer, it will have to make arrangements for its own infrastructure including facilities for shipping, transit storage, transportation and distribution. Revenues from the export of this type of facilities might contribute to maintaining a certain equilibrium on the balance of payments.

CONCLUSIONS

The energy situation in the Netherlands will be characterized by considerable shifts in the supply pattern of primary energy carriers and the relating diversification of central and regional energy-supply systems.

A most likely breakdown of the energy consumption pattern of the Netherlands at the beginning of the next century may be gas: 35 Mtoe, oil: 50 Mtoe, coal: 20 Mtoe, uranium: 10 Mtoe and other energy sources especially solar and wind energy and possibly also underground coal gasification: 5 Mtoe (Table 3).

Since the most preferred consumers of a clean fuel such as natural gas are evidently households and small- and medium-sized industrial users in urban areas, any national gas reserve still available after the year 2000 should be earmarked exclusively for this group of users.

An active policy on both energy saving and acquisition of foreign gas are to be seen as the most effective ways to preserve indigenous natural-gas resources.

Another element of active energy policy would be to envisage the first steps towards the construction of coal gasification plants in order to increase knowledge and experience on all aspects of this technology.

Availability and environmental impact will limit the use of coal in the Netherlands in the first decades of the next century: 30 million tons a year (20 Mtoe) seems to be a maximum (Table 3).

The pollution caused by the use of coal can only be abated effectively in large-scale installations.

Gas-fired heat pumps are very promising as energy-saving devices in home heating. The conditions for introduction of this type of heat pumps especially should be studied.

REFERENCES

1. June 1978. Report TNO/ECN: *Gas en Elektriciteit in Nederland, een toekomstverkenning.* Centrum voor Energievraagstukken TNO, Apeldoorn; Holland; Energie Studie Centrum ECN, Petten, Holland.
2. June 1978. Report N.V. Nederlandse Gasunie. *Plan van Gasafzet 1978, de plaats van aardgas in de toekomstige Nederlandse energievoorziening.*
3. A. J. Bogers and J. Quakernaat, September 1977, "The possible role of hydrogen in the energy supply sector of the Netherlands". *Papers of the Tenth World Energy Conference,* No. 3, 5-3, pp. 1-17.

United States' Gas Supplies: An Overview

L. D'Andrea

United States of America

INTRODUCTION

Natural gas currently represents approximately 27% of all energy consumed in the United States. Each sector of the economy is vitally dependent upon this clean-burning fuel. Over half of all residences and commercial establishments utilize gas for space heating and/or cooling. Gas comprises 40% of industrial energy demand. Despite the heavy dependency on gas as an energy source, the United States has experienced a pervasive gas shortage for the past several years. Domestic production rose from 9.4 Tcf* (trillion cubic feet) in 1955 to 18.4 Tcf in 1967 to a peak of 22.6 Tcf in 1973. Since that time, production has declined steadily to 19.4 Tcf in 1977. The United States has basically been self-sufficient in natural gas supply, with natural gas imports, primarily from Canada, providing about 3.5 to 4% of total consumption. Imports are likely to increase, however, in order to close the gap between supply and demand.

The gas-pricing policies of the United States were developed at a time when there was an over-abundant supply of natural gas. The increased demand on clean fuels due to environmental considerations were not evident at the time critical pricing policies were being considered and implemented. Market growth was the primary thrust of the industry, with little regard placed on how or where the gas was being used.

The administered low prices for natural gas relative to other energy sources have led to a highly uneconomic situation and poor utilization of this valuable energy resource. The pricing policies resulting from the implementation of federal controls on wellhead prices for gas sold and delivered in interstate commerce has promoted high consumption while discouraging needed incentives to seek out new sources of gas supplies. In time these policies led to curtailments of gas deliveries to low-priority consumers, and moratoria on new service in the regulated interstate market. The distortion created by federal pricing policies between the unregulated intrastate and regulated interstate markets have further accentuated the supply problem because new supplies of gas are flowing to the unregulated intrastate market at the expense of interstate consumers.

This market disparity and pricing consideration prompted the President to include in his April 1977 National Energy Plan a comprehensive programme for natural gas.

---

*To convert cubic feet of gas to cubic metres, multiply by 2.835×10^{-2}.

This plan has been subjected to intense national debate, with the Congress presently having accepted certain provisions but having reached no firm solution to the issue surrounding wellhead gas prices.

THE NATIONAL ENERGY PLAN

The National Energy Plan deals with oil and gas by recommending a series of policy initiatives that include higher prices to encourage exploration and development of new gas sources, and a more efficient utilization of the available supplies. The plan also prevents windfall profits. In the case of gas, the plan makes a series of specific recommendations as to prices, identity of gas supply sources, extension of federal wellhead price jurisdiction, and taxation on certain users, all of which give rise to certain estimates of future gas production and consumption.

Aside from the relationship that these considerations have on production, two major factors determine the domestic supply of natural gas. One of these is the extent of the gas resource base. Given a certain resource potential, actual production will also depend on the quality of the resource base and the price of the gas compared to the costs of locating, developing and producing these deposits. With higher prices, more of the resource base becomes economically attractive to produce. Other factors that influence gas production are accessability to so-called offshore frontier areas, much of which is controlled by the rate of leasing, and the lead time needed to build required facilities to bring the gas supply to markets. The construction of a pipeline to deliver gas available in the North Slope of Alaska is a good illustration of this factor.

Given a series of assumptions, the plan forecasted the following gas supply, expressed in trillions of cubic feet (Tcf):

Domestic Supply (Tcf)

	Associated — dissolved gas	Non-associated	Import (Tcf)	Consumption (Tcf)
1978	2.3	14.6	1.0	17.9
1980	2.3	14.3	1.0	17.6
1985	2.0	15.0	1.2	18.2

The consumption is projected to maintain a low rate of growth because of the provisions contained in the plan which will constrain growth of the low priority gas markets. This in turn would free up gas for higher priority residential/commercial markets. The fuel-conversion provision of the plan designed to shift low-priority oil and gas consumption to coal is one which is most complex and highly optimistic in its goals. None the less, any shift in gas loads will increase the availability of gas for other worked at the expense, some argue, of additional oil imports other than a shift to coal.

Because of the general belief that the demand for gas will continue to rise beyond the levels of supplies that may be forthcoming from the known conventional sources, additional emphasis is being placed on the availability of supplemental and non-conventional sources of gas. Greater emphasis is being placed in the development of new technologies which will increase the production of energy before the turn of the century. In the case of natural gas, new supply sources that are likely to emerge during the mid-1980s and the 1990s are several, including unconventional gas, Alaskan gas, gas imports from Canada and Mexico, LNG imports and high-Btu gas from coal.

UNCONVENTIONAL GAS

Unconventional sources of gas (such as tightsands, geopressured methane and Devonian shale) have the potential to produce substantial quantities of gas, but at costs higher than the prevailing, free-market price for new natural gas. Aside from the economic factors impeding the rapid development of these supply sources, there is the need for additional requirements for the development of new technologies to bring such resources into commercial markets.

The general location of these unconventional gas resources is shown in Fig. 1.

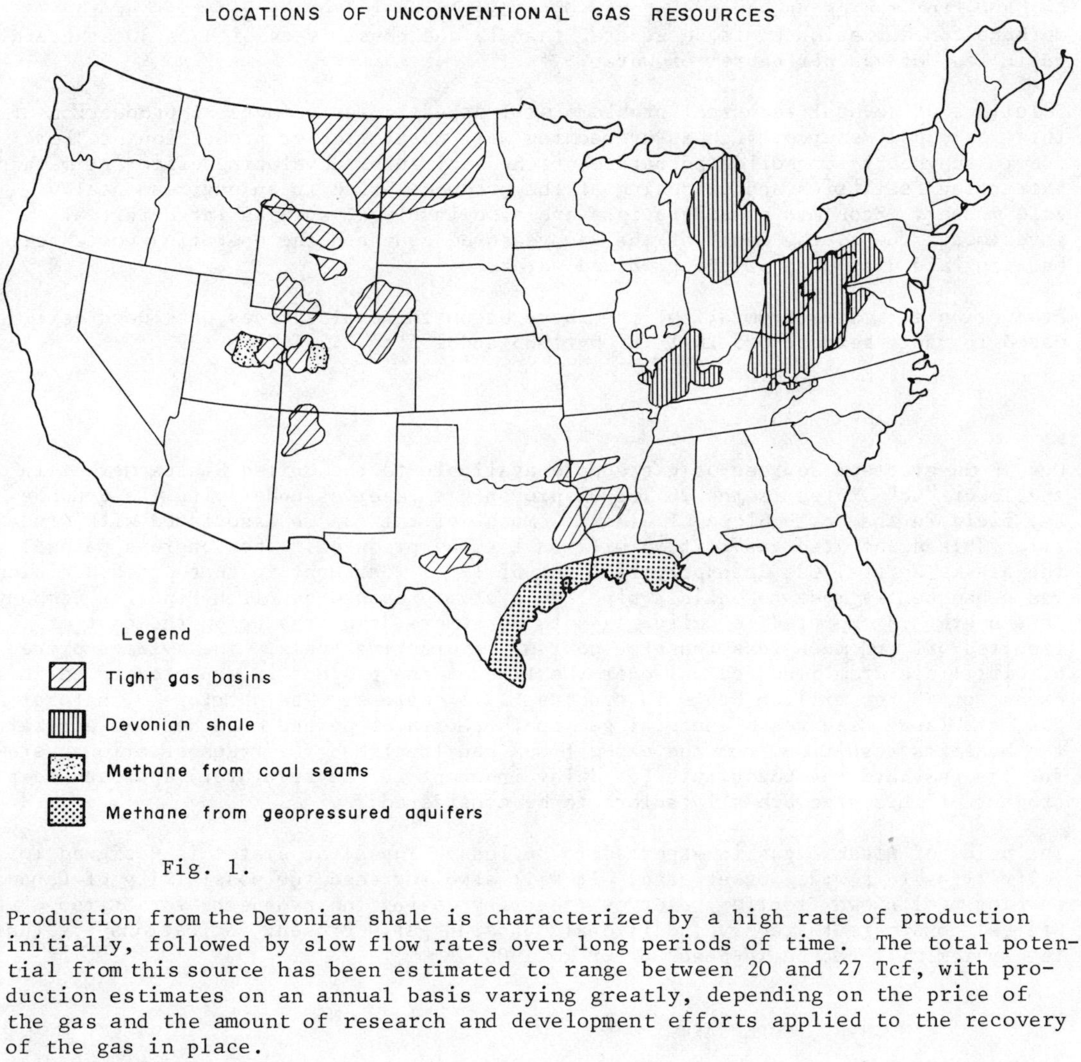

Fig. 1.

Production from the Devonian shale is characterized by a high rate of production initially, followed by slow flow rates over long periods of time. The total potential from this source has been estimated to range between 20 and 27 Tcf, with production estimates on an annual basis varying greatly, depending on the price of the gas and the amount of research and development efforts applied to the recovery of the gas in place.

Production from tightsands could be brought on stream between 1980 and 1985 in significant quantities. Production from tightsands is not new. These sources currently provide about 1 Tcf per year. The difficulty is the very low rate of production existing for many of these sands. The gas-in-place resource for tightsands has been estimated to be about 400 Tcf. Advanced technology in massive hydraulic fracturing (MHF) and chemical explosion are being actively tested, with prevailing results showing increased flow rates. Significant additions to the recovery and acceleration by production/commercialization could be achieved through greater government/industry research and development of a commercialization programme.

Underlying a large portion of the northern shoreline of the Gulf of Mexico are clastic sedimentary deposits of great thickness. Certain areas contain sedimentary deposits which have trapped water at higher than normal pressures (geopressured). This trapped water is estimated to contain significant quantities of dissolved methane. A great deal of interest has been developed in the recovery of this methane from the geopressured water. Data developed to date suggests that the methane concentration could be greater than 15 and possibly as high as 50 standard cubic feet of gas per barrel of water.

Solutions of several technical problems will expedite the commercial production of this gas-supply source: such as overcoming well completion or production problems due to the poorly consolidated nature of the sediments, developing efficient methane extraction facilities and disposing of the produced brine in an environmentally safe manner. Economic considerations are also involved, such as large initial investments due to the depth of the geopressured zone and the operating costs for handling and disposing of the produced water.

Production estimates from all of the above unconventional sources have been estimated to range between 0.5 to 1 Tcf by 1985 and 1.5 to 2.5 Tcf by 1990.

ALASKAN GAS

One of the greatest sources of gas supply available to the United States market in the lower "48" states is the 26 Tcf of proven gas reserves underlying the Prudhoe Bay field in the North Slope of Alaska. Much of the gas is associated with crude oil. This associated gas in time will have to be produced. The Congress passed the Alaska Natural Gas Transportation Act of 1976. Pursuant to that Act a decision was subsequently made to build a pipeline system proposed by Alcan Pipeline Company. This system is expected to deliver gas to consumers at prices below the cost of imported oil and much less than the cost of alternative fuels. The average price of distillate from imported oil over the life of the project is expected to be in excess of $3 per million Btu's in constant 1975 dollars. The shortage of natural gas, the large reserves of natural gas in the North Slope and elsewhere in Alaska, the benefits resulting from the expeditious construction of a transportation system for the gas, and the potentials for delay inherent in the regulatory approach to a project of this size are all factors to be considered.

The price of Alaskan gas is expected to be less. The Alcan system is designed to deliver gas to a wide market area. It will also increase the possibility of Canada developing its own frontier reserves, thus increasing the prospects for United States' consumers to receive additional Canadian gas. Present indications are that the system will not be in place prior to 1985.

HIGH BTU COAL GAS

The United States has an abundant supply of coal, constituting about 90 per cent of its conventional energy reserves. It currently supplies, however, only 18 per cent of energy consumption. An expanded utilization of coal is a major component of the United States energy strategy.

Among the options available for this purpose is the production of high Btu coal gas. Not only is this a long-term, assured supply of domestic gas, but it also appears to be more favourable than the costs of imported gas, which are subject to price escalation and other uncertainties.

Technically, coal gasification plants can be built and a number of gas utilities are seeking to develop commercial facilities. The problem inhibiting their development centres is the need for assured marketability of the gas and assurance that the debt will be repaid to lenders in the event of non-completion. Capital requirements for a full scale, 250 million ft^3/day plant can range from \$1.1 to 1.6 billion (in 1978 dollars), depending on how the project will be financed. Several initiatives considered important for the rapid development of high-Btu gas includes the ability to "roll-in" the price of the coal gas with low-cost domestic conventional gas, elimination of the non-completion risk, and loan guarantees. Production estimates for coal gas have been at 0.2 Tcf by 1980 and 0.9 Tcf by 1990, with price estimated at \$4.20 to 4.60 per MMBtu, based on average prices over the lifetime of a project.

CANADIAN AND MEXICAN IMPORTS

Canada has been supplying the United States' gas markets for several years. These imports total about 950 Bcf a year, representing approximately 4% of total US gas consumption. Volumes of Canadian exports under current licences are on a steady decline and will cease by 1992. Canada had been experiencing a growth in demand for gas, causing it to curtail certain volumes to utilities on the West Coast. More recently, however, Canadian gas supply has improved considerably and domestic demand has declined due to increased prices, causing Canada to make additional volumes of gas available to US importers. In terms of total volumes delivered to US consumers, Canadian supplies currently represent our most expensive source of gas, with the import price at the border set at \$2.16 per MM Btu's. Canada's gas resource potential is substantial, but Canadian policies are directed toward an orderly development of its resources. Undoubtedly, US markets are the most promising outlet for disposing surplus gas supplies. It is conceivable that the construction of a pipeline through the McKenzie Delta, as part of the pipeline system to deliver North Slope gas, will increase the prospect of continuing the current level of exports for years beyond the 1985 period.

One of the most promising prospects for increasing gas supplies to the US is the availability of significant volumes of gas from Mexico. Six gas-transmission companies have signed an agreement with PEMEX to purchase natural gas available in the Reforma area of south-eastern Mexico. Plans call for the construction of a 750-mile pipeline capable of ultimately delivering about 2 to 2.5 Bcf per day by the mid-1980s. A border price of \$2.60 to \$2.80 per MM Btu's has been proposed, which is equivalent to that of imported no. 2 fuel oil at the New York Harbor. Because of the impact that the proposed price would have on US consumers and the national policy objective of maximizing domestic energy resource development, no decision has been rendered on this project at this time. There is every reason to believe that these differences will be resolved and in time Mexican gas will begin to find its way in United States' markets.

LIQUEFIED NATURAL GAS (LNG)

Among the various gas supply options being considered by gas utilities, the importation of liquefied natural gas (LNG) is one which is quite uncertain to quantify or estimate, because of the active role the government has played in this area. Up to March 1978, the United States was a net exporter of LNG. LNG shipments from

Kenai, Alaska to Japan are averaging about 50 Bcf a year. Imports, on the other hand, are about 15 Bcf, all from Algeria. In March 1978, the first large, base-load import project from Algeria commenced initial deliveries. This project, often referred as the El Paso I project, involves the importation of about 1 Bcf per day. This LNG will be supplying primarily East Coast consumers. Several other projects have been authorized or are currently pending before the Economic Regulatory Administration for authorization. See Table 1 for a list of these projects.

TABLE 1

LNG Projects Approved or Awaiting Approval

Country	Project	Volume per year/ schedule date	Entry points
Algeria	Distrigas I*	15 Bcf/1971	Everett, Mass.
	El Paso I*	388 Bcf/1977	Cove Point, Md. Savannah, Ga.
	El Paso II**	365 Bcf/1980	Matagorda Bay, Tx.
	Distrigas IV*	43 Bcf/1977 (Volume includes Distrigas I above)	Everett, Mass.
	Trunkline*	153 Bcf/1980	Lake Charles, La.
	Tenneco Atlantic**	416 Bcf/1985	St. John, New Brunswick, Canada
	Total for Algeria	1603 Bcf	
Indonesia	Pacific Light*	197 Bcf/1981	Oxnard, Calif.
	Subtotal	197 Bcf	
	Total	1800 Bcf	

*Approved.
**Pending before ERA.

Because LNG imports represent high-cost gas supplies and involve significant safety issues, there has been a great deal of controversy and concern surrounding the desirability of increased levels of imports. Such issues as technical and political reliability of exporting country, balance-of-payment impacts, price-escalation provisions, proper siting of terminal facilities, and safety have prompted the Federal government to conduct a policy review on LNG imports. This policy review will establish broad guidelines and criteria for consideration in the acceptance or rejection of particular LNG import projects.

There is no doubt that LNG will play a role in supplying gas to utilities seeking to supplement existing sources of gas supplies. A critical factor that will control the rate of growth of LNG imports is the pricing arrangement at the marketing level. If LNG is permitted to be "rolled-in", as urged by parties sponsoring LNG projects, the marketing of the LNG, which currently ranges from $3.37 to 4.50 per MM Btu on a c.i.f. basis, will be more acceptable to consumers. On the other hand, if it is required to be sold incrementally or at its margin, LNG will not be easy to market because of its high cost, risk of interruption in deliveries caused by curtailments of deliveries and difficulty in securing financing. Current inclination is toward some form of incremental pricing.

Supply projections indicate that by 1985 there may be as much as 1.8 Tcf of imported LNG, primarily from Algeria. On the other hand, there could be as little as 0.780 Tcf if no further LNG projects are authorized.

CONCLUSIONS

The shortfall in domestic natural-gas supply is highlighted by six salient conditions. First, total proved reserves have declined by 24 per cent since 1970, from 283 Tcf to 216 Tcf (32 Tcf are in Alaska). Second, the ratio of proved reserves to production is at an all-time low, less than 11 to 1 (about 9½ to 1 when Alaska is excluded). Third, new additions to supply, from new discoveries and extensions of known fields, have replaced only 1 Tcf for every 2.5 Tcf consumed over the last 7 years. Fourth, while exploration in 1976 was three times that of 1970, largely as a result of significant price increases, reserve additions per completed exploratory well declined 4 billion cubic feet (Bcf) per well in 1970 and 1 Bcf per well in 1976. Fifth, new frontiers (Alaska, deeper waters, deeper wells) could contribute significantly, but finding, producing and delivering this gas will be costly and will require long lead times. Sixth, development drilling, which supplies 80 per cent of additions to reserves, has more than doubled in the past 7 years. However, the base for future developmental drilling is not being renewed because of low exploratory success.

While the gas reserve base has dwindled during the past decade, demand has remained high, hovering around 20 Tcf per year. Some observations about future demand can be made despite the difficulties caused by uncertainty over the degree of supply regulation and future constraints and incentives (negative or positive) aimed at gas usage. First, the National Energy Plan forecast of domestic gas consumption in 1985 is about 18.2 Tcf per year. This forecast is based upon the assumption that all the provisions of the plan are implemented — not only those that address gas supply and usage, but those that deal with other fuel forms as well. Second, it appears likely that this forecast of demand is low, given the changes the energy legislation has undergone in Congress. Finally, gas demand could stay level (20 Tcf) or even increase if difficulties are encountered in increasing coal production, implementing fuel switching on a large scale or encouraging significant conservation measures. Emphasis for accelerating unconventional gas resources will be expedited through improvement in technology and economic incentives.

The Place of Liquefied Petroleum Gases in the Energy Situation Around 1990

The European Liquefied Petroleum Gas Association

It is thought appropriate to begin with a few words about the Association itself. The Association Européenne des Gaz de Pétrole Liquéfiés (AEGPL) is a non-profit-making organization set up in Paris in 1968 by the national LPG (butane and propane) associations of six West European countries. Its membership is now 16, with Turkey also having recently applied to join.

The AEGPL co-ordinates on the European level the work of its member associations and provides a means by which the LPG industry can make its views known in official and legislative circles in an authoritative manner. Thus it plays an active part in the creation of the appropriate international and European standards and the Directives of the Council of the European Economic Communities (EEC), particularly in relation to the elimination of technical trade barriers.

The AEGPL operates through three commissions, "Common Market", "Utilizations/Normalization" and "Transportation/Distribution/Storage". The first of these is responsible for liaison with the Brussels Commission of the EEC. The other two commissions are responsible for the technical work of AEGPL, for preparing advice to other organizations and drawing up rules and codes of safe practice in the industry.

In addition to the standards organizations and the Brussels commission mentioned above, AEGPL works closely with a number of other international organizations, in its own and allied fields, in particular with the International Gas Union in whose committees and triennial world Gas Congress it takes an active part. AEGPL is represented at meetings of the RID/ADR Joint Committee, at Bern and Geneva.

AEGPL organizes its European LPG Congress every 2 years, marking the culminating point of the period in office of each president, which is open to all persons interested in the LPG industry. The next of these is being held in Madrid, in April 1979.

AEGPL has kept in touch for many years by correspondence with the Committee on Gas. The increasing interest in LPG on the part of this committee is gratifying to the industry, providing as it does a welcome opportunity for improving knowledge of this useful, versatile and environment-conscious fuel. It was a particular pleasure and honour when Mr. Edmond Janssens, Director of the Energy Division of the Economic Commission for Europe, addressed AEGPL's Congress at Florence in 1977, in a well-received presentation on the consumption of LPG in the ECE Region.

We have been pleased to note that this study has been repeated by the Economic Commission for Europe recently in a comprehensive manner, and the paper which now follows is intended to provide more information and throw more light on the general perspective in the context of energy as a whole.

By the nature of their derivation liquefied petroleum gases (LPG) are, and will always be, a very small constituent of the complex mix which forms the world's energy supply. It is very easy, at present, to overlook their significance in any consideration of total energy problems.

Historically, these gases have been by-products, derived from the manufacture of the main petroleum fuels. As such they have acquired added value in many applications — usually in competition with main fuels — but until comparatively recently their intrinsic worth as hydrocarbons has been regarded as low.

This situation is now changing. New sources of LPG are being specifically created, at enormously high cost, whilst the current sources will continue to produce. Between now and 1990 LPG will become, at the same time, more abundant and much more expensive. Long before 1990 it will be necessary to find ways of using this material. Failure to do so will lead to conflict between the energy-producing and the energy-consuming nations. The object of this paper is to try and indicate where, by 1990, the new supplies of LPG might find outlets which will represent their true value to both producers and consumers. It is also to stress the need for thought and action which already may be overdue.

In a paper of this length it is not possible to provide more than broad indications of trends in the development of energy demand and provision. The figures used to support our arguments may be challenged but they are as good as any others which might be offered in their place. The logic they illustrate, we believe, is sound and we offer it in the hope that it will provide food for thought.

DEVELOPMENT OF THE ENERGY SITUATION TO 1990

Figures la and lb give a forecast of the energy situation in the ECE Region in the year 1990. They show the increases over the year 1976 and indicate, so far as information is available, two aspects of the situation:

The sectors of usage in which increased demand will fall.
The amount of the different types of energy which it is forecast will be available.

The total energy demand in 1990 is shown under each area heading.

Although the growth of demand in North America in the coming decade will be at a lower rate than in either Western or Eastern Europe, in terms of total energy demand in 1990 it will still be highest of the three. The growth rate in Eastern Europe will be highest of all.

What is more interesting than rates of overall growth, however, is to try and identify the sectors of energy usage in which this growth will occur and to assess the types of energy which will be required as well as the quantities. We are, in this endeavour, hampered by the incomplete nature of our available information but there is sufficient to enable us to draw some conclusions.

One thing is quite certain. From now on governments will want to exercise more control over the energy situations in their countries. The nature and intensity of the forms of control applied will differ in the three areas under discussion, but the basic objectives will be the same, namely to minimize both the costs of energy

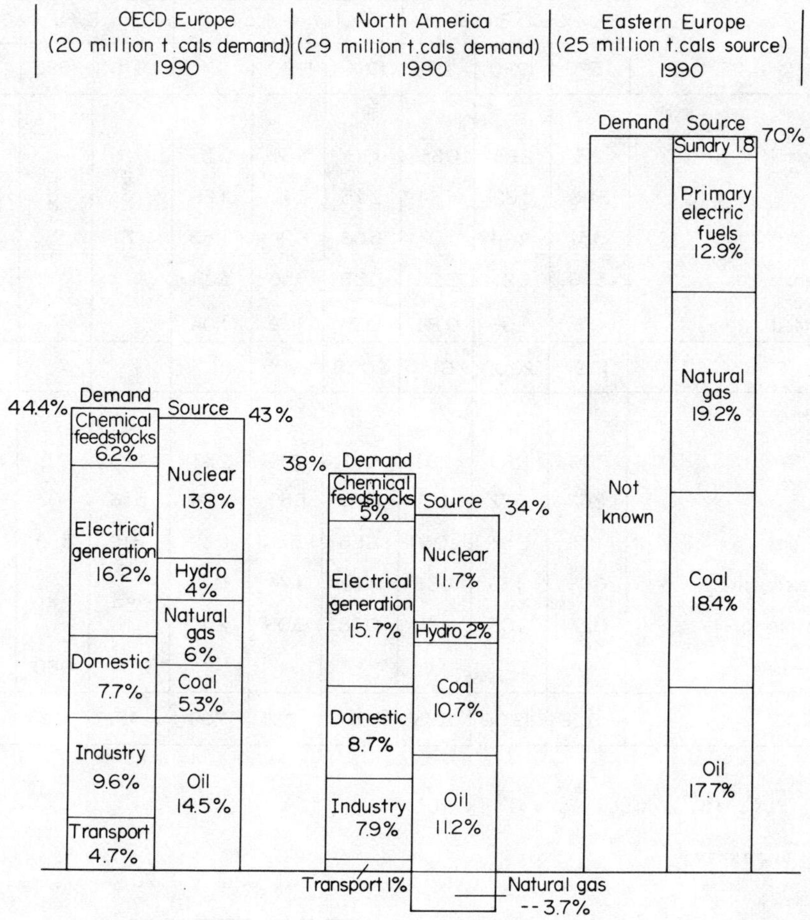

Fig. 1a. The energy situation, 1990. Percentage increase over 1976.

and dependence on foreign energy sources. National energy policies will be formulated on the principle of optimizing the use of locally available sources within the patterns of demand which develop. In some cases, the approach will be from the opposite direction and demand will be tailored to fit supply.

Thus, assuming that governments will try to optimize local supply to fit the quality as well as the quantity requirements, we can draw up the supply/demand balances below, given in T cals × 10^6 of incremental energy.

Western Europe

Demand	Source		
Electricity generation 2.25	Nuclear	1.76	
	Hydro	0.53	
		+ 0.04	+ 0.04

	OECD Europe			N. America			E. Europe		
	1976	1990	Diff.	1976	1990	Diff.	1976	1990	Diff.
Demand							Not known	Not known	
Transport	2.3	2.95	0.65	5.13	5.38	0.25			
Industry	3.48	4.82	1.34	3.75	5.41	1.66			
Domestic	3.37	4.44	1.07	5.08	6.91	1.83			
Elec. gen.	3.96	6.21	2.25	6.27	9.56	3.29			
Chem. feed	0.8	1.66	0.86	0.75	1.79	1.04			
Total	13.91	20.08	6.17	20.38	29.05	8.07			
Source									
Oil	7.76	9.69	1.93	9.77	12.10	2.33	4.93	7.5	2.57
Coal	2.62	3.32	0.7	3.65	5.89	2.24	5.36	8.03	2.67
Natural gas	1.77	2.57	0.8	6.56	5.19	-0.37	3.11	5.89	2.78
Hydro electricity	0.86	1.39	0.53	1.32	1.74	0.42	0.54	2.41	1.87
Nuclear energy	0.27	2.03	1.76	0.58	3.03	2.45			
Sundry fuels							0.54	0.80	0.26
Total	13.28	19.00	5.73	20.85	27.95	7.06	14.51	24.63	10.12

Fig. 1b. Units T. cals x 10^6.

Domestic 1.07	Natural Gas	0.80	
	Oil	0.27	
		–	
Industry 1.34	Coal	0.70	
	Oil	0.64	
		–	
Transport 0.65	Oil	0.65	
		–	
Chemical feedstocks 0.86	Oil	0.37	
		(0.49)	(0.49)
			(0.45)

There will thus be a shortage of oil-based energy which will create problems in the oil using sectors.

North America

Demand	Source			
Electricity generation 3.29	Nuclear	2.45		
	Hydro	0.42		
	Coal	0.42		
		–		
Domestic 1.83	Natural Gas	(0.37)		
	Oil	2.04		
	Coal	0.16		
		–		
Industry 1.66	Coal	1.66		
		–		
Transport 0.25	Oil	0.25		
		–		
Chemical feedstocks 1.04	Oil	0.04		
		(1.00)	(1.00)	
			(1.00)	

The problem in North America will be concerned with finding the optimum substitute for diminishing quantities of indigenous natural gas. The locally available alternative is coal. Synthesis natural gas can be produced from coal and whilst this is a more expensive solution *at the present time*, its employment in the domestic sphere may well be regarded as preferable to increased dependence on imported oil. For this reason we believe that the energy shortage will fall most heavily in the transport and chemical feedstocks sectors and that it will have to be made up from imported oil.

We have no figures for energy demand by sector in the Eastern Europe area. We believe, however, that by 1990 the supply of indigenous energy in that area will still enable government to maintain growth without recourse to any kind of imported energy. Eastern Europe may still, in fact, be a net energy exporter.

Our conclusion with regard to the development of the energy situation up to 1990 is that both North America and Western Europe will not only require to import energy but will increasingly require that energy to be of specific type. In both areas coal will tend to replace heavy fuel oil while the energy shortage will be concentrated in applications demanding the lighter petroleum products. The continued importation of crude oils — particularly the heavier crudes — will only increase a situation of imbalance between product yield and product demand. The solution of building conversion facilities to up-grade unwanted heavy products is, itself, energy intensive.

LIQUEFIED PETROLEUM GASES

These are the fractions of petroleum-derived hydrocarbons which lie within the molecular range of C3 and C4 and comprise the products propane, propylene, normal butane and iso-butane. Their physical properties are such that they can be liquefied at comparatively low pressures and light refrigeration. They can be converted back to the gaseous state by simple release of pressure and/or heating. The following is an indication of their volumetric efficiency:

1 m^3 of liquefied propane yields 274 m^3 of gaseous propane,
1 m^3 of liquefied butane yields 233 m^3 of gaseous butane.

LPG is thus capable of being stored and transported in ways that enable it to be used anywhere and in any quantity. The gas cigarette lighter and the domestic cylinder are well known to everyone but the range of LPG uses made possible by its relative ease of storage and transportation extends from the smallest scale applications right through the gamut of energy provision up to continuous burning under industrial boilers.

In terms of calorific value it is superior to natural gas on a volume basis:

Natural Gas	9590 kcal/m^3	at normal temperature and pressure.
Propane	23400 kcal/m^3	
Butane	30700 kcal/m^3	

It shares with natural gas the characteristics of clean, low-pollution combustion.

LPGs are found in both natural gas and crude oil, of which they form varying proportions depending on the compositions of the host substance. Currently they are obtained either from natural gas or during the process of refining crude oil. In either case, they represent less than 3% of the products derived from petroleum sources and an infinitesimal proportion of the total primary energy supply.

Historical Availability and Consumption of LPG in the ECE Region

Figure 2 shows analysis of LPG within the ECE Region which has been compiled from the ECE *Annual Bulletin of Gas Statistics for 1976*.

It is notable that for the years 1970 and 1976 all three areas were capable of meeting better than 97% of their consumption by internal production. The net import/export requirements were marginal. All groups showed a growth in availability but in Western Europe and North America this was at a slowing rate.

What is of particular significance in the picture shown is the range of values which can be ascribed to the various uses of LPG and this is linked to the close correlation between availability and supply. As noted in the introduction, LPG has developed as a by-product; in North America from the treatment of natural gas and in Eastern Europe from both natural gas and refineries. Whilst in every case its original market was for household use – cooking – it has had to find new additional outlets where it could and at whatever values it could obtain. As a last resort there was always a "sink", which in Western Europe was refinery fuel and in North America the chemical feedstocks pool.

The Changing Situation

Throughout the world the nature of LPG business has developed on strictly localized lines. Markets have tended to be sought close to the sources of supply and any international trading has been of a balancing nature, e.g. exports of butane surpluses and imports of propane deficits. There has been, and is, one exception to this general rule. Japan began some years ago to import LPG from the Persian Gulf oil-producing states. This was a matter of environmental necessity, the air-

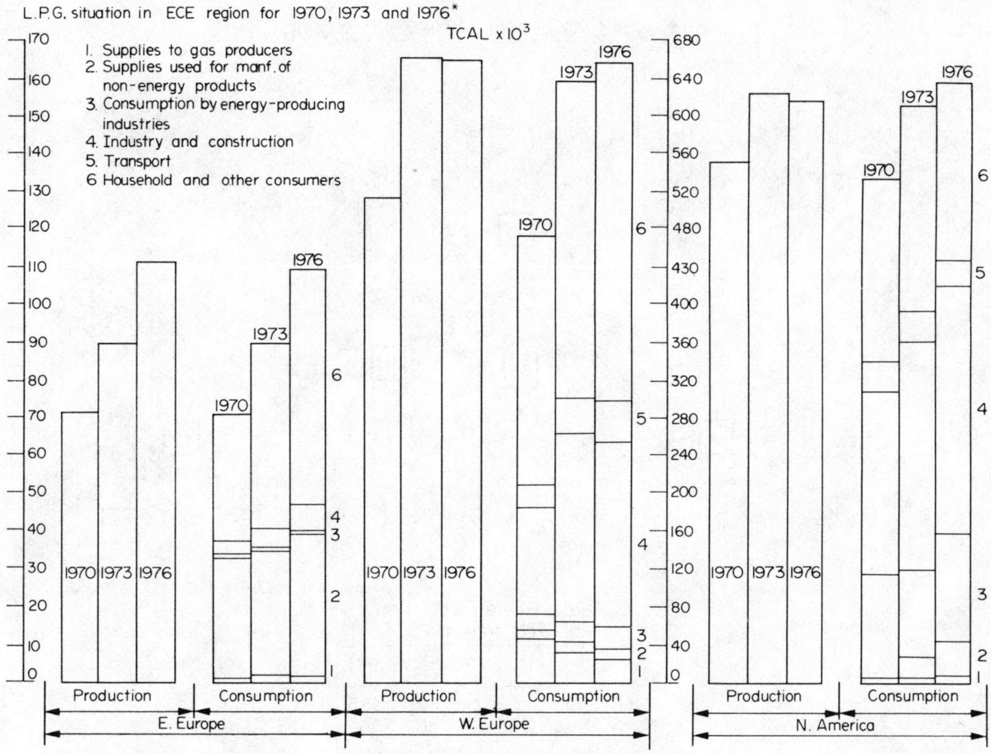

Fig. 2. *The ECE Annual Bulletin of Gas Statistics for 1976.*

pollution problem being such that Japan was prepared to pay a high premium (much higher than that offered by the calorific advantages) to replace some fuel oil with an environmentally acceptable fuel. Even so this trade had only reached 7 million tons per year by 1976 and formed only a small part of international energy traffic.

The situation has already begun to change. Since the oil industry began, and particularly since the discovery of the huge Middle Eastern oilfields, gas has been flared off at the well heads in order to stabilize the crude and ensure the quality of natural gas. Many millions of tons of these valuable fuels, mainly LPGs, have been and are still being wasted each year. In the days when oil energy was cheap and appeared to be everlasting, it was simply not considered economically viable to incur the high costs of collection, separation, transportation and storage necessary to get them to where they could be used.

Energy is no longer cheap and the world is gradually beginning to realize that its resources, particularly its petroleum resources, are limited. In particular, the oil-producing states are sensitive to the fact that their most valuable sources of wealth — in some cases their only sources of wealth — are limited. The flares on their oil fields are particularly irksome and no one can fault their desire to use this gas if only in order to conserve the oil for a little longer.

There are now projects and plans, in various stages of development, for recovery of the well-head gases, the larger part of which will be collected as LPG. Estimates of the amounts which will ultimately be available are numerous and are to be found

World wide LPG availability 1978 through 1985

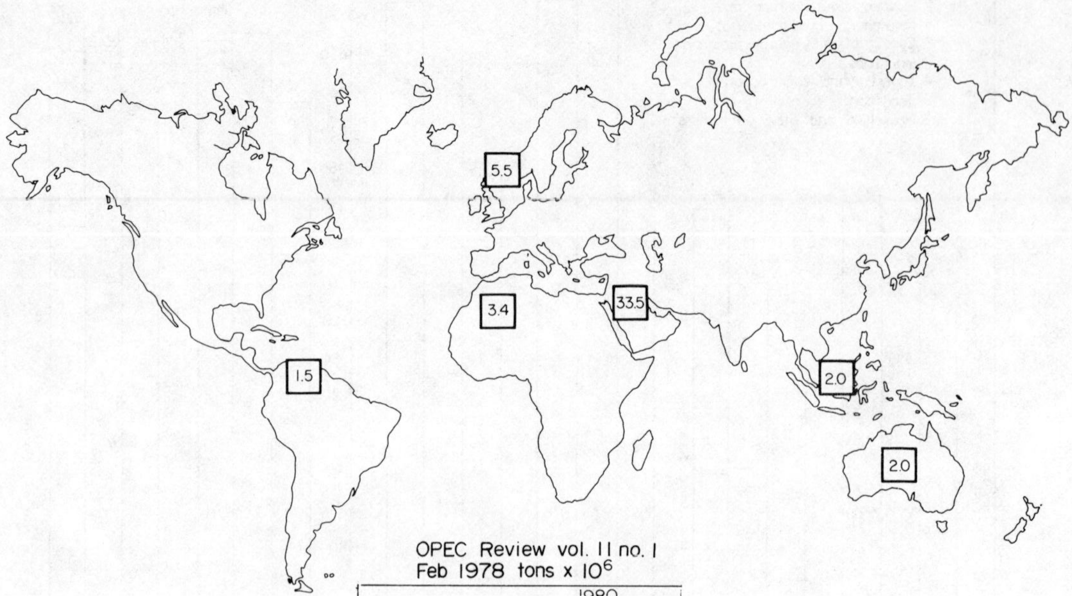

OPEC Review vol. 11 no. 1
Feb 1978 tons x 10^6

Source	1978	1980 1982	1985
Middle East	10.3	22.5	33.5
North Africa	1.2	3.0	3.4
Indonesia	0.7	1.5	2.0
Venezuela	1.7	1.7	1.5
OPEC total	13.9	28.7	40.4
Australia	1.6	2.0	2.0
North Sea	1.2	4.2	5.5
World total	16.7	34.9	47.9

Fig. 3.

in every issue of the petroleum press. Figure 3 shows one such set of estimates — a conservative one based on known projects considered to have a good chance of achievement. They may well be understated.

We have hinted at the danger of conflict arising over these new volumes of LPG. It is a real one and its roots are to be found in the confrontation of two sets of logic, that of the producer and that of the consumer.

The producer logic looks upon this new LPG as a means of conserving crude oil by replacing it. It will be enormously expensive to make available (the Saudi Arabian Mastergas plan is estimated in billions of dollars) and, once available, it should be made to pay for its costs and earn a reasonable return on the money spent to collect it. At the very least it should be sold at a value equivalent to the crude oil it replaces if the producing country's overall revenue is not to fall. Thus it will be expensive at the loading port.

The consumers' logic is also related to cost. The LPG will have to be transported over very long distances to reach its markets. This will involve expensive, large-scale, refrigerated vessels with high freight rates. Also, it will be necessary to build large-scale terminals to receive the products, store them and in some cases reheat them for inland distribution. Finally, the existing LPG markets are not capable of considerable expansion in the short term. Those which have high values and are able to pay high prices are saturated. Those which can be expanded will require the LPG to be at costs close to those of existing fuels — most notably industrial fuels. The conflict which could arise from this situation is most likely to take the form of enforced liftings of LPG linked to liftings of crude oil. Producing countries will be compelled to dispose of this LPG somehow, which for example could be linked to crude offtake. Obviously they will be looking for relatively high prices to give acceptable returns on the investments involved in producing LPG. Consumer countries, on the other hand, may be forced to receive this material and most probably develop new markets to absorb it at prices which will also give suitable returns on investment, without disrupting existing market patterns.

It is our belief that a situation of conflict need not arise at all if sufficient thought and planning is given to this problem immediately.

THE PLACE OF LPG IN THE 1980s

We can predict, with some certainty, that the new supply situation will not result in major geographical changes in LPG markets. We have already pointed out that LPGs cannot be cheap and only the highest industrialized areas will be able to use the greatly increased volumes. Nor is it likely that Eastern Europe will be directly affected by them.

The main markets will continue to be Japan, Western Europe and North America (in this case the USA — not Canada). Japan will continue to increase the amounts of LPG it takes, so long as it considers that the price justifies the clean-air premium it currently pays for LPG over fuel oil. There will be a physical limit imposed by Japan's capacity to import and use the material and this will be well below the total volume available. Nevertheless Japan will continue to be an important market.

Potentially the USA might be thought to be capable of absorbing all the forecast incremental volumes that will become available. Its natural gas stream is diminishing and its huge LPG market is beginning to feel the effects of tightening supply. We have tried to show, however, that the USA is concerned about both dependence on foreign energy and reduction of energy costs and that it will seek to use its large reserves of indigenous coal. There will be strong government pressure to encourage LPG users to switch to less expensive and/or more widely available forms of energy. Some LPG will certainly go to USA but this will probably not be as much as many current forecasts indicate. There is, as Fig. 2 shows, enormous scope within North America to convert many existing LPG applications to other fuels and the domestic market, which is the only really sensitive one from a political point of view, offers better prices than do any of the others.

We have already expressed the opinion that the USA will choose coal and not LPG for synthesized natural gas feedstock. This view is not universally shared but it fits best with our belief that government controls over energy policy will become stronger. We cannot envisage that the USA government will continue to allow imported energy to be subsidized in the price of declining supplies of natural gas. Over the course of the 1980s we foresee the value of LPG in the USA becoming higher as it gradually moves to the markets which can afford high prices. There will be some importation of LPG but we do not foresee a large-scale expansion of LPG terminal facilities to handle imports unless there are guarantees of price stability and government non-intervention. Overall LPG usage in the USA may even decline over the next decade.

The only remaining market will be Western Europe which, predictably, has not received a great deal of attention from forecasters of this subject. It is very possible, however, that LPG usage in Europe could expand very rapidly indeed by 1990.

At the present time this market is around 14 million tonnes based virtually entirely on refinery-produced gas and is characterized by small-scale and diversified sectors. In point of fact Western Europe has less need of LPG than either of the other two areas and could face static or even declining volumes without serious disruption of economics or standards of living.

On the other hand, Western Europe is very heavily dependent on imported energy, particularly oil. The North Sea production of both oil and natural gas is solving the problem for part of the area but the dependence on imports from Middle East and Africa will remain at least till the end of the century.

At the present time one of the greatest problems facing Western Europe is an over-capacity in refineries and a growing imbalance in the out-turn of those refineries. In particular there is a need for gasoline which continues to grow despite enormous price inreases, but industrial demand for fuel oil has fallen with industrial recession.

We have shown that this area's needs will be felt principally in a requirement for lighter products. There will be demand for industrial fuel oil; however, it will be primarily in the automotive, domestic and feedstocks sectors that the deficits will fall most heavily. In each of these sectors LPG could play a very important role although it will not necessarily find direct outlets in all of them.

Recalling the elements of the confrontation situation described earlier, we spoke of the need to find outlets for LPG which would provide both high netbacks to producers and large volumes to provide the scale required to justify tankers, terminals, etc. Where are these to be found in Western Europe? In industry — large volumes, but the competing fuels are low in value. Domestic use — large volumes could be possible — whole towns could be fuelled with LPG — but natural gas is also becoming available (in many cases as the mother-lode of the LPG) and is cheaper. Feedstocks — LPG is a good substitute for naphtha, the main petrochemical feedstock in Europe, and very attractive quantities could be absorbed into the European chemical industry without a lot of additional capital cost. This is an outlet which is expandable or retractable within reasonable limits and some development will almost certainly take place.

The remaining potential market is the automotive sector and this could be the key to the solution of a number of problems. LPG is an excellent automotive fuel which is already widely used in many countries throughout the world. Its technology is established and accepted; costs of vehicle conversion are low in comparison to existing vehicle cost; a distribution system for LPG can be quickly established within those already developed for gasoline and diesel fuel. Additionally, it is virtually pollution free and more than meets emission standards — even some not yet published. Finally, LPG can be taken straight from an ocean tanker to a vehicle, it does not require refining or other energy-demanding processing on the way.

Where LPG has been allowed to compete with gasoline it has found ready acceptance within the commercial (mainly urban) fleets which are vital to the smooth running of a country's transportation services. In Holland and Italy, for example, LPG vehicles are now commonplace. In Tokyo almost every taxi runs on LPG. But it must be allowed to compete with other automotive fuels.

The only obstacle to the growth of automotive LPG markets is its price compared with that of gasoline. Where fiscal systems allow a differential in the pump price of fuel which is favourable to LPG, use of that product flourishes. The reluctance of governments to allow this development is based on a fear that gasoline revenues will fall, but the point has not been studied with a view to seeing how the country might benefit by allowing it. LPG is a more efficient automotive fuel than gasoline but it is inconceivable that all the LPG available world-wide will ever amount to more than a small fraction of Western Europe's gasoline demand. It could provide an acceptable solution to problems on both sides of the producer/consumer dilemma. Certainly, no other outlet offers the kind of "mass market with high value" solution which will have to be found by this, the area most heavily dependent upon imported energy supply.

CONCLUSIONS

Our studies lead us to one main conclusion, which we have already expressed in the introduction; by 1990 LPG will be both more plentiful and more expensive than it is today. By 1990 no longer will it be treated as a by-product but as a high-value energy form which will justify its costs only by use in applications which save even more expensive energy.

Current supply facilities will continue to operate, the refineries and local natural-gas plants will continue to make LPG but the value of their production will reflect that of imports and will be enhanced. Locally produced LPG will have the cost advantage of freight and terminal penalties incurred by the imported material.

Current markets will change as costs change. The "sinks" will be other mass markets which can act as buffers to maintain the smooth operation of production — loading — transportation — reception, which will be needed to ensure that all parties enjoy a harmonious relationship. Values will thus range from a high-level, close to gasoline retail prices, to a low-level equivalent to naphtha for chemical feedstocks.

Western Europe could well become the most important LPG market in the world, with North America gradually declining as it reduces its dependence on foreign energy supply and develops new techniques to use local resources.

The next decade will see many changes, many problems and many crises. We offer the foregoing in the hope that it may provide at least a starting-point for those who will have to face the problems we are sure will be associated with these hitherto unregarded products.

The American Point of View on the Future of the Gas Industry

The American Gas Association

UNITED STATES ENERGY DEMAND

Phenomenal consumption levels have characterized United States energy development. In 1950, following the rapid post-war development after World War II, and entry into the conflict in Korea, the USA consumed 34 quadrillion British thermal units of energy (quads), equivalent to 35.9 Exajoules (EJ). By the middle of the present decade consumption passed 70 quads (73.9 EJ) and it is projected to reach a national level of 100 quads (105.5 EJ) by 1985. Estimates indicate as much as 150 quads (158.3 EJ) may be involved at the turn of the century.

Because of this trend and concerns about energy supply, the Carter Administration's proposed National Energy Plan focused on conservation as a mainstay of future energy policy for the United States. An annual energy-consumption growth rate of only 2.5 per cent between now and 1985 was assumed despite previous increases of about 4 per cent per year. Many US energy analysts consider conservation benefits over-estimated by the Carter Administration planners. Careful examination leads us to the conclusion that overall energy usage will be reduced by no more than 5 per cent from the predicted totals noted above. Thus, US energy imports will not decline substantially from present high levels. To the extent the Carter plan fails to meet ambitious conservation goals, larger levels of oil imports are mandated throughout the rest of this century to sustain economic health.

The 100-quad (105.5 EJ) level in 1985 can be analysed to provide rough percentages of the expected energy mix. Although coal is America's most abundant fossil fuel, development of the vast domestic reserves is constrained by technical, environmental and political problems. Coal's share of the 1985 total is expected to be 17.0 quads (17.9 EJ). Domestic crude oil production is estimated at 27.1 quads (28.6 EJ). Natural gas liquid projections are dependent upon overall gas production but are assumed to provide 2.2 quads of energy (2.3 EJ). Technical, financial and environmental delays have scaled down previous optimistic United States government predictions for nuclear power development to 11.8 quads (12.5 EJ). With the exception of natural gas, all other energy sources, including hydropower, geothermal, oil shale and solar, are forecast by the US Department of the Interior at a level of 4.8 quads (5.1 EJ).

Assuming removal of Federal field price controls which would eliminate three decades of inhibited development for dry natural gas, and development of supplemental supplies not presently on stream, the gas contribution for 1985 could be 25.3 quads

(26.7 EJ). All of the above-noted projected contributions from domestic sources total only 88.2 (93.1 EJ) of the 100 quads (105.5 EJ) required, leaving 11.8 quads (12.5 EJ) to be supplied by imported oil. Shortfalls in any of the domestic energy sources would increase the demand for oil imports, with attendant exacerbated economic consequences for the United States.

Clearly, in planning to meet energy needs for 1990 and the remainder of this century, the United States does not have the luxury of debating which energy source should be developed first. All sources must be developed as rapidly as possible to meet increasing demand required for a growing, healthy economy.

ENERGY SUPPLY

The factors and components of gas supply are of special interest. United States' proved reserves are reported currently at more than 200 trillion cubic feet [Tcf ($5.7 \times 10^{12} m^3$)] and estimates of potential reserves vary from 700 ($19.8 \times 10^{12} m^3$) to 1200 Tcf ($34 \times 10^{12} m^3$). This represents approximately 60 years of conventional supply at present consumption rates of about 20 Tcf ($565 \times 10^9 m^3$) per year. Plainly, the United States has vast resources of natural gas waiting to be produced but development has been hampered by unproductive Federal price controls on sale of gas from producing states to areas of consumption across state boundary lines.

Even assuming near-future relaxation of Federal governmental restraints on production of natural gas, it would be around 1990 before US conventional gas production reaches maximum level. However, only gradual decline in production rates would then occur over a period of decades. Concurrently, supplemental gas sources will increase their share of the gas mix. About half the gas sold by utility companies in the year 2000 could come from supplemental sources.

The forecasts of gas supply which are presented in this paper are, of course, not guaranteed. However, we believe they are reasonable and logical if proper incentives are provided. Government inaction, on the one hand, could reduce supply; but research breakthroughs or a prolific gas discovery, on the other hand, could increase supply. The outlook should be treated for what it is — a reasonable scenario based on today's knowledge.

Examination of the possible gas-supply components in 2000 indicates that conventional natural gas from the contiguous forty-eight states (excluding Alaska and Hawaii) will continue to be the major source of gas energy. Of a total 33 Tcf ($935 \times 10^9 m^3$) of gas consumed, 18.1 Tcf ($513 \times 10^9 m^3$) will be from historic conventional sources augmented by offshore development.

Proved reserves in Alaska are about 32 Tcf ($906 \times 10^9 m^3$), but potential reserves in the range of 200 Tcf ($5.7 \times 10^{12} m^3$) are expected to come on stream with construction of a pipeline to market so that Alaska will yield approximately 3.6 Tcf ($102 \times 10^9 m^3$) annually long before the turn of the century.

Imports from Canada of gas in excess of Canadian needs will probably decline to amounts of about 0.6 Tcf ($17 \times 10^9 m^3$) per year. However, negotiations with Mexico (currently stalemated by US government concerns regarding price levels) are expected to result in an import level of 1 Tcf ($28 \times 10^9 m^3$) annually long before 2000.

Production of synthetic natural gas (SNG) from liquid hydrocarbons will have expanded, with twenty-eight plants providing 1.2 Tcf ($34 \times 10^9 m^3$) per year. Coal gasification will yield 4.0 Tcf ($113 \times 10^9 m^3$) of SNG, and other sources such as geopressurized-geothermal zones in the Gulf of Mexico area, new technology in biomass conversion and hydrogen developments should provide 1.5 Tcf ($42 \times 10^9 m^3$).

These North American sources listed above total 30 Tcf ($850 \times 10^9 m^3$), half again as much as present consumption levels. However, the US gas industry will also be looking to other gas-production areas around the world for liquefied natural gas (LNG) imports amounting to 3.0 Tcf ($85 \times 10^9 m^3$), about 10 per cent of total supply, as the remainder of supply in this turn-of-the-century scenario.

Liquefied natural gas will serve a vital role in maintaining and eventually increasing US levels of gas supply, thus preventing escalating costs to the consumer because of under-use of plant and distribution systems. Using imported LNG will ease pressures for greater crude-oil imports. A growing trade is projected to reach a level of 1.6 Tcf ($45.3 \times 10^9 m^3$) by 1985.

Dependence on overseas sources of petroleum is a relatively new experience for the United States which now imports almost half its oil supplies. Natural gas, however, currently is about 95 per cent domestically produced and, even with growing LNG importation, will remain a secure and reliable energy source.

LNG import projects approved by and pending before the newly created US Department of Energy provide opportunity to replace 10 per cent of oil imports with LNG. This plan is superior to use of oil and other energy alternatives because of factors relating to consumer cost, balance of trade, employment, environment, national security and capital investment.

Here are the supporting facts for this contention.

ARGUMENTS FOR LNG

On a cost basis to the end-user, LNG from pending projects would be less than one-third the cost of electricity and about equal to number two fuel oil refined from imported crude oil at world price levels. Recent US government regulatory decisions have approved imported LNG pricing on a rolled-in basis so that cost to consumers is minimal when the higher-cost supplemental fuel is averaged into the distribution system.

LNG importing is superior to oil importing because of greater capital intensiveness, a much larger portion of cargo which would move in US flag carriers and lower landed cost. Plans to import 1.6 quads ($45 \times 10^9 m^3$) of LNG annually by 1985 in lieu of an equivalent amount of energy from oil would reduce net dollar outflow for energy by \$1.5 to \$2.5 billion per year depending on the extent of US financing, capital goods and construction labour used. Terminal and ship construction for LNG development could create almost 20 000 jobs for American workers over a 5-year construction period. This would result in an annual payroll of \$310 million and addition of \$900 million to the nation's gross national product. Permanent employment would increase by 2900 jobs.

Air and water pollution and solid wastes from LNG operations are negligible after initial coastal zone impact during port and regasification terminal construction. Compared with increased generation of electricity from coal or more crude-oil-tanker traffic, LNG clearly represents a superior environmental alternative.

LNG importation diminishes US dependence on other nations by diversifying imports more equally among supplying nations. Such nations would have substantial investment in liquefaction facilities and considerable foreign debt requiring servicing. They would also have fewer market alternatives than crude-oil-exporting nations have.

United States capital investment required to provide an additional unit of end-use energy from LNG is only one-ninth that of electricity. The investment required for LNG receiving terminal and regasification facilities is slightly less than for oil-refining capability of the equivalent end-use energy output.

THE WORLD OUTLOOK

Can the world gas-resource base sustain increased production and movement from production areas to consuming nations? The answer is clearly, yes. In contrast to the United States supply situation where conventional production is expected to peak about 1990, resources of natural gas in the rest of the world have hardly been tapped. Because world gas reserves are being depleted far less rapidly than oil reserves, gas can be expected to remain available in substantial amounts even though oil production may peak in the 1980s according to US government studies.

As the major consuming nations of the world begin to deplete their own indigenous natural-gas resources, opportunities for intercontinental trade in LNG will increase. Long-term availability of natural gas throughout the world enhances the attractiveness of world trade in this premium energy commodity.

The American Gas Association analysed future world gas supply and reported last year to the Tenth World Energy Conference in Istanbul on natural-gas resources stating that there is a worldwide capability for substantially increasing production of conventional natural gas in the coming decade and sustaining steadily increasing levels of consumption far higher than currently at least through the first quarter of the twenty-first century.

World-proved reserves are estimated at about 2200 Tcf ($62 \times 10^{12} m^3$) and remaining undiscovered resources are estimated at about 7500 Tcf ($212 \times 10^{12} m^3$). Should the present world production rate of 48 Tcf ($1.4 \times 10^{12} m^3$) per year double to 96 Tcf ($2.7 \times 10^{12} m^3$) annually, the estimated remaining conventional natural-gas resource base would be sufficient to sustain production at or near that level for at least another 50 years.

UNITED STATES LNG PLANS

Despite this optimistic outlook, the history of US LNG import development is a story of frustration because of uncertain government policy and regulatory delay. Early in the 1970s American gas-industry planners considered some 2.2 quads (2.3 EJ) annually or 6 billion cubic feet ($.17 \times 10^9 m^3$) per day as a likely rate of LNG importing by 1976, drawing on a half-dozen sources and projecting a dozen terminals in use. Actual imports totalled only 0.01 quad (.01 EJ), from Algeria to a single regasification terminal, in Boston harbour. Industrial supply curtailments during cold winter months would hardly be a problem for the United States to surmount if planned projects had come to fruition. But 1978 marks the start of the first large-scale shipments of a project designated El Paso I, from Algeria to receiving terminals at Cove Point, Maryland, and near Savannah, Georgia. A second large-scale project, Trunkline, has received US Department of Energy approval to import LNG from Algeria to Lake Charles, Louisiana, beginning about 1981.

Four long-term import projects are presently pending before the Department of Energy which is still trying to determine just what its policy is regarding this vital fuel. Conditional approval has been expressed for one plan to bring LNG from Indonesia to southern California. If all are approved, the total level of LNG imports could rise to a modest 1.68 quads (1.8 EJ) by 1985 with plans being formed for an additional 3.2 quads (3.4 EJ), but all of these proposals are not yet filed with the government.

These prospective developments include trade with Australia, Trinidad, Colombia, Iran, Nigeria and the USSR. The latter involves both a trans-Pacific Nahodka-to-US West Coast link and a Murmansk-to-Atlantic Ocean route.

MARITIME CONSIDERATIONS

Development of LNG importation offers several advantages to the ailing US maritime industry. LNG tankers are central to attaining the US Merchant Marine Legislation objectives of assuring a maritime capability to construct and repair a US fleet, maintain a skilled maritime labour force, preserve and develop technological skills and induce purchase of modern capital equipment and facilities.

About 55 per cent of LNG imports to the US will move in US flag tankers compared with about 4 per cent for crude-oil imports. Eighteen of these LNG tankers have already been ordered, built, or are under construction. Total orders add $3 billion gross revenue to America's ailing shipbuilding industry.

Should the US government fail to encourage LNG imports as it considers its policy on projects now pending, hesitation would constitute a major policy inconsistency. Delay would result in lay-up of ships in service or cancellation of construction. These ships have been granted Maritime Administration loan guarantees and eleven of them are receiving construction subsidies. Several nations entered LNG tanker construction prior to the United States and this has created a world surplus of such ships through 1985. As a result, US flag tankers not employed in the project for which they were intended would be unlikely to find alternative employment.

CONCLUSION

The United States has experienced growing curtailments of natural-gas supply for industry since 1972. This is an avoidable impact upon industrial users since large quantities of gas are being flared around the world. Total US curtailed volumes since 1972 amount to 1.6 Tcf ($45.3 \times 10^9 m^3$) while more than 5 Tcf ($140 \times 10^9 m^3$) of a premium commodity are being flared annually in producer nations. Large non-associated world gas reserves exist as well. LNG importation could help meet US needs for clean, efficient energy in coming decades and at the same time help gas-producing nations utilize their gas resources to mutual advantage.

If rational men are still able to function in a rational manner, an adventure in technological trade is only beginning. The inherent advantages of natural gas will continue to be appreciated and used by the United States in coming decades and with appropriate technological development as far into coming centuries as can be foreseen by scientists and engineers.

High Depth Underground Coal Gasification: Potential Energy Source for the Future

M. Ledent and J. Ribesse

Belgium

INTRODUCTION

Coal layers are by far the largest fossil energy reserve of the earth's crust, especially if one takes into account the very deep layers that are not productive by conventional methods.

Extracting energy from these layers by underground gasification would free man from the arduous work of mining, putting this high depth enormous energy resource at his disposal in a clean fluid form and at economically acceptable conditions.

The first development tests that have been done up to now at shallow depth have come against two difficulties: gas leakage and water penetration in the underground gas-producing layers. Performances were poor.

New developments aim at producing high depth gasification leading to three advantages: no leakage, no water and high-pressure operating level, favourable factors for performance and better possibilities.

This study underlines certain theoretical aspects of underground gasification such as the influence of operational factors and describes some technical systems giving evaluations of gas-production costs.

Important experimental research programmes on high depth underground gasification are being undertaken in different countries. This is the case in Belgium with the INIEX* Project being undertaken in co-operation with the German Federal Republic for a period of 6 years ending in 1984.

The success of *in situ* high-pressure gasification would cover a large part of our countries' energy demands in the form of transport gas, for instance, in addition, or as substitute for natural gas.

Many have tried to assess the importance of world reserves and of local reserves of fossil energy, in order to direct long-term energy options.

The study of the various assessments brings out the following considerations:

---

*Institut National des Industries Extractives.

Uranium cannot cover an important part of the energy needs before advanced reactors are developed.

Coal reserves are considerable: they alone represent 4 times more energy than all other non-nuclear fuels taken together, and their geographical repartition is rather diversified. However, the estimated reserves do not include the coal resources lower than 1200 m as they are considered up to now as being unexploitable.

For Europe, the amount of these deep-lying resources is tremendous (Fig. 1); their estimate has not yet been made but some indications enable us to assess their importance. German geologists think that the exploitable Ruhr reserves down to a depth of 1200 m amount to 10 billion tons and that the total reserves down to a depth of 5000 m reach 1000 billion tons. In Belgium, the probable reserves reach 15 to 20 billion tons if one includes the northern extension of the Campine deposits, i.e. 8 times the total quantity that has been won during a 150 years' period. Dutch geologists reckon that the underground of the Netherlands conceals about 1200 billion tons of coal, the major part of which would lie between 1500 and 3000 m. But conventional coal extraction is only possible where the seams are thick and not at great depths.

From a social point of view, the profession of coal miner is a very hard one requiring considerable physical effort and people are clearly leaving it. From an economic point of view, European coal has become, except for some exceptions, one of the most expensive sources of energy (see Fig. 8).

Conventional mining techniques must be excluded for both the deep-lying deposits and the more frequent partly covered seams. New techniques will be required for the exploitation of these substantial deep-lying deposits. These techniques must meet three requirements:

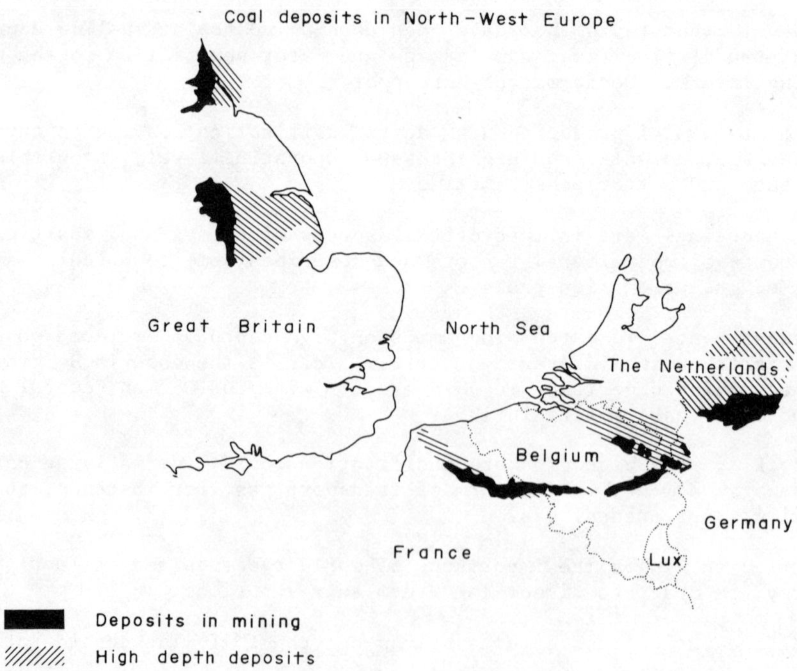

Fig. 1. Coal deposits in North-west Europe.

relieve the human element of underground mining work,
enable the exploitation of deep-lying deposits,
extract the coal's energy in fluid form.

The introduction of high-pressure underground coal gasification satisfying the above conditions would fill an important part of the energy requirements of our countries while simultaneously reducing our dependency.

UNDERGROUND GASIFICATION

Underground Gasification Experiments Carried Out[1]

The underground coal gasification by air has been experienced in the USSR for over 40 years at depths between 100 and 300 m. This method is suitable for seams with a minimal opening of 2 m and gas with a calorific value between 3700 and 4200 kJ/m^3 (900-1000 kcal/m^3) was obtained.

This gasification technique in shallow depths has to cope with difficulties as to the tightness of the underground gas generator and its interference in water-bearing beds.

The results recently obtained in Wyoming, USA, are much better: gas with 5400-6300 kJ/m^3 (1.300-1.500 kcal/m^3) thanks to greater tightness of the gas generator. It is, however, too soon to draw any conclusion as the mining work is still recent and done on a small basis, and water-infiltration problems could arise later on.

Considering all the information presently available, it can be inferred that the major obstacle to the development of gasification at shallow depths is the untightness of gas generators. In addition to this, the pressures which might be used are limited and it is impossible to benefit from the advantages given by using high pressures, as, for example, much better transportation of the fluids in the gas generator.

Objectives of Underground Gasification at High Pressure

The experience acquired in Belgium on extraction of deep-lying coal has shown that the shale covering coal layers manifest a plastic behaviour which makes up a tight barrier blocking the passage of gas and liquids. Thus their deformation plugs the holes resulting from the operation. The depth at which the coal shales behave so depends to a great extent on the geological conditions of the deposit and on the composition of the beds. In Belgium, it can be reached towards 500 to 600 m in the north and can move towards 1000 m and even more in the south.

Going down to great depths — which is an important handicap in the classical mining industry — can bring three advantages to underground gasification: no water, tightness of the rocks, suitability for resistance against high pressures which is a favourable factor for the gasification performance; the pressure level should, however, remain lower than the hydrostatic pressure of the environment.

On the basis of these considerations and of the preliminary studies made in Belgium by INIEX, it has been decided to start research tending to develop deep underground gasification and to work out the technical and economic conditions of its industrialization.

Principle of Underground Gasification

Underground gasification consists of injecting a reactive gas (air, oxygen and steam or hydrogen) in the boreholes drilled down to the coal seams. The quantity of reactive gas must allow an exothermic reaction to be maintained in the underground gas generator.

Before the reaction is started up, a passage should be made for the gases between the boreholes in the coal seam through an operation called linking. Experiments showed that this operation can be realized at distances of up to 80 m.

Figure 2 shows a principle scheme of an installation for underground coal gasification under pressure according to which coal seams will be valorized ever deeper; this operation will be experimented with in Belgium.

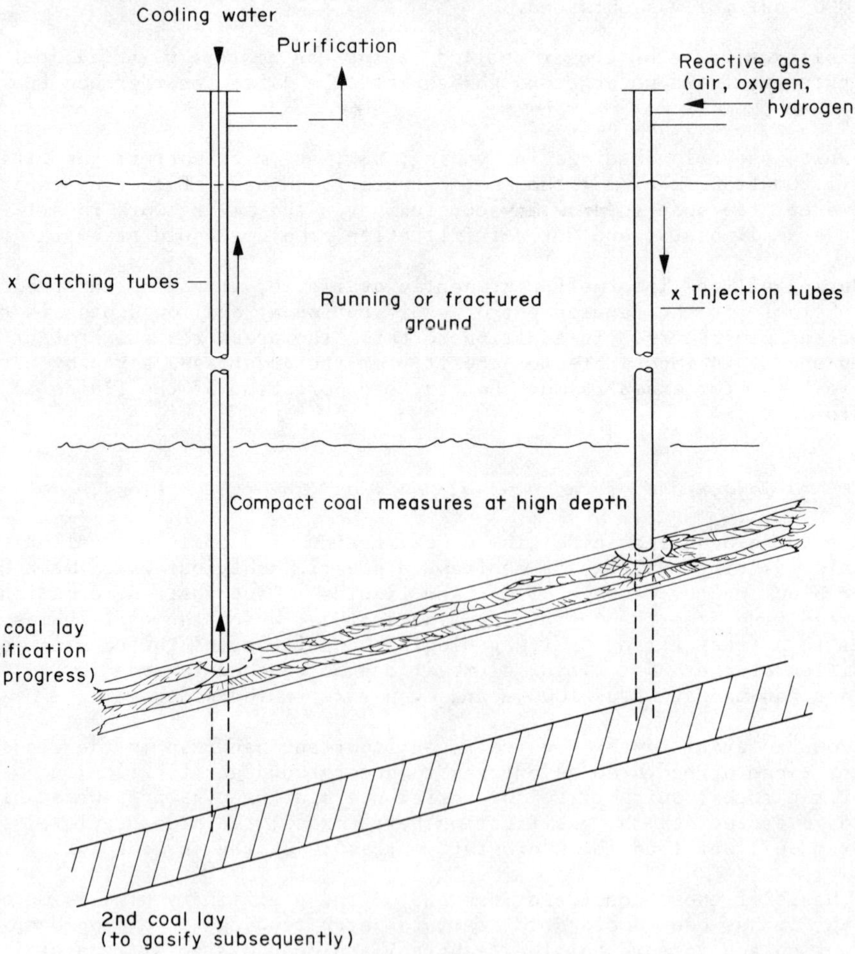

Fig. 2. Underground coal gasification. Principle of high-pressure gasification

The reactive gas used for underground gasification is either air under high pressure and then the gas produced from the gasification is lean, i.e. 4000 to 5000 kJ/m^3 and can only be used for the local electric power generation, or oxygen and steam, which considerably improves the quality of the gas, i.e. 10 000 to 11 000 kJ/m^3, and this enables a diversified valorization of the gas that will be submitted to a further treatment with a view to producing ammonia or methanol synthetic gas or synthetic natural gas.

The energy needed to produce oxygen absorbs part of the gas produced by the gas generator.

Finally, if one uses hydrogen as the gasifying agent, i.e. $C + 2H_2 \rightleftharpoons CH_4$, one obtains a gas with a higher calorific value and the coal can be valorized under the form of gas similar to natural gas which is able to be transported over long distances and used in all energy activities: gas-distribution system for domestic use, thermal use in industry, chemical industry and power plants. The hydrogen needed to gasify can be produced by reforming with steam part of the gas produced by the gas generator

$$\tfrac{1}{2}CH_4 + H_2O \rightleftharpoons \tfrac{1}{2}CO_2 + 2H_2$$

This technique requires a tight underground gas generator operating under high pressure and this can be realized at great depths.

Figure 3 shows a principle scheme of ground installations for the injection of the gas intended for underground gasification and of utilization of the gas leaving the gas generator.

Theoretical Aspects of Underground Gasification at Great Depth[2]

Basic principle for an underground gasification reactor able to be used in a mathematical model (Fig. 4). The reactive agent, either air or oxygen or hydrogen plus steam, heats up when entering into contact with hot solid materials before starting to burn (air or oxygen) or to producing methane (hydrogen) with hot coke.

The gases produced by the combustion of the coke then cool down and balance the watergas, reforming and Boudouard reactions that are endothermic. Then the gas transmits its sensible heat to the coal thus taking part in its distillation and heating.

Among the basic parameters that characterize the reaction, one should mention the coal reactivity. This defines the reaction's temperature level as little reactive coal requires a higher gasification temperature which means that more reactive gas and less steam are needed, and that the loss of thermic energy is high.

A second important parameter is the level of *thermic losses* of the gas generator into the environment. These losses are linked with the gas generator temperature and with the specific features of the coal seam.

A third important parameter is the gasification *pressure* that has a considerable influence on the operation characteristics of the gas generator, and more particularly in the case of hydrogasification. Its impact on the cost of gas is very important and shows the unquestionable advantage of using high pressures.

As to the composition of the coal to be gasified, its influence mainly determines its ability to gasify at low temperatures, i.e. the reactivity mentioned above.

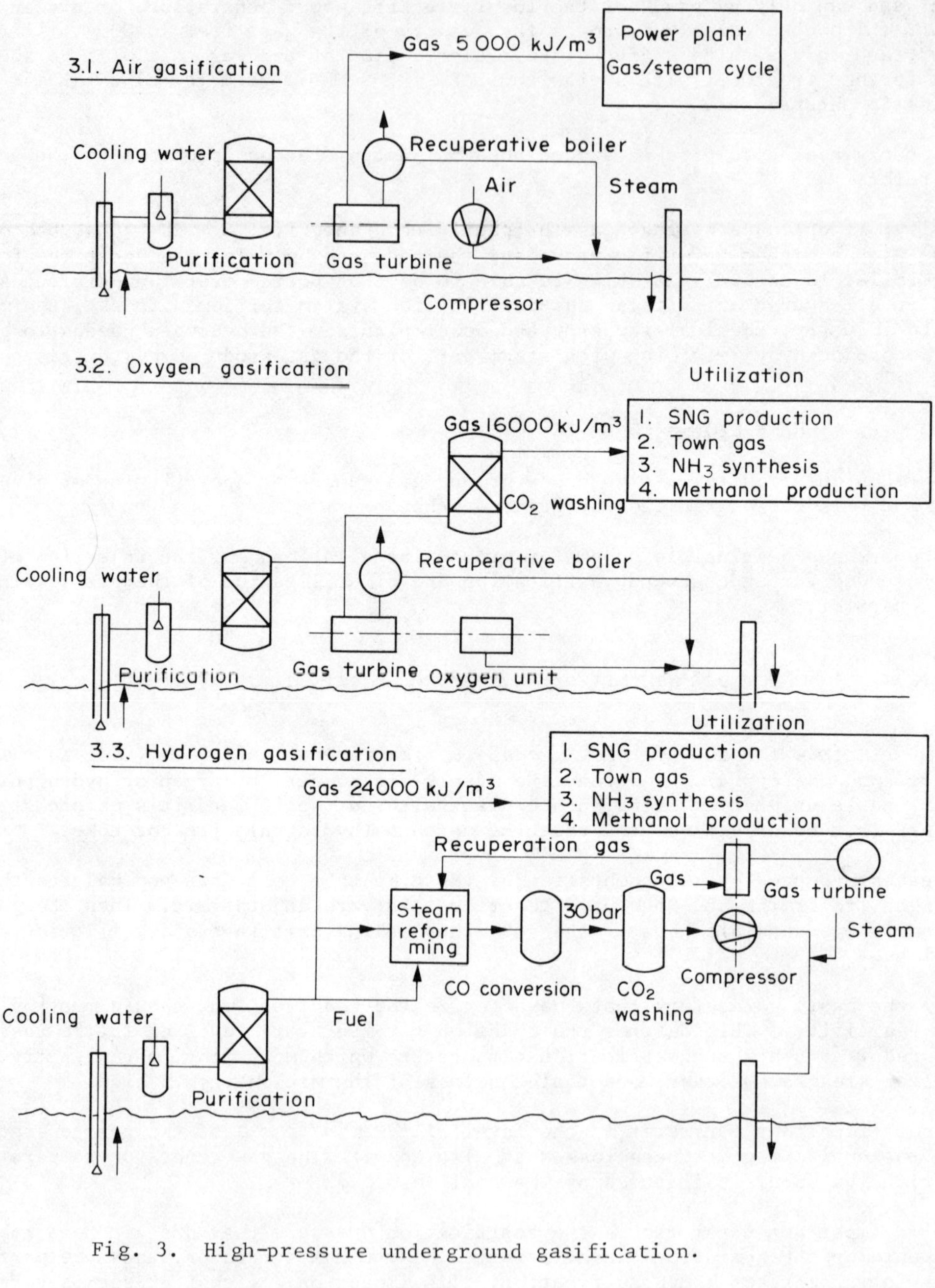

Fig. 3. High-pressure underground gasification.

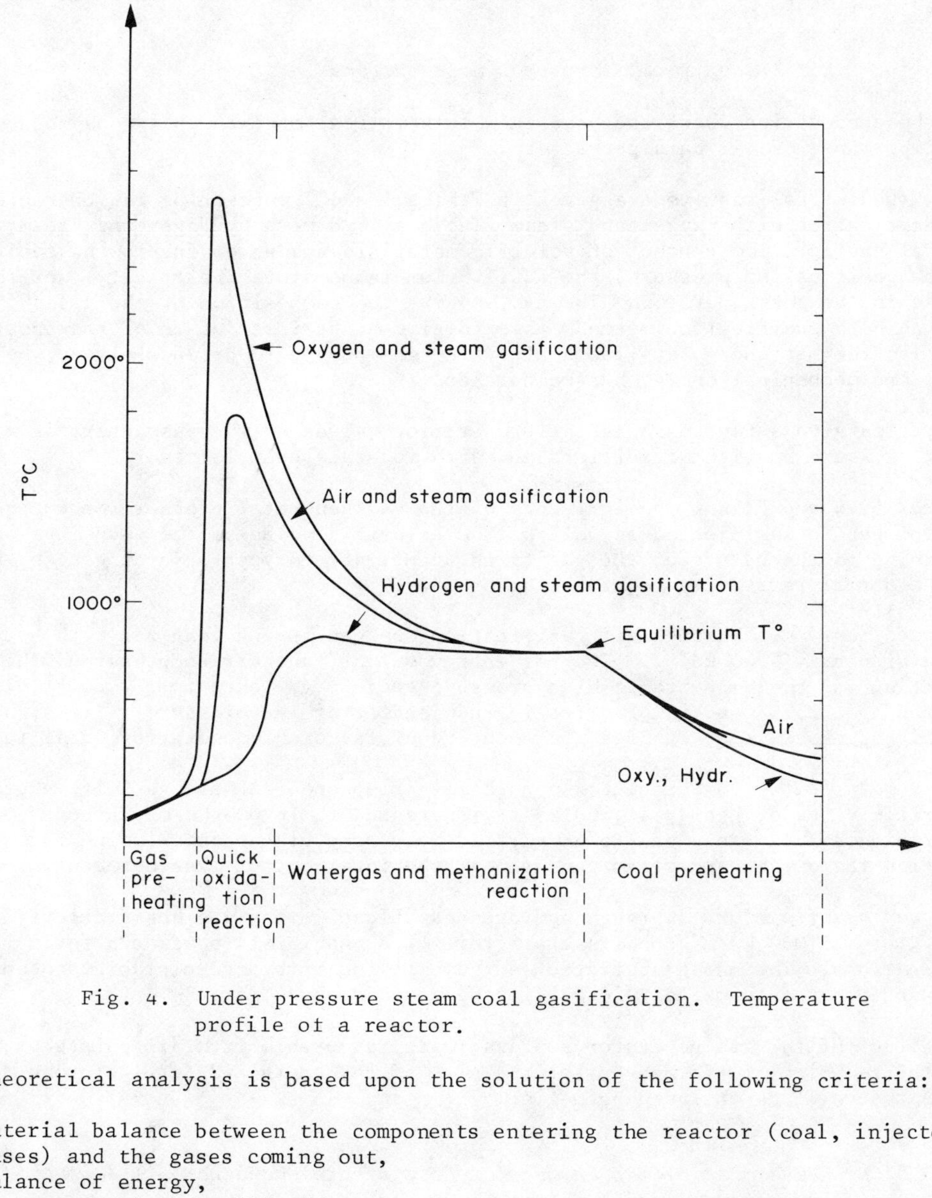

Fig. 4. Under pressure steam coal gasification. Temperature profile of a reactor.

Our theoretical analysis is based upon the solution of the following criteria:

a material balance between the components entering the reactor (coal, injected gases) and the gases coming out,
a balance of energy,
the thermochemical equilibriums.

The balance of energy takes into consideration the calorific value of the coal, of the gas produced by the reaction, of the sensible heat of the gases, the enthalpy of the preheated coal against the gas flow after the reaction, the heat content of the injected gases, the losses in the ground.

The thermochemical equilibriums mean that the following reaction equilibrium constants are nearly reached depending on the equilibrium temperature and of the pressure in the gas generator.

$$2CO \rightleftharpoons CO_2 + C \qquad CO + H_2O \rightleftharpoons CO_2 + H_2 \qquad CH_4 + H_2O \rightleftharpoons CO + 3H_2$$

Results of the Theoretical Calculations

All these conditions were included in a mathematical model enabling us to assess the importance of all said parameters.

A few theoretical results are given in Tables 1 and 2 mentioning the characteristics of gasification with oxygen and steam (Table 1) and with hydrogen and steam (Table 2) of coal having a low content of volatile materials and according to the evolution of the gasification pressure, the equilibrium temperature (reactivity) and the heat losses in the gas generator. The tables give the composition of the gases obtained through both gasification methods able to give a gas intended to be transported, and also the net energy available after the gas needed to produce oxygen, hydrogen, steam and mechanical energy has been taken.

The tables give a few characteristic operation values of the gas generator working under pressure in tight conditions and with moderate heat losses.

Figures 5, 6 and 7 show the influence of the gas generator pressure on the quantities of gas to be injected as well as the calorific value of the crude gas obtained according to the nature of the gasifying agent: air, oxygen or hydrogen for coal with a normal reactivity (equilibrium at 750°C).

Air gasification. Through air gasification one can have a lean gas with a calorific value of nearly 5000 kJ/m^3. The net energy output, the air compression, the steam injection and the recoverable high-pressure gas energy taken into account, lies between 55 and 70% and is practically independent of the pressure. It is, however, the high pressure that enables the securing of favourable operating conditions.

Oxygen gasification. Through oxygen gasification one can have gas with a gross calorific value of nearly 11.000 kJ/m^3, increased to 16.000 kJ/m^3 through CO_2 washing, which enables this gas to be carried over long distances. Taking into consideration the oxygen and steam generation, the total output lies between 40 and 60%.

Hydrogen gasification. Through hydrogen gasification the gas has a calorific value from 21 to 25 000 kJ/m^3 and its characteristics enable it to be used for numerous applications. The net gasification output, taking into account the hydrogen generation and other internal services, lies between 50 and 60%.

The effect of the gas generator's pressure is favourable from all points of view: for the technical performance, for the size of the units and for the production costs that will be analysed hereafter.

Assessment of the Production Cost of the Gas Generated Through Underground Gasification

The production cost of gas generated through *in situ* coal gasification is linked with the financial costs of the investments to reach above-ground level, for drilling equipment and the operation costs.

The main items are:

Drilling — parameters of depth, mass of the fluids, permeability — compression of the fluids (rate of flow, pressure) — oxygen and hydrogen generation — cleaning and treatment of the gas.

TABLE 1

Underground Gasification of Coal with Oxygen and Steam
Low volatile composition
Theoretical characteristics

Ref.	Press., bar	T° reactivity balance °C	Losses gas gener., % Ch.	Gas injected moles/kg ch.		Composition gas out, %						Volume gas, mole/kg	NCV dry gas, kJ/m³ × 10³	Energy gas kJ/kg × 10³	Gas treatment Energy consumed			NCV clean gas (CO_2), kJ/m³	Total energy output
				Oxyg.	Steam	CO	CO_2	H_2O	CH_4	H_2	N_2				Oxyg.	Misc.	Avail.		
1	3	750	12	31.5	16.3	38	15	10	3	33	1	140	10.4	29.1	5.25	1.2	22.72	12.2	0.63
2	3	750	21	38.5	33.5	40	17	10	2	29	2	130	9.9	26	6.42	1	19.7	12	0.55
3	10	750	12	26.8	70.1	24	20	19	6	29	1	152	10.5	28.7	4.9	3.7	20.1	13.7	0.56
4	10	750	21	36.2	57.4	26	23	19	5	26	2	144	9.8	25.6	6.6	2.1	16.9	13.3	0.47
5	30	750	12	25.4	88.3	15	24	27	11	22	1	155	11.3	28.5	5.1	5.3	18.1	16.4	0.50
6	30	750	21	33.3	76.3	16	26	27	9	20	1	149	10.4	25.4	6.7	3.7	15	15.8	0.42
7	100	750	12	22.2	100.6	9	26	34	16	14	1	153	12.6	28.4	4.9	6.4	17.3	20.3	0.48
8	100	750	21	30.3	89.7	9	29	34	14	13	1	150	11.4	25.3	6.7	5	13.6	19.5	0.38
9	3	850	17	37.5	13.5	60	6	3	1	27	2	113	11.2	27.4	6.3	-	21.2	11.9	0.59
10	6	850	31	47	1.6	62	13	4	1	18	3	102	10.4	22.8	8.4	-	14.5	11.8	0.40
11	10	850	17	36.6	28.7	47	13	9	2	27	2	124	10.7	27	6.7	-	20.3	12.2	0.57
12	10	850	31	46.8	9.8	53	16	7	1	20	3	109	9.9	22.7	8.6	-	14.1	11.9	0.39
13	30	850	17	34.6	48.5	33	19	16	6	25	2	135	10.6	26.6	6.9	0.6	19.1	13.4	0.53
14	30	850	31	45.5	29.5	36	23	15	4	20	2	122	9.5	22.2	9.1	-	13	12.7	0.36
15	100	850	17	31.4	69.4	20	24	25	10	19	1	141	11.1	26.3	6.9	2.7	16.7	16	0.47
16	100	850	31	42.9	51.5	22	28	24	8	16	2	132	9.6	21.8	9.4	0.4	11.9	14.9	0.33

TABLE 2

Underground Gasification of Coal with Hydrogen and Steam

Theoretical characteristics for 1 kg of coal with low volatile composition 35 881 kJ

Technical hydrogen: CO_2 = 1 CO = 3 H_2 = 88.5 CH_4 = 4.5 N_2 = 3

Ref.	Press., bar	T° balance, °C	Losses gas gener., %	Gas injected, moles/kg ch.		Composition gas out, %						Volume gas mole/kg	NCV dry gas, kJ/m³ × 10³	Energy gas kJ × 10³	Gas treatment, 10³ kJ			Total energy output, %
				Hydrog.	Steam	CO	CO_2	H_2O	CH_4	H_2	N_2				Prod. H_2	Misc.	Gas avail.	
1	30	700	4	187	57	4	4	21	37	31	3	207	21.8	79.4	53.7	3.9	21.8	60.7
2	30	700	8	221	35	3	2	17	42	33	4	203	22.9	86.6	63.5	2.9	20.2	56.3
3	100	700	4	137	71	3	7	29	41	18	3	176	23.9	66.7	39.4	4.9	22.4	62.4
4	100	700	8	171	48	2	4	24	47	19	3	171	25.5	73.9	49.1	3.9	20.9	58.2
5	300	700	4	115	78	2	9	34	43	10	2	163	25.3	61	33	5.8	22.2	61.9
6	300	700	8	148	55	1	5	29	50	11	3	158	27.2	68.3	42.5	5	20.8	58
7	30	800	6	321	5	3	0	6	37	51	4	262	20.1	111.5	92.2	1.7	17.6	49
8	30	800	12	368	-23	-	-	-	42	54	4	259	20.8	121.3	105.7	0.5	15.1	42.1
9	100	800	6	209	29	4	2	15	45	31	4	187	23.5	83.7	60.1	3.2	20.4	56.9
10	100	800	12	248	-4	1	0	5	55	34	5	173	24.8	91.8	71.3	1.8	18.7	52.1
11	300	800	6	166	43	3	4	22	49	19	3	163	25.7	72.9	47.7	4.6	20.6	57.4
12	300	800	12	206	6	1	1	10	62	21	5	147	27.6	81.6	59.2	3.2	19.2	53.5

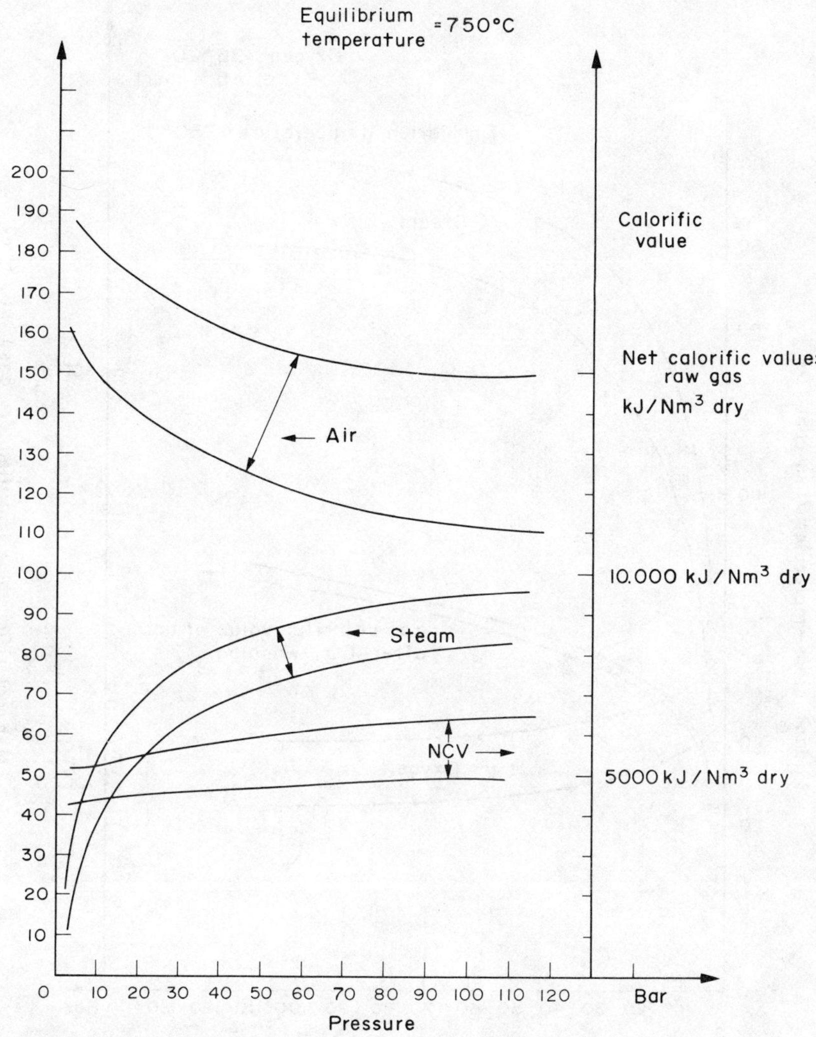

Fig. 5. Under pressure air & steam coal gasification.

It is possible to assess, on the basis of the theoretical evaluations mentioned in the previous chapter and in spite of some uncertainties inherent in the development of new systems, the cost of the gas in given conditions (satisfactory coal reactivity, limited heat loss in the gas generator, satisfactory total depth of the coal seams (2 m), tightness of the structure), as the techniques to be used are based to a great extent on well-known installations.

Our assessment of the cost price of the gas is given in Fig. 8 and refers to a yearly operation of 7000 h, the yearly cost of financial and working expenses being estimated at 20% of the investments.

Figure 8 shows the influence of pressure on the net production cost of the gas generated through gasification with air, with oxygen and with hydrogen, and gives, as

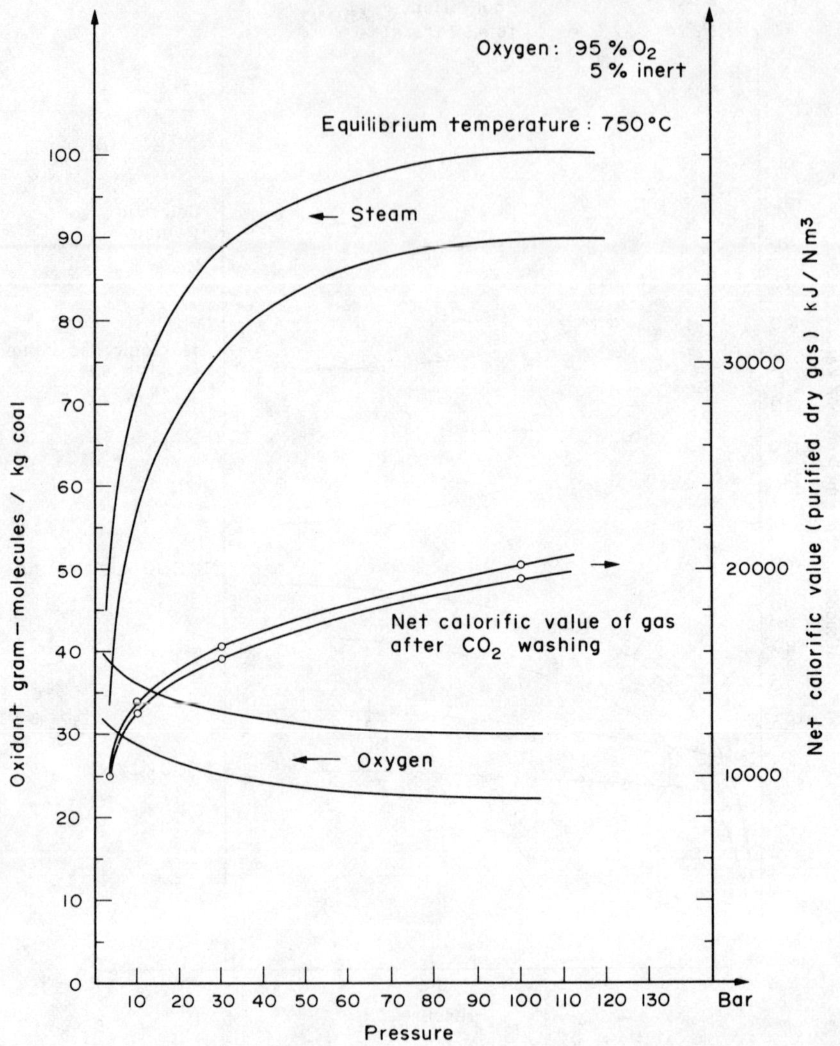

Fig. 6. Under pressure oxygen & steam coal gasification.

a comparison, the reference costs of coals won all over the world (best open mines: USA, South Africa, UK, Ruhr; coal imported in Belgium and cost price of the Campine coal).

This diagram shows that there is a possibility to produce pipeline gas for a price of 75-90 BF/GJ, i.e. half of the production cost of SNG through gasification of cheaply won coal, and that the price of said pipeline gas is competitive with the import prices of natural gas.

It should be noticed that hydrogen gasification implies that pressures above 50 bars will be used.

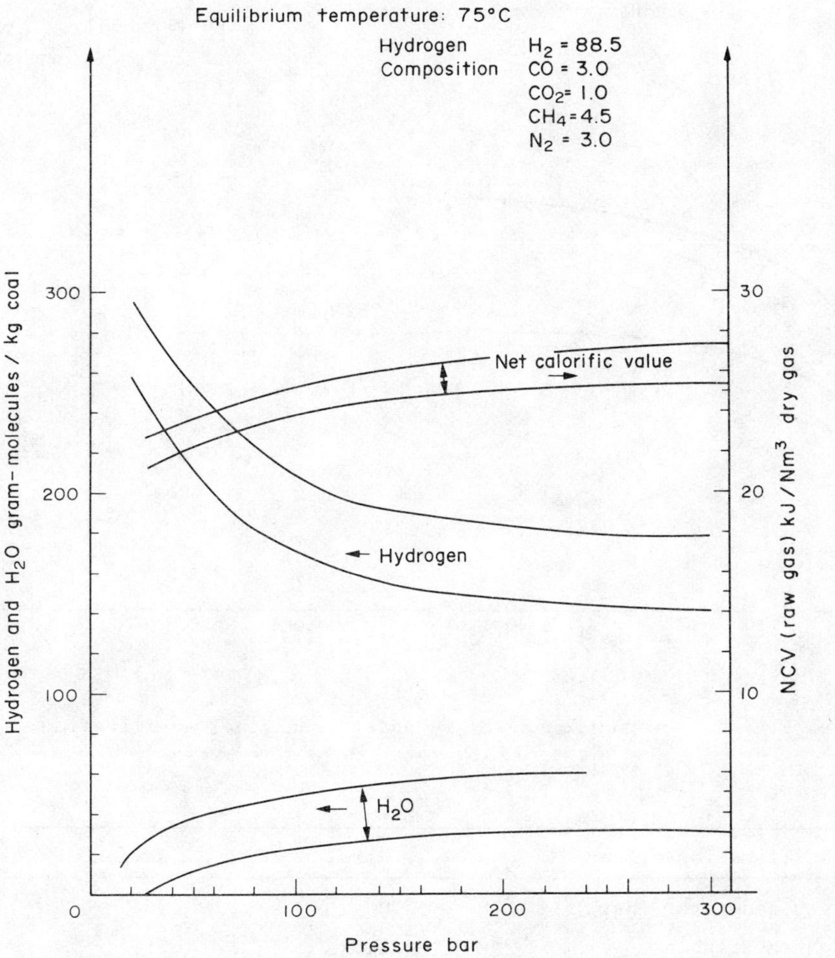

Fig. 7. Under pressure hydrogen & steam coal gasification.

Field of Application of Gas Generated Through Underground Gasification

We will not consider here the air gasification, the application of which is limited on account of power generation, as lean gas (5000 kJ/mole) is not suited to transport over distances longer than a few miles and as the treatment to make it into pipeline gas is too expensive (e.g. separation of inerts).

Oxygen gasification. The gas must be cleared by absorption from the CO_2 ballast; the composition of the cleaned gas is given in Table 3 and can be compared to that of an old manufactured gas.

$$GCV = 16\ 000\ kJ/m^3\ (3850\ kcal\ NCV),\ specific\ density \sim 0.5\ (W = 6000).$$

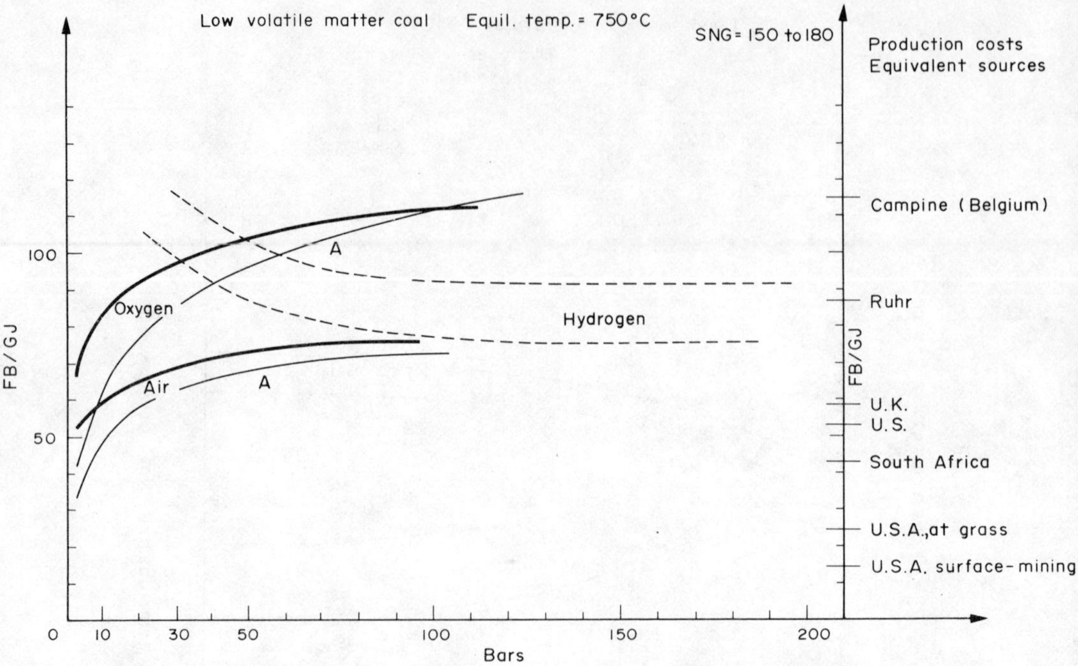

Fig. 8. Production costs by underground coal gasification.

TABLE 3

Gasification method	Air	Oxygen	Hydrogen
Pressure gas generator (bar)	30	30	100
Pressure gas available	21.5	21.5	71.5
Composition cleaned gas		(CO_2 washing)	
CO	13	31.7	3.4
CO_2	15.4	2	3.4
CH_4	4.90	20.8	58.4
H_2	15.50	42.8	30.5
N_2	51.20	2.6	4.4
NCV of gas (kJ/m^3)	5.100	16.100	24.600
Gross output	0.73	0.75	2.2
Net output (incl. own consumption)	0.59 (with recovery of high-pressure gas)	0.47	0.58

Rational and massive use of the gas can be as follows:

1. Direct use as pipeline gas, type town gas.
2. Adaptation to production of SNG through partial conversion of the CO, CO_2 washing and methanization, either gas of the L type (equivalent to the Slochteren gas), or gas of the H type (rich natural gas). This adaptation to SNG entails an increase of costs of about 10%.
3. Direct raw material for the production of ammonia or of methanol or of reducing gas. The crude gas can be transformed into synthesis gas by catalytic oxidation with air or with oxygen, a method that can replace the more expensive steam-reforming technique.

Hydrogen gasification. The gas obtained has a composition which is more like that of natural gases CV = 24 600 kJ/m^3 (5900 kcal NCV) specific gravity = 0.47 Wobbe index = 9600 -- and it can be valorized with less costs for the uses mentioned above.

SNG production brings about additional costs reduced to 3% as its transformation through methanization is rather limited.

It must be noticed that the gas will be available under high pressure (gas generator 100 bars, usable gas 70 bars).

Experimental Research Programme in Belgium for Underground Gasification at Great Depth

On the basis of the results of preliminary research works made by the INIEX and at the Aachen University on the advantages of using a high pressure in underground gasification, the Belgian and German governments reached an agreement on carrying out a common research programme aimed at developing the technical and economical conditions of this gasification. This programme is directed towards a widening of the field of application of the existing underground gasification techniques. It tests the operation under high pressure in a tight, dry, deep-lying gas generator, which should make it possible to produce either lean gas and electrical power through air gasification, or pipeline gas or synthetic natural gas and various synthesis gases through oxygen or hydrogen gasification.

The starting-up period of this programme includes air gasification trials under high pressure spread over three stages:

First stage: gasification trials of a first coal seam with four boreholes moderately distant from one another.
Second stage: gasification trials with three boreholes very distant from one another.
Third stage: trial of deepening the drillings with a view to the gasification of a second coal seam.

This research programme is spread over 6 years and will be completed by the end of 1984. It aims at developing the techniques of firing the gas generator, the underground equipment, the control of the reactions under high and changing pressure, the verification of the parameters, the operation of the gas generator and the evolution of the gas-production cost.

As soon as the operation conditions are mastered well enough, gasification trials will be made with oxygen and hydrogen.

The total budget for this starting-up period of underground gasification amounts to about 1 billion Belgian francs.

CONCLUSION

Experiments must now give proof of the possibility to exploit deep-lying coalfields by underground gasification under changing high pressure. Preliminary studies show that the process is economically attractive. The industrial success of a process of exploiting deep-lying coalfields through underground gasification would limit to a great extent our countries' dependence upon energy imports by putting at our disposal, in the form of gas, the tremendous quantities of coal that make up the main source of energy in our soil.

As far as Belgium is concerned, producing yearly 10 billion m^3 of gas for distribution is theoretically possible, and we do hope that the experiments initiated by the INIEX will lead to favourable results within the next few years.

BIBLIOGRAPHY

1. "Gazéification souterraine de charbon" par P. Ledent, Directeur de l'INIEX, Liège, 1er septembre 1977.
2. "Modèle mathématique de gazéification souterraine" par J. Ribesse, Distrigaz, Bruxelles, février 1978 (non publié).
3. "Evaluation du coût de fabrication de gaz destiné aux réseaux de distribution" par J. Ribesse, Distrigaz, Bruxelles, mai 1978 (non publié).
4. *Moniteur Belge* no. 74 du 15/4/78. Loi approuvant l'accord entre la Belgique et la République Fédérale d'Allemagne pour la réalisation en commun d'un programme de recherches sur la gazéification souterraine de houille et de lignite, signée le 1/10/77.

Economic Aspects of Gas Gathering in the North Sea

R. Beggs and M. Beale

Gas Gathering Pipelines [North Sea] Ltd. and Scicon [United Kingdom] Ltd.

INTRODUCTION

All North Sea oilfields produce some associated gas. Although few are likely to have enough gas to justify a separate pipeline to the shore, the total amount of gas involved could be very large. The United Kingdom Government is concerned to minimize the quantity of gas to be flared and to maximize the benefit from this indigenous energy resource.

After a preliminary study had established a prima facie case for gas gathering, Gas Gathering Pipelines (North Sea) Limited, a joint public sector/private sector company, was set up to establish the viability (or otherwise) of gas-gathering schemes in the UK sector of the northern North Sea. A final report was submitted in April 1978.

GGP's work programme was divided into four areas of specialization:

Gas Supply and Availability

Estimation of reserves for over eighty fields and discoveries.

Engineering

Assessment of the costs and technical feasibility of the various elements of gas-gathering systems.

Marketing

Study of the values and market opportunities for the gas collected.

Optimization and Evaluation

Selection and evaluation of the most promising potential systems.

This paper discusses very briefly the various specialist studies and then describes in more detail the evaluation and optimization models developed for GGP by Scientific Control Systems Limited and used to assess the economic significance of all the data.

The Evaluation Model evaluates a prescribed configuration, defined by the location of each pipeline link, its internal diameter, and its commissioning date. The model checks that there is enough capacity in each year to transmit all the gas available at all the fields connected to the systems and, if not, it rejects some gas. It then accumulates all gas flows, costs and revenues and prints tables summarizing the physical and financial consequences of the system.

The Optimization Model uses integer programming to find the configuration that maximizes the net present value of the system at a chosen test discount rate. Some minor simplifications are made for computational convenience. If many alternative possible connections are specified for each node, the model may not find a guaranteed global optimum, but it always produces a good solution. It has proved very useful when examining complex economic trade-offs. Part of the output from this model is an input file for the Evaluation Model, so detailed reports on the solution can be produced from this model.

PART I. BACKGROUND TO THE ECONOMIC APPRAISAL

Early Proposals for Gas-gathering Systems

Several companies have been active over the past few years in promoting interest in the development of UK offshore gas-gathering systems which would take as their sources of supply:

associated gas from oil-production platforms,
gas production from condensate fields,
gas from small gasfields which could not justify independent development.

Up to now, no such system has been completed in the North Sea.

Initiative by Government

In September 1975 the UK Government commissioned Williams-Merz (consulting engineers) to see if there was a prima facie case for a pipeline system to gather gas, especially that produced in association with oil, in the UK sector of the northern North Sea.

Their report, published in May 1976, proposed two new pipeline systems to bring gas ashore, and estimated that, excluding the output of the Brent and Frigg fields, 28.3 to 42.5 million cubic metres a day of specification gas could be available and that there might be, in addition, 6 to 9 million tonnes a year of heavier gases. However, in the interest of speed, the scope of the study had been limited by simplifying assumptions and was based on immediately available data of uncertain quality. Because of this, the Secretary of State for Energy decided that more work would be needed before any decision could be made to go ahead with a pipeline project.

In December 1976 Gas Gathering Pipelines (North Sea) Limited (GGP) was established to plan and commission the further studies needed. GGP is a joint public sector/private sector company with the British Gas Corporation (BGC), the British National Oil Corporation (BNOC), British Petroleum (BP), Imperial Chemical Industries (ICI), Rio-Tinto-Zinc Corporation (RTZ) and Total and Elf/Aquitaine in consortium (Total) as participants. The objectives of the company were to identify and carry out the studies required and, if appropriate, make proposals for the development of any recommended gas-gathering scheme and for the formation, structure and financing of a company to undertake the recommended venture.

Scope of the Work

The programme developed in three phases: the first was concerned with the collection of basic data and constructing the means to evaluate it; the second with the evaluation of possible gas-gathering systems and the third phase where the more rigorous study of certain specific aspects led to the further optimization and evaluation of possible systems.

The work programme was divided into four areas of specialization:

Gas supply and availability,
Engineering,
Marketing,
Optimization and evaluation.

Within each area certain specific subjects were entrusted to outside consultants. A list of the consultants employed and the subjects studied is shown in Table 1.

Gas Supply and Availability

To carry out this part of the study, GGP and its reservoir consultants needed to have access to confidential information, particularly on the subject of gas availability and interpretation of discoveries. To satisfy licensees that any sensitive information provided would be fully safeguarded, GGP agreed with the Department of Energy detailed confidentiality undertakings acceptable to all the operators concerned. As a result, adequate co-operation was received from most of the North Sea licensees.

An evaluation was carried out of more than eighty fields and discoveries in the area under review. The evaluation covered in-place reserves, ranges of recovery efficiency, oil- and gas-production rates, platform fuel use and gas availability. Although the amount of data varied, in most cases, sufficient information was obtained to provide a reasonable estimate of gas quantities.

Special attention was paid to condensate discoveries which, between them, could contribute the major part of the gas likely to be available.

A separate study was carried out of the potential prospectivity of blocks and part-blocks not covered in the reservoir studies in order to assess whether their distribution was likely to change the overall pattern of a gathering system.

Engineering

Studies were based on established technology, although consideration was given to the effects of technological changes which may be introduced during the next 5 years.

Engineering consultants were commissioned to assess the costs and technical feasibilities of alternative facilities for off-shore separation of oil and gas, pipelines, booster platforms, compressors and shore terminal facilities.

The main effort was devoted to estimating the capital and operating costs of the

TABLE 1

Consultants to GGP

Area of study	Consultants appointed
1. Gas supply and availability	
(a) Reservoir analysis	Gaffney Cline & Associates
(b) Prospective gas reserves	The British National Oil Corporation
2. Engineering	
(a) Field platforms	Humphreys & Glasgow Ltd. Redpath Dorman Long (North Sea) Ltd. Sea Tank Co.
(b) Booster platforms	Humphreys & Glasgow Ltd. Redpath, Dorman Long (North Sea) Ltd. Sea Tank Co.
(c) Shore terminals	Humphreys & Glasgow Ltd.
(d) Ports, terminal sites and pipeline routes	Scottish Development Department
(e) Submarine pipelines	Pipeline Technologists (Offshore) Ltd. Brown & Scot (UK) Ltd.
(f) Land pipelines	Constructors John Brown Offshore Ltd.
(g) Gas-flow and phase calculations	London Research Station of British Gas Corporation
(h) Offshore gas processing	D.U.T. Pty. Ltd.
(i) Floating production facility	Vickers Offshore (Projects and Developments) Ltd.
(j) Offshore operating costs	P. A. Management Consultants Ltd.
(k) UK capacity of project management and equipment supply	Taywood-Santa Fe Ltd.
(1) Condensate field development	Humphreys & Glasgow Ltd. Gaffney, Cline & Associates
(m) Sub-sea repairs	H. O. Mohr
3. Marketing	
(a) Dry gas values in Europe	Societe Nationale Elf Aquitaine
(b) Market for natural gas liquids in Europe	Arthur D. Little Ltd.
(c) Market for imported NGLs in US	The Pace Company Consultants & Engineers, Inc.
(d) Existing LPG market in Europe	Totalgaz BP Trading Ltd.
(e) Economics of LNG handling	BP Trading Ltd.
(f) Transportation costs	H. Clarkson & Co. Ltd.
(g) Forecasts of crude oil price	Arthur D. Little Ltd. The Pace Company Consultants & Engineers, Inc.
4. Optimization and evaluation	
(a) Construction and operation of computer models	Scientific Control Systems Ltd.
(b) Applicability of decision-analysis techniques	Buchanan & Clacher Ltd.

elements (see Fig. 1) of gas-gathering systems. A "base case" was selected for evaluation purposes with high and low cases for use in sensitivity analyses.

Marketing

The object of the marketing studies was to forecast the values and market opportunities for gas collected over the life of a gas-gathering system. This gas can be divided into four basic product groups which were studied in detail:

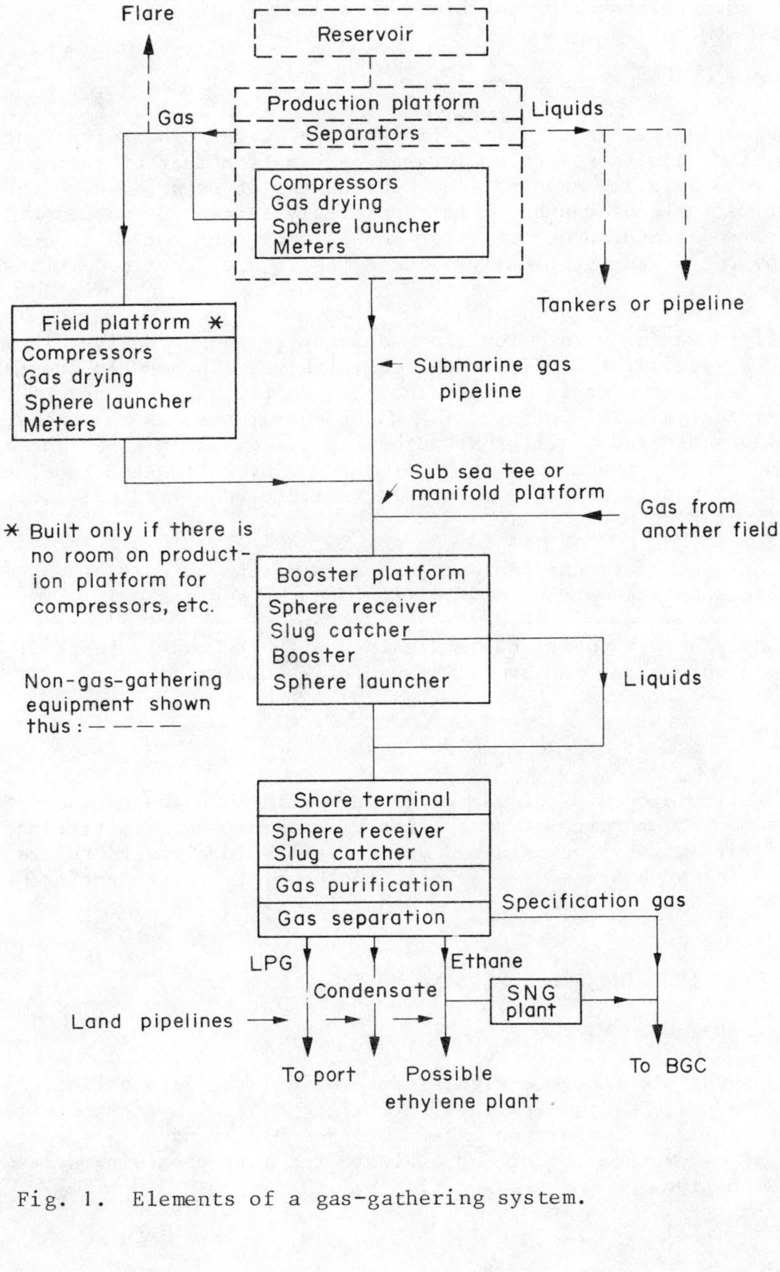

Fig. 1. Elements of a gas-gathering system.

Specification gas to BGC specification
Ethane (both as fuel and petrochemical feedstock) } Natural gas liquids (NGL)
LPG (propane and butane) }
Condensate (pentanes and higher hydrocarbons). }

The market potential for specification gas in Great Britain and continental Europe was examined. In the case of NGL, markets in Great Britain, continental Europe and the USA were considered.

An important part of the market studies was to forecast future crude-oil price changes in real terms and the effect of such movement on product prices which are directly or indirectly related to them.

Optimization and Evaluation

The basic objective was to compare efficiently a large number of potential gas-gathering systems and to select and evaluate a small number of the most promising systems for more detailed examination. The number of ways in which the many fields and discoveries could be connected is exceedingly large and two computer models, an Evaluation Model and an Optimization Model, were constructed in order that this task might be performed efficiently. These models are described in the second part of this paper.

The factual information with which the models operated was derived from the results of the various specialist studies referred to above. The models are capable of operating in real terms or in money values but do not take account of taxation or financing strategies. The Optimization Model operates so as to choose, out of all the alternatives offered to it, that gathering system which gives the highest net present value at the chosen test discount rate. The Evaluation Model cannot consider alternatives but it operates in more detail and requires less computing time than the Optimization Model.

Sensitivity analyses were carried out to determine the effects of changes in the major variables and assumptions on the final chosen systems.

Overall, a satisfactory position was attained with the models responding realistically to the problems set and producing useful solutions.

Conclusions and Recommendations

GGP found that the economics of gas-gathering systems in the northern North Sea were subject to wide margins of uncertainty. Within these limitations, GGP recommended the early action it considered was necessary to minimize future flaring as well as later action designed to maximize the benefits to the nation of its indigenous natural gas.

PART II. THE GGP SELECTION MODELS

Introduction

In order to assess the economic significance of all the data collected against a tight time schedule, GGP commissioned Scientific Control Systems Limited (Scicon) to develop computerized selection models which could be used to compare potential gas-gathering systems and select and evaluate the more promising systems for further detailed examination.

The models were developed while the other consultants gathered their data. Indeed, once outline specifications for the models had been agreed between GGP and Scicon, discussions were held with the other consultants to agree the logical structure of the data they would supply. For example, it was agreed that the costs of each facility would be defined by six components:

A Capital Cost incurred in the first year of operation.
A Capital Cost incurred in the year before commissioning.
A Capital Cost incurred 2 years before commissioning.
A Capital Cost incurred 3 years before commissioning.
A Fixed Annual Cost in every year of operation.
A Variable Operating Cost proportional to throughput.

These agreements formed an essential step in co-ordinating the work of the consultants. The initial versions of the models were developed over the first 3 months of the study period, by which time GGP had a comprehensive set of data and the models started to contribute to GGP's analyses. But revisions continued in both the data and the models. Some features of the models proved not to be useful, and others were added.

To evaluate the overall economics of any gas-gathering system it is necessary to:

- select the fields to be considered;
- define the availability of gas from each field in terms of quantity, composition and time profile;
- decide precisely how the field will be connected, including the route of the main trunk line ashore;
- choose the sizes and commissioning dates for each pipeline link;
- check that the lines can carry all the gas available at every field connected to the system in each year, and if not to decide which gas should be rejected;
- calculate the amounts of compressor power needed at each platform to pump the gas ashore;
- define the type and size of shore terminal needed to receive the gas stream and separate it into specification gas and other components. Specification gas is assumed to consist of methane enriched with other components in specified proportions;
- calculate the capital costs of connecting to the production platforms, installing compressors, laying the pipes and constructing the shore terminal;
- calculate the annual operating costs of the system;
- calculate the revenues earned from the system through the sale of marketable products;
- calculate the internal rate of return and the net present value (NPV) discounted at the test discount rate;
- report in detail on the gas flows in each pipeline leg, on the estimated cash flows in each year, and on the capital and operating costs of each field, of each pipeline section, and of the system as a whole.

The data for both models are divided into two parts: Databank Data, describing relevant facts about the real-world situation, and Run Time Data, describing the possible courses of action to be considered. The data from the other consultants are all incorporated in the Databank. The Run Time Data for each analysis were selected by GGP, often in consultation with Scicon.

The Evaluation Model

The Evaluation Model evaluates a prescribed system quickly and cheaply by performing the steps listed above. The Run Time Data for the model consist of a defined network of fields and pipelines connections, with sizes and commissioning dates for each pipeline link. Pressures for the gas arriving at the shore terminal and at each offshore platform are prescribed.

Gas is rejected at any platform where either the output pressure or the compressor power would otherwise be excessive. In principle one could make an elaborate calculation about which gas to reject. But the present version of the model does not do this, because the practical solution is usually to choose a larger pipe.

Having checked the system for feasibility in each year, rejecting gas if necessary, and having computed the input pressures to each pipeline in each year, the model costs the entire system using data from the Databank. This analysis includes the cost of the shore terminal needed to handle the volumes of each gas component landed.

The landed gas is separated into specification gas and other saleable components, and the resulting revenues are calculated. Detailed reports are given for each pipeline section, for each field, and for the system as a whole. The NPV and internal rate of return are computed for a range of assumed prices for specification gas.

The Evaluation Model has an additional optional facility whereby it considers removing the pipeline connecting any end field (i.e. any field with no incoming pipeline). If this increases the NPV, then the pipeline is removed, otherwise it is restored. This process continues until there is a system whose NPV cannot be improved by removing any single end field. The improve option was originally intended to consist of a more elaborate search for a local optimum solution. In particular consideration was given to testing a range of diameters and commissioning dates for each pipeline link. But there is no guarantee that a local optimum solution to such a problem will be a global optimum. And the Optimization Model was so much more effective than originally expected that a sophisticated Improve Option did not seem necessary.

The Optimization Model

In principle, the Optimization Model performs the evaluations following on from field selection. It considers a given number of fields connected by a large number of alternative routes each of which can consist of a number of different sized pipes and can be built in any one of up to nine time periods spanning the 30-year period under examination. The alternative routes can also include a number of different routes ashore to the same terminal from different fields or to different shore terminals from the same field or combination of both.

A typical problem can have millions of possible solutions and the model examines a substantial number of these in order to choose the particular system which gives the maximum NPV discounted at the given test rate.

The model can take several hours of computer time to search exhaustively all possible combinations and arrive at the best. It usually finds a "good" solution fairly quickly but there is no guarantee that this is the best. In practice this model has been used to narrow the number of possibilities which can then be further examined using the Evaluation Model.

The Optimization Model has proved particularly useful at examining complex economic trade-offs. Examples of this include:

- early connection of a field with additional or larger pipes laid versus a later connection with less or smaller pipes;
- temporary flaring from some fields caused by downstream capacity constraints versus an increase in capacity;

whether or not it is worth while connecting a field in its first years of low gas production;
whether it is worth while connecting a field at all;
the early connection of producing fields whose start-up cannot be deferred versus rejection of that field's gas until later fields come on stream.

The Optimization Model will always act so as to maximize its objective function which generally has been to maximize the NPV of the system discounted at the given test rate. A particular field will only be connected to the system if it gives a marginal increase in the system's total revenues which is at least equal to the marginal increase in the system's total costs needed to connect the field when discounted at the test rate. This model thus provides the means of establishing whether or not a particular field or indeed a whole series of fields can be economically connected in the context of the given costs, gas volumes, prices and test discounted rate. The model makes two simplifying assumptions in addition to those of the Evaluation Model. The more important of these is that the total time horizon is divided into a moderate number of time phases, e.g. 1979-1981, 1982-1983, 1984-1985, 1986-1990, 1991-2006. Installations can only start to operate at the start of a phase. And in each phase the gas production is treated as constant, if the phase is defined as a constant phase, or else is assumed to vary linearly with time, if the phase is defined as a variable phase. The actual rates are chosen by the model, to agree with the annual data in the Databank as closely as possible consistent with these simplifications. The number of constant phases plus twice the number of variable phases must not exceed 9. The other simplification is that the input and output pressures on most of the pipeline links are treated as constants defined by the input data. This simplifies the calculation of pipeline capacities at the expense of exaggerating the compressor power requirements on platforms. This exaggeration rarely has much effect on the overall economics of the system.

Optimization could be performed by evaluating all possible schemes and selecting the best. But the number of possible schemes may be very large when we consider alternative configurations, and all possible pipe types and commissioning dates for each link together with all possible starting dates for the production from each field. Fortunately, a better computing strategy is possible. This is known as Branch and Bound. In general terms it consists of finding a reasonable trial scheme and computing its NPV, and then selecting key components from this scheme, such as the commissioning date and pipe type for a particular link and showing that if a wrong decision is made the NPV would certainly be worse than for the trial scheme, even under simplifying assumptions about the rest of the scheme that exaggerate its NPV. No further analysis is then needed of any scheme that includes this wrong decision.

Even more fortunately, this computing strategy can be implemented by formulating the problem in a standard mathematical form for which there are ready-made computer programmes to carry out the optimization. Except for one extension used on some runs, the GGP Optimization Model is formulated as a mixed integer linear programming problem. Most of the integer variables must be either zero or one, since each represents the "probability" that some particular investment — such as a pipeline of a specified type over a specified link — is commissioned in a specified year. In fact most of the integer variables are grouped into special ordered sets of type one to speed up the solution. The extension arises when the pressures at some nodes must also be chosen by the Optimization Model, and in practice this extension has only been used when pipelines are linked at sub-sea tees. The problem is then one of non-linear integer programming, but it can be solved by similar means using a new technique known as linked ordered sets introduced by Beale and Forrest (1978).

The Use of the Models

Although the Evaluation Model is run alone when specific possibilities need to be studied, the models were designed to complement each other. The main output from the Optimization Model consists of an Input File to the Evaluation Model defining the run time data corresponding to the optimum solution found. The Evaluation Model is then run to give a detailed report on the solution, and it is easy to check that the simplifying assumptions made in the Optimization Model do not seriously distort the solution.

The following diagram outlines the information in the Databank and the run time data, and the output from each component of the two models.

The final output tables consist of reports on:

Gas flows for each pipeline.
Compression needed.
Capital and operating costs for each pipeline and platform.
Sales revenues by market and product.
Total cash flows.
Economic data on each field, each pipeline, the shore terminal and the whole system.

The Computer Programmes

Computer models intended to analyze complex sets of data are of little use unless they provide good facilities for editing the data, for checking them for consistency and for printing them neatly so that the user can check them for accuracy. They must also produce detailed reports designed to meet the user's needs, if only because data errors often remain undetected until their detailed consequences are analysed.

The GGP Selection Models use five computer programmes that were developed specially for the project.

One programme updates and tabulates the Databank.

A second programme is a matrix generator for the Optimization Model. This takes the Databank and run time data and creates an input file defining the problem in a form suitable for the mathematical programming system (SCICONIC) which is an existing commercial package used to carry out the optimization.

A third programme is an output analyser for the Optimization Model. This takes the output from SCICONIC, and, using the Databank and run time data, prints a brief report on the solution and creates the appropriate run time data for a detailed analysis via the Evaluation Model.

A fourth programme carried out the evaluation, checks whether gas must be rejected. It prints a one-page summary of the system, including a calculation of the NPV.

The improve routine is an option provided as part of the evaluation programme.

The output routine is another separate programme. It prints 10 sets of tables occupying about 100 pages.

It would obviously have been possible to combine these last three programmes into a single large programme, but the overhead cost in running them as separate programmes is small, and the fact that some calculations made in the evaluation routine are repeated in the output routine has led to the detection and correction of some errors and anomalies.

The five programmes have been written in Univac Fortran V. Together they contain about 30 000 Fortran statements. As noted, the Optimization Model also uses the SCICONIC System which can only be run on a Univac 1100 Series computer.

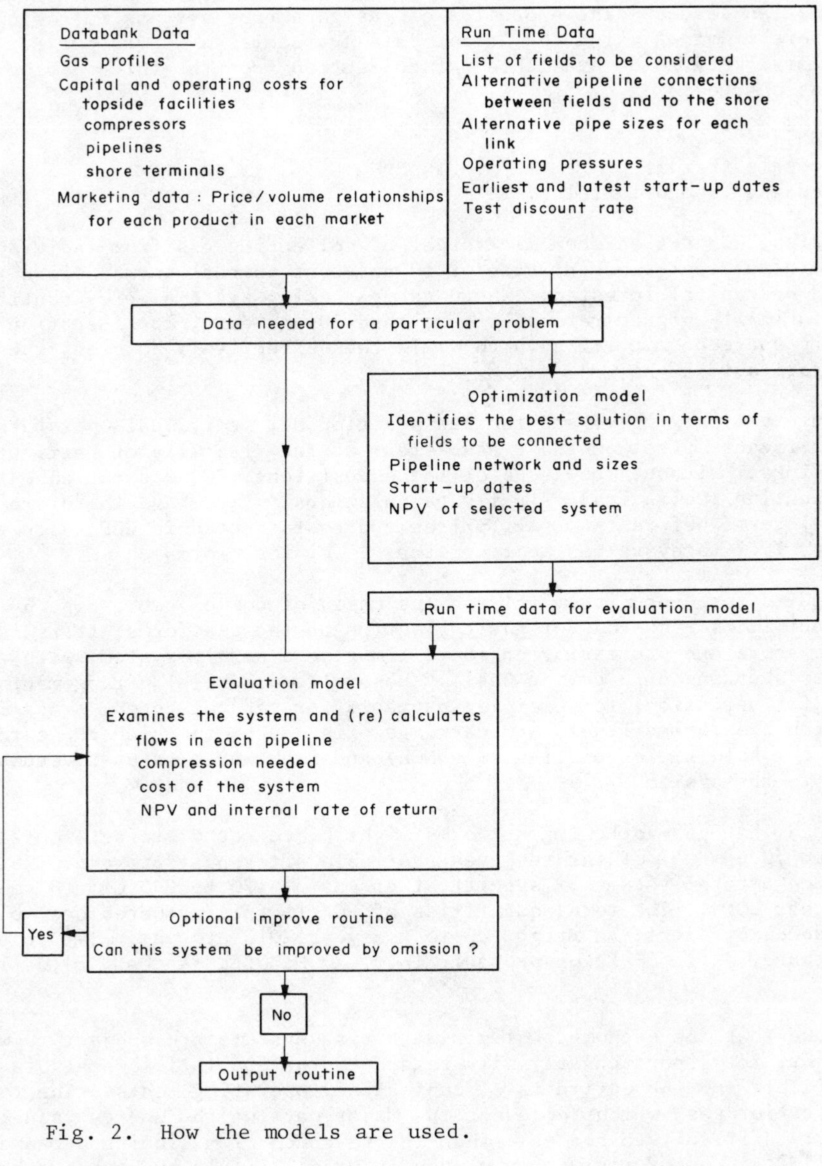

Fig. 2. How the models are used.

REFERENCE

1. E. M. L. Beale and J. J. H. Forrest (1978) "Global optimisation as an extension of integer programming", in *Towards Global Optimisation 2*, ed. L. C. W. Dixon and G. Szego (To be published).

ADDENDUM

GGP prepared its reports to meet the requirements outlined by the Secretary of State for Energy in his statement on 9 December 1976. GGP was required to look in detail at the possibilities for gas-gathering within the UK sector of the northern North Sea between latitudes 56°N and 62°N. No account was to be taken of reserves lying in the Norwegian sector. The economics of gas-gathering systems in the northern North Sea were examined using the latest data available to GGP to project the quantities of gas likely to be available for collection from the three categories of offshore gas sources:

gas fields,
gas condensate fields,
associated gas from oil fields.

In its studies, GGP set an economic target of collecting gas from individual fields only if a minimum test discount rate of 10 per cent in real terms before tax could be achieved on capital invested. Where gas-gathering systems were identified which had a substantially higher rate of return than 10 per cent, consideration was given to including sub-economic links which would then effectively be cross-subsidized by the more profitable links.

The discount rate chosen by, and the tax position of, individual operators could have a significant effect on their assessment of the viability of particular gas-gathering links. Without knowledge of these positions, GGP was not able to reflect the tax situation realistically in its calculations, which have therefore all been made in real terms before tax. All prices and costs quoted in GGP's study are based on 1 January 1977 money values kept constant (i.e. real terms).

GGP has studied the economics mainly on the basis of common ownership, by assuming that all facilities required for gas collection on the platforms, transmission from platform to shore and processing on shore comprise a single system, with all costs and revenues accruing on a common basis. Possible commercial arrangements provide wide scope for variation, for example, payments for gas collected, transmission charges through existing lines, and carriage of products to which the gathering system may not hold title (particularly NGL) and from which sales revenues might not accrue to the system.

GGP concluded that gas-gathering networks might be connected to between 214 and 302 Gm^3 (7.5 and 10.6 Tft^3) of the reserves available after platform use. This quantity could produce a total volume of specification gas of 170 to 240 Gm^3 (6 to 8.5 Tft^3) up to the year 2015. The total quantities of NGL in this gathered gas, plus quantities produced by Brent and Frigg, up to the year 2015 are estimated to be 48 to 59 Mt of ethane, 29 to 39 Mt of propane, 14 to 20 Mt of butane and 8 to 11 Mt of condensate.

GGP has found that the economics of gas-gathering systems are subject to wide margins of uncertainty, particularly with regard to the quantities, analysis and timing of gas available and the estimates of capital and operating costs. The revenues for specification gas, which represent the major part of the energy value of the gas, are more difficult to forecast than the revenues from liquids because of the statutory limitations on its use and supply in Great Britain by anyone other than BGC. Revenues are also dependent upon energy prices in the future, particularly crude-oil prices. However, GGP has tried to identify gas-gathering schemes which are sufficiently attractive to offer the prospect of their fulfilment by normal commercial initiatives. GGP has also reported systems where the economic merits are commercially doubtful, but which the Secretary of State may wish to consider on the grounds of national interest and conservation policy. These possible gathering schemes are shown on the reference map (Fig. 3).

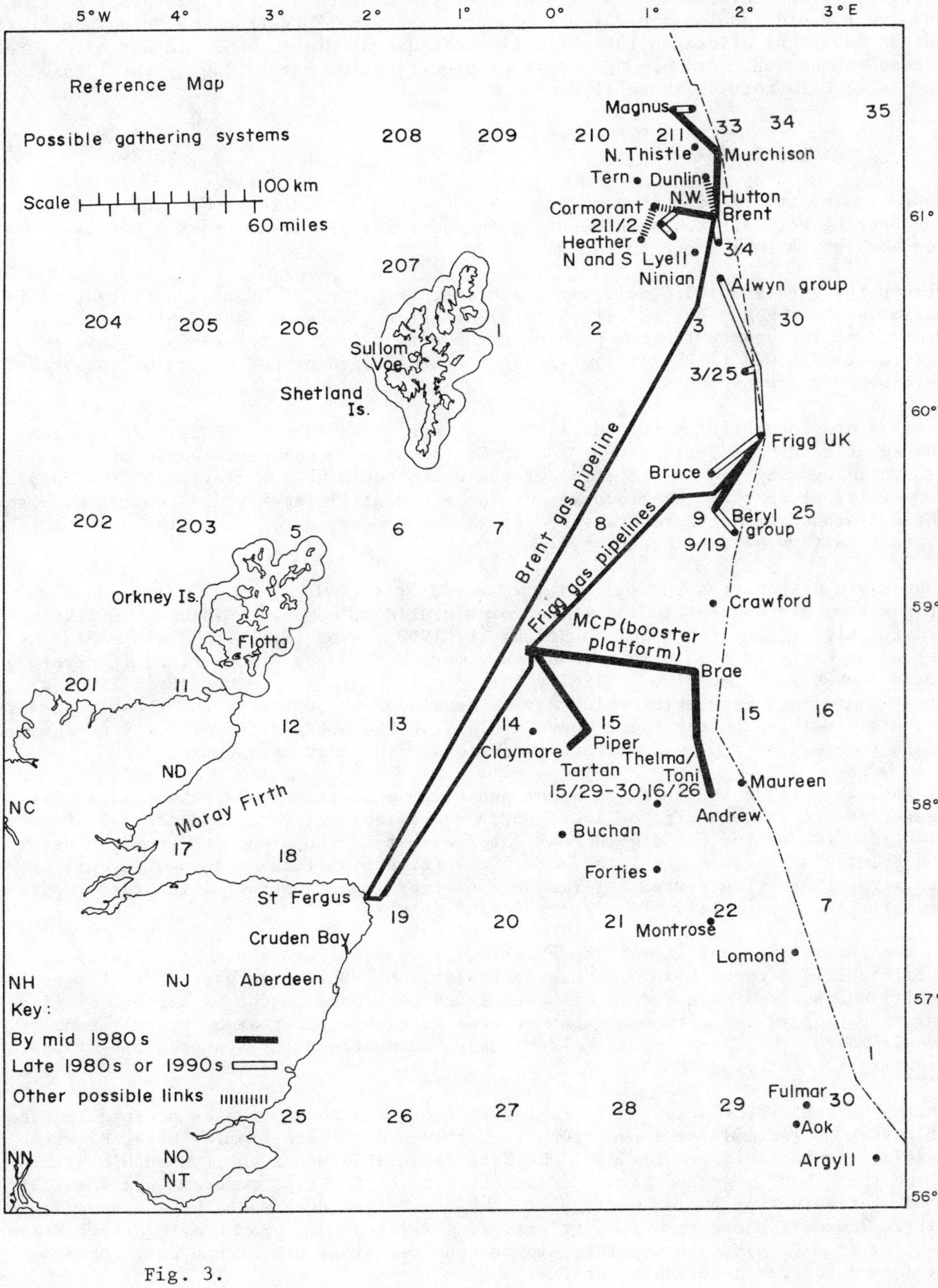
5°W
4°
3°
2°
1°
0°
1°
2°
3°E
Reference Map
Possible gathering systems
Scale
100 km
60 miles
208
209
210
211
33
34
35
Magnus
N. Thistle
Murchison
Tern
Dunlin
N.W. Hutton
Cormorant
Brent
211/2
Heather
N and S Lyell
3/4
61°N
207
Ninian
Alwyn group
204
205
206
1
2
3
30
Sullom Voe
Shetland Is.
3/25
60°
Frigg UK
Bruce
Brent gas pipeline
Frigg gas pipelines
202
203
5
6
7
8
9
Beryl group
25
9/19
Orkney Is.
Flotta
Crawford
59°
MCP (booster platform)
Brae
201
11
12
13
14
15
15
16
Claymore
Piper
Tartan
Thelma/Toni
ND
Maureen
Moray Firth
15/29-30, 16/26
NC
Andrew
58°
17
18
Buchan
Forties
St Fergus
19
20
21
22
7
Montrose
Cruden Bay
Lomond
NH
NJ
Aberdeen
57°
Key:
By mid 1980s
Late 1980s or 1990s
Other possible links
1
25
26
27
28
29
Fulmar
30
Aok
NN
NO
NT
Argyll
56°N

Fig. 3.

No single UK discovery has been identified which could itself sustain a new gas pipeline to the UK coast. The existence of the 36-inch diameter gas pipeline from the Brent Field and the two 32-inch diameter gas pipelines from the Frigg gas field has an important effect on the evaluation of gas-gathering. GGP did not have access to the contracts for the supply of specification gas to BCC by Shell/Esso and Total/Elf, through these lines.

Conclusions

Based on the information it has received and appraised on reserves, marketing and engineering work and taking account of the existing gas trunk lines, GGP has reached the following conclusions:

Within the area studied, excluding the Brent, Frigg and UK Statfjord fields, it is estimated that about 442 Gm^3 (15.6 Tft^3) of gas could become available for collection up to the year 2015 from present discoveries, after platform use. This is reduced to 334 Gm^3 (11.8 Tft^3) when discoveries judged by GGP not to be commercial developments are excluded.

A study of prospectivity in undrilled areas suggested that there is a 25 per cent chance of finding a further 580 Gm^3 (20 Tft^3) but a 50 per cent chance of this figure being only 140 Gm^3 (5 Tft^3) of gas which could become available from future commercial prospects. The location of these potential reserves is consistent with the pattern of existing reserves and is therefore considered unlikely to change the overall pattern of a gathering system.

Condensate discoveries in aggregate represent about half of the 334 Gm^3 (11.8 Tft^3) of gas from discoveries judged to be commercial developments. They are unlikely to provide significant gas supplies before the 1990s, even if their development commenced as early as 1980. This is because most condensate discoveries, if developed, would probably be recycled initially for 5 to 10 years to recover their liquids first. Smaller discoveries which may be developed by depletion and whose reserves are also included in the total above would require technical innovation to reduce costs in order to induce operators to proceed with their development.

By the early 1990s the existing Brent and Frigg pipelines could have substantial spare capacity. Depending on location, the available gas arising before 1990 from oil field production could be accommodated either in the Brent pipeline, by using the Brent gas cap as a regulator, or in the Frigg pipelines by a small amount of boosting. The Frigg system intermediate platform was constructed in order to provide for the possibility of boosting in the future.

In the 1980s, the capacity of the existing Frigg pipelines would be unlikely to be sufficient if account had to be taken of additional Norwegian gas, or if large additional sources of gas were discovered and developed in the UK sector, or if for any reason spare capacity could not be made available. For example, the priority provisions of the UK/Norwegian Frigg Treaty could affect the availability of spare capacity for carrying UK gas.

On the present information there is insufficient gas to provide an acceptable rate of return on capital for a new trunk line to shore. The economics of such a line based on present discoveries would be less favourable under any reasonable assumptions than those of connecting the same fields to the Frigg or Brent systems. If a new gas-gathering system were built, the St. Fergus area would be the best landfall. However, there is only a relatively small cost difference between the sites ranging from Cromarty to Tayside, and so considerations other than cost could be used to determine a suitable landfall.

The transmission of gas as a two-phase fluid is technically feasible, but there are advantages in reducing the liquid dropout by operating the pipelines at a high pressure. The minimum operating pressure of the gas-gathering system will need to be greater than about 110 bar. Where gas is to be injected into existing trunklines (which operate at maximum pressures of 140 to 145 bar) without compression at the point of injection, parts of the gas-gathering system may need to operate at pressures up to 170 bar.

The additional gas which it is proposed should be carried in the Frigg pipelines differs substantially in composition from Frigg gas. This could lead to technical problems both in the pipelines and at the shore terminal. These problems should be investigated in the near future. The investigations should include the cost and timing of any modification required to the existing installation and the range of gas composition that could be accepted into the Frigg system. These costs should be compared with those of offshore processing to achieve an acceptable composition.

Associated with the above, there is a need for further study of the possibilities and consequences of redefining the gas purchase specification of BGC in order for it to accept a significantly greater NGL content.

Gas-gathering systems designed to reduce flaring to a minimum would result in much higher collection costs than systems meeting commercial criteria. However, there would only be a small difference in the amount of gas which could be collected from present discoveries by the two different types of systems.

Gas from several fields is grossly uneconomic to collect and therefore some flaring must be accepted as inevitable.

In the Brent area there is a choice between a system which would be started almost immediately at an estimated total cost of £283M and a smaller system starting later at an estimated total cost of £182M. Both these systems would be developed by the mid-1980s. The first would enable gas to be collected from Heather, Cormorant and Dunlin and thereby minimize flaring, but this system would be sub-economic and would only increase the total amount of gas collected by a relatively small amount — from 20.6 Gm^3 (0.73 Tft^3) to 24.3 Gm^3 (0.85 Tft^3). Either system could be further extended to collect additional gas available later in the area.

In the Frigg and Central areas the operators of the fields which are in production or at an advanced stage of development and whose gas is economic to collect (Piper, Beryl, Tartan) are already taking steps to avoid flaring. There are two fields in production (Forties and Montrose) and a further two (Fulmer and Buchan) which are scheduled to come into production within the next few years, whose gas cannot be collected economically into the Frigg system. However, Fulmar gas might be economically transmitted to Ekofisk and the Emden pipeline. Other discoveries in these areas are still being appraised but it appears that the gas from most of these could be economically connected into the Frigg system. Such a gathering system, collecting all of the available gas from discoveries judged to be commercial, could land about 160 Gm^3 (5.6 Tft^3). This is equivalent to rather more than half of the present total Frigg field reserves. The overall capital cost of a system to collect this additional gas is estimated to be £657M, excluding the shore terminal. The cost of a shore terminal could add a further £200M to £300M.

BGC is considered to be in a position to absorb any specification gas arising in the 1980s from GGP's proposals and would no doubt welcome the additional supplies that would arise in the 1990s. A very large market which could bid competitively also exists on the European mainland where this gas would be a welcome addition to declining resources.

There are a number of options available for disposing of the natural gas liquids that could arise. It is possible by suitable treatment to accommodate the surplus ethane, and the LPG if necessary, in specification gas. On present information, until the early 1990s, ethane supplies alone will not build up to an adequate level to support another ethylene plant in addition to the one planned for Mossmorran, unless a way is found of drawing on the ethane used to maintain the calorific value of specification gas. There is, however, sufficient NGL (ethane plus LPG) to supply a new plant from the mid-1980s, although GGP is doubtful whether the ethylene market could sustain such a plant at that time. The LPG could be sold on the world market, probably to the USA, although a premium market in Europe could also develop.

So long as available gas in the area is to be collected into the existing Brent and Frigg trunk lines, there is no present requirement to stimulate additional exploration or appraisal drilling for further gas in the northern North Sea to support the overall economics of these gas-gathering systems. There may, however, be a case for an exploration programme to discover sufficient gas to make viable a southern extension of a Frigg or Central collection system.

GGP has concluded that BGC and Shell/Esso must play a key role in a Brent area gas-gathering system. The design and construction of this system should be initiated forthwith if flaring is to be minimized.

There is no immediate need to set up an organization for building a gas-gathering system in the Frigg and Central areas. There are several different types of corporate structure which could in due course be set up to build such a system. The selection should be made in time to meet the requirements of any development plans for discoveries in the area.

A successor to GGP seems desirable to assist the Department of Energy in the whole field of activities associated with gas-gathering.

RECOMMENDATIONS

Based on the above conclusions, GGP makes the following recommendations.

At this stage no gas-gathering system requiring the construction of a new trunk pipeline to the shore is necessary. Maximum use should be made of the existing pipelines. This strategy would give the necessary flexibility in dealing with the very uncertain time-scale of development of oil discoveries.

In future all platforms in the UK North Sea should, if appropriate, have risers for gas collection installed at the outset. Their design should also allow for the installation of other equipment necessary for collection and/or onward transmission of gas for delivery, within a controlled specification, to a future high-pressure gas-gathering system. These requirements should be subject to a waiver in certain special cases (e.g. for very small quantities or inconveniently located gas). Consideration should be given to the funding of such advance investment.

BGC and/or Shell/Esso should be invited to construct and operate a Brent area gas-collection system.

In the Frigg area, the discussions now proceeding between the Frigg pipeline operator and the operators of Tartan, Piper and Beryl should be encouraged. At the same time adequate consideration should be given to the collection of other gas and to the contractual arrangements to accommodate future developments in the area.

A successor organization should be nominated as soon as possible to assist the

Department of Energy in the whole field of activities associated with gas-gathering. This organization should cover the whole UK Continental Shelf. The functions of this successor organization would include *inter alia:*

recommending the type and timing of any organization that would be necessary to construct and finance a gas-gathering system, and
preparing contingency plans which could be put into effect rapidly if the matters referred to above do not achieve the desired results.

Arrangements should be made to investigate in the near future the technical and commercial problems created by the carriage of a gas of a different composition in the Frigg pipelines and its treatment at the St. Fergus terminal. At the same time, the consequences of changing the sales gas specification should be examined.

There is no immediate requirement to initiate any exploration or appraisal drilling to support the overall economics of gas-gathering systems.

Associated gas which cannot reasonably be collected, reinjected or stored should be flared.

Prospects for the Gas Industry in Great Britain from 1990 to 2000

S. A. Price

United Kingdom

INTRODUCTION

The UK gas industry received two major stimuli during the 1960s. The first was the introduction of manufactured gas from petroleum at a cost significantly below that from the then principal source, coal. Sales, which had been stagnating at around 2½ billion therms per annum during the 1950s, rose to nearly 4 billion therms by 1967, when almost 40% of manufactured gas was petroleum based. Of greater significance was the discovery of natural gas in the Southern Basin of the North Sea which followed the discovery of the giant Groningen field in the Netherlands. The first licensing round had been held in September 1964 with the first contract, for West Sole gas, being signed the following year. Other discoveries quickly followed West Sole such that by 1970 some 60% (see Table 2) of gas supplies to the UK was coming from the Southern Basin gasfields. Currently, natural gas sales account for a substantial proportion of the UK's energy requirements. Over 6½ billion therms of gas were sold in the domestic market in 1977, some 44% of the total domestic market. In the other major market, industry, the corresponding figures are almost 6 billion therms and 26% respectively. Manufactured gas is now an insignificant percentage of total gas supplies and is confined to a few areas remote from the national pipeline system or to gas which is the by-product of industrial processes.

This paper looks at the industry in the longer term and follows some of its possible developments. The industry's freedom to vary its depletion profile is subject to certain constraints, both physical and contractual, but some flexibility is available. On top of this are the inevitable uncertainties surrounding timing and amount of future supplies. However, in general we can quite confidently expect the period around 1990 to be the time when natural-gas supplies reach a plateau, although to present a fuller picture this paper does not confine itself strictly to that date. By looking at a wider timescale, the implications of possible developments such as, say, large new gas discoveries, or the long-term role of substitute natural gas, can more readily be seen in context.

The paper falls conceptually into three sections. The first section is concerned with the likely size and potential recovery of reserves. The second section deals with the future supply profile, while the third looks at gas demands in each of the major markets.

UK CONTINENTAL SHELF GAS RESERVES

Estimating the size of the total gas reserves available to the UK is a hazardous operation. Even for known fields there is a good deal of uncertainty as to the total quantities of gas available, the composition of the gas and the economics of winning and bringing the gas ashore. For areas not fully surveyed the uncertainties are, of course, very much greater. For gas reserves located in relatively small fields in difficult waters their development may well depend on the economic merits of constructing offshore pipeline gathering systems.

The Department of Energy's publication *Development of the Oil and Gas Resources of the United Kingdom 1978* (the "Brown Book") gives estimates of gas reserves remaining at 31 December 1977 in known discoveries in the UK Continental Shelf. The estimates are shown in Table 1.

TABLE 1

Estimated UK Continental Shelf Gas Remaining in Known Discoveries at 31 December 1977*

	Proven	Probable	Possible	Total
Southern Basin				
Under contract to British Gas	14.9	0.5	0.9	16.3
Other discoveries believed commercial	1.8	2.3	-	4.1
Other discoveries	-	1.1	1.4	2.5
Total Southern Basin	16.7	3.9	2.3	22.9
Other Areas				
Under contract to British Gas	6.3	-	-	6.3
Other dry and condensate fields	1.4	6.9	11.1	19.4
Other oil associated (under development or possible development)	2.0	1.8	2.3	6.1
Total other areas	9.7	8.7	13.4	31.8
Total UK Continental Shelf	26.4	12.6	15.7	54.7

At the end of 1977, the remaining proven reserves were estimated to be 744 billion m^3 (26.4 trillion cubic feet), which is slightly lower than the estimates produced in 1977, although total known reserves are now put at 1546 billion m^3 (54.7 tcf) compared with 1443 billion m^3 (51.0 tcf) last year.

*The conversion factor assumed is 1 tcf (10^{12} cubic feet) = 28.317 billion m^3 (10^9 m^3). The terms "proven", "probable" and "possible" are given the internationally accepted meanings in the context of:

Proven those which on available evidence are virtually certain to be technically and economically producible.

Probable those which are estimated to have better than a 50% chance of being technically and economically producible.

Possible those which at present are estimated to have a significant but less than 50% chance of being technically and economically producible.

In the Department of Energy's recent Green Paper, "Energy Policy: A Consultative Document" two assumptions on reserves were used in examining the energy situation that could face the UK in the period up to the year 2000. A prudent case put total remaining reserves at between 50 and 60 tcf; this figure includes the contracted reserves from the Norwegian part of the Frigg field, plus reserves in fields which could be connected in one way or another to the Frigg field or which have an option on transmission through the Frigg line. A higher figure of 75 tcf was also examined to test the effects of higher gas reserves on the preferred depletion profile and to consider the consequences for other fuels, coal, electricity, etc. The Green Paper assumptions are consistent with the Brown Book estimates; the figures differ because in forecasting an allowance is made to allow for the possibility that only part of the "probable" and "possible" reserves referred to in Table 1 will be developed. The reserves estimates used in forecasting also include an assumption about the commercial reserves likely to be discovered after 31 December 1977, during the period to which the forecasts relate.

FUTURE GAS SUPPLIES

Table 2 shows the development of natural-gas production in the UK over the important period 1967-1976. The table shows the importance of the Southern Basin gasfields in the development of the UK gas industry. In 1967 gas from the Southern Basin accounted for only 4% of total UK public gas supply, but by 1976 gas from this source had reached 14.2b therms (3800 million cubic feet per day) accounting for over 96% of total supply. The largest Southern Basin gasfield, Leman Bank, was itself in 1976 accounting for some 40% of total supply.

Since 1976 the major development has been the introduction of UK and Norwegian gas from the large northern North Sea Frigg gasfield. Gas from this field first reached the UK in the autumn of 1977: supplies are, on current projections, expected to grow rapidly and would reach their plateau by the early 1980s by which time they will be accounting for a significant proportion of the total supplies available to the UK. Other gas is expected to come ashore through the Frigg gas-line system before 1980, such as gas from the Piper oil field, but before 1980 quantities are likely to remain small. By about 1980 the first gas from the Brent field is expected to come ashore at St. Fergus in Scotland. Supplies from all these Northern Basin sources could reach some 8 billion therms per annum (some 2160 mcfd) by the early 1980s. After the mid-1980s the supply position starts to become less certain. An important part of expected future supplies of gas not yet under contract is expected to be in the form of gas found in association with oil in the North Basin of the North Sea. Unlike the dry gas found in the Southern Basin, this gas contains substantial quantities of the heavier natural gases — ethane, butanes and propane — which are potentially valuable as feedstocks for the petrochemical industry as well as for fuel uses. Unlike Brent and Frigg, the quantities of gas associated with each field so far known are too small to justify the construction of individual pipelines to shore. While some of this potential supply is from proved oilfields, the major part appears to lie in fields which are still under assessment. For such sources, the uncertainty as to whether the field will be developed has to be added to the uncertainty whether the gas from it can viably be collected. At present, however, we cannot say for certain to what extent exploitation will be practicable. Further study of the characteristics of the oil reservoirs and of the physical, engineering and economic constraints is needed. A joint public sector (British Gas Corporation and British National Oil Corporation) and private sector company, Gas Gathering Pipelines (North Sea) Ltd., was set up to carry out the detailed studies necessary to appraise the viability of various gas-gathering schemes. The company submitted a preliminary report in December 1977 and a full report in April 1978. The results of this study are being examined by the Department. A further possibility is collaboration with Norway on a joint gas-gathering system. If such a system were

TABLE 2

UK Public Supply and Disposal of Gas 1967-1976[a] Million Therms

Year	West Sole	Leman	Hewett	Indefati-gable	Viking	Rough	Total Southern Basin	Onshore (Lockton)	UK Frigg	Imports	Other gas supplied to gasworks[b]	Manu-factured gas	Total supply
1967	166	–	–	–	–	–	166	–	–	305	1218	2691	4380
1968	472	274	–	–	–	–	746	–	–	396	1044	2815	5001
1969	579	1088	197	–	–	–	1864	–	–	415	952	2609	5840
1970	432	2928	722	–	–	–	4032	–	–	333	743	1693	6851
1971	671	4780	1248	63	–	–	6762	60	–	332	561	849	8564
1972	800	4939	1911	1670	503	–	9823	37	–	306	398	609	11,173
1973	645	4934	2110	1691	1295	–	10,675	9	–	293	335	728	12,040
1974	616	5878	2587	2066	1756	–	12,903	4	–	243	157	417	13,724
1975	616	5668	2774	2324	2031	4	13,417	–	–	335	69	184	14,005
1976	681	5773	2951	2379	2237	188	14,209	–	–	384	49	67	14,709

[a]Excludes a small quantity of gas supplied direct to the petrochemical industry (from West Sole).
[b]Includes celliery methane, coke oven gas and petroleum gases.

constructed it could mean that significant quantities of additional gas would reach the UK possibly in the 1980s or 1990s; but the Norwegians would decide the ultimate destination of any Norwegian gas which might be landed in the UK as a result of such a new joint trunk pipeline and it would not necessarily be available for UK consumption. It is possible to speculate only in the most general terms about further sources of supply from additional finds in other areas of the UK Continental Shelf such as in the Southern Basin and Irish Sea; even speculation becomes difficult for areas not yet tested by the drill where gas has been found, such as the South Western Approaches. In the Green Paper forecasts, crude ranges were put on such estimates, but it was felt that it was unlikely that such "speculative" gas would have any significant impact on UK gas supplies before the mid-1990s.

As the above description makes clear, the exact sources of all likely UK gas supplies after the mid-1980s are at present not certain. We have prudently assumed available reserves of 55 tcf. These could be supplemented, if we are fortunate, by discoveries in Britain and its Continental Shelf above expectations (but of course 55 tcf could, on the other hand, turn out to be an overestimate). There is the possibility of further imports either by pipeline or as LNG. On this prudent view of gas reserves it becomes increasingly likely that substitute natural gas (SNG) produced from oil and coal will be needed to meet the market in the late nineties.

DEMAND FOR GAS AND MARKETING STRATEGY

Gas is a clean high-grade fuel presenting no storage or transport problems to the user. Consequently it is attractive in high value, "premium", markets, particularly in the domestic sector where it competes principally with electricity and in the commercial and industrial sectors where it competes with the LPG and distillate oils used for process purposes and for space heating. As can be seen from Table 3 these characteristics together with competitive prices following the development of North Sea gas resulted in a rapid growth in gas sales. In the domestic market, sales of gas have over the past 10 years increased from 2 billion therms to some 6½ billion therms by 1977, an increase of over 11% per annum. High rates of growth have taken place in other markets: over the 10-year period sales have increased by 20% per annum in the industrial market and by 10% per annum in the commercial and public services market.

TABLE 3

UK Gas Consumption by Markets 1970-1976

	1966	1967	1968	1969	1970	1971	1972	1973	1974	1975	1976	1977
Domestic	2177	2473	2829	3211	3542	3930	4509	4815	5384	5891	6194	6590
Industrial*	932	902	951	1047	1416	2474	3688	4597	5087	5072	5679	5883
Commercial	419	448	479	514	561	632	634	571	729	750	834	1587
Nat. & Local Government	157	161	175	190	201	234	362	512	540	597	687	
Total final users	3685	3984	4434	4962	5720	7270	9193	10495	11740	12310	13394	14060
Power stations	-	2	9	38	60	263	630	285	985	858	662	519
Total sales	3685	3986	4443	5000	5780	7533	9823	10780	12725	13168	14056	14579

*Including non-energy uses.

At the expense of some over-simplification, one could say that over the past 10 years the marketing strategy of natural gas has been "supply led" with markets being developed in consequence of the development of the offshore gasfields. In the future this is likely to change in the realization that the "stock" of gas available to Britain is limited and most probably no greater than will be required to meet the likely demands over the next 20 years or so. In forecasting demand, we have therefore not limited ourselves to extrapolating from the past the growth of consumption in the various sectors of consumption and of gas's share of that consumption. We have also taken account of the possible movements in competitiveness of the various fuels. We have thirdly taken account of the different values gas has in different markets since, in view of its limited availability, it becomes worth while to switch gas from an early but less valuable market to a later premium market which, in the absence of natural gas, would have to pay the much higher costs of SNG. The full calculation of these effects is complex. But, in short, the expansion of market share up to about the mid-1980s to assimilate already contracted supplies from the Brent and Frigg fields will be concentrated into the premium rather than non-premium markets. Beyond 1990 we expect that the gas industry would probably gradually withdraw from some of the markets which, if the present price relationships between fuels were preserved, it would retain.

Any forecast of energy demand beyond 1980, especially a forecast which distinguishes particular markets, is highly speculative. In making its forecasts the Department of Energy examines a wide range of possible futures which take account of different patterns of economic growth, possible movements in energy prices and changing supply conditions in the individual fuel industries. The estimates shown in Table 4 are a forecast produced by the Department and based on the following main assumptions: economic growth in the UK is projected to continue to the end of the century at an average rate of just under 3 per cent per annum, broadly in line with past trends but with some allowance made for the effects of North Sea oil; energy prices related to the world price of oil which is assumed to double in real terms from its present level by the end of the century; and total UK "expected" available gas reserves of 55 tcf. To put the gas demand estimates into perspective, Table 5 provides an energy balance over the forecast period showing the position of natural gas in relation to other fuels.

TABLE 4

Forecast of UK Gas Sales by Market 1980-2000 (for Energy Uses), Billion Therms

(Figures shown in brackets are the proportion of gas in each market in relation to total energy consumption in the market)

	1980	1985	1990	2000
Domestic	7.7(53%)	8.7(59%)	9.6(64%)	10.0(63%)
Iron and steel	0.6 (9%)	0.6 (8%)	0.7 (9%)	1.0(10%)
Other industry	4.7(26%)	5.8(30%)	6.4(32%)	3.5(14%)
Commercial	2.3(32%)	2.6(35%)	2.8(35%)	3.2(36%)
Transport	-	-	-	-
Total	15.3(25%)	17.7(28%)	19.5(29%)	17.7(23%)

*Non-energy uses of natural gas, in petrochemical plants, could account for perhaps 1000 m therms per annum over the next 15 or more years.

TABLE 5

Energy Balance, UK Primary Energy Consumption*
(Million tons of coal equivalent)

	1975	1985	2000
Coal	118	130	172
Oil	154	175	239
Nuclear/hydro	13	27	93
Natural gas	52	78	70
Total	337	410	574

*Including non-energy use and bunkers.

Table 4 shows that in the domestic market sales of natural gas are expected to continue growing up to 1990 by which time gas will account for some two-thirds of the total energy supplied to domestic consumers. After 1990 sales will plateau out at a level of about 10 billion therms per annum and it is expected that sales at this level will continue until the end of the century by which time natural gas will be progressively replaced by substitute gas based on oil and coal as a feed-stock. Similarly in the industrial and commercial markets, gas sales are expected to grow up to 1990 by which time, very broadly, gas will be providing about one-third of the energy requirements of these markets. After 1990 gas to the industrial market is expected to fall and by the turn of the century sales are expected to be less than 5 billion therms — less than 15% of the market.

The demand forecast reflects an expectation that interruptible sales will not be expanded greatly beyond their present levels and that they will be gradually reduced in the longer term. Of course some such sales are unavoidable as an insurance against emergencies. But, in spite of the higher load factor, since the next tranche of supplies is likely to be from associated and other Northern Basin gas, we expect the role of storage to be expanded rather than to increase the use of interruptibles as a means of matching contracted supplies to the seasonal demand of the premium and particularly of the domestic and commercial markets. This change in emphasis is likely to lead to a search for new forms of flexibility in supply such as the storage of gas in salt cavities, which is currently being examined by the BGC. Other forms could involve the more extensive use of existing "take-or-pay" clauses in the Southern Basin contracts where gas paid for now can be taken later; and the tailoring of wholly British Gas Corporation-owned fields, such as Morecambe, to seasonal demands. This reduction in gas sales to the industrial markets after 1990 reflects the expectations discussed earlier that will be increasingly concentrated in the most premium markets.

No allowance in the above forecast is made for possible sales of UK natural gas to the Continent. On present assessment of our reserves it seems very unlikely that the gas found in the UK Continental Shelf will exceed the country's foreseeable needs, although further large discoveries could change the position. If the UK and Norway decide to collaborate in the development of gas resources in the Northern Basin of the North Sea, it might be feasible, if the Norwegians so wished, to re-export their portion of the gas across the Channel to the Continent.

CONCLUSION

In presenting a picture of the UK gas market up to the turn of the century, emphasis has been placed on the considerable uncertainties involved in putting forward such a picture. In particular we are uncertain about the ultimate size of UK reserves and their commercial viability and of the future energy demands and developments in alternative fuels. However, the paper shows with some confidence the growing importance of natural gas up to 1990. The period around 1990 is of some interest as it is likely with current reserve estimates to be then that production is at its plateau level.

After 1990 the position depends crucially on the size of reserves. On the 55 tcf estimate, supply of natural gas falls away and SNG production begins around the turn of the century. Should reserves prove larger than this or imports became available in greater quantities than assumed, the principal effect is more to extend production of natural gas at the plateau level, delaying the need for SNG, than actually to increase the plateau level. With the development in the 1980s of sources of seasonal supplies of gas (e.g. storage or fields operated at low load factor) the industry is likely to have the flexibility to match the supplies coming forward to the desired pattern of premium markets outlined here.

Consumption

B. Consumption

N. Cojan

France

Under the general topic of the prospective development of gas consumption up to 1990, three main questions have been studied:

- forecasts of gas demand in aggregate terms and by major sector of use — industrial, household, commercial and public;
- energy-saving measures and their impact on total and sectoral gas consumption;
- the place of gas in energy consumption.

Of the reports following, three are mainly concerned with the techniques, means of information and regulations to be applied with a view to saving energy, and with the effects they are likely to have. They are as follows:

- The contribution of natural gas to energy-saving in industry;
- Natural-gas savings in the household and tertiary sector in France, and
- Predicted consequences of measures on gas conservation and efficient utilization in various industries.

Four papers deal more particularly with forecasts of energy consumption. They are:

- Sectoral approach to the development of gas consumption in France up to 1990;
- The place of gas in the energy supplies of France;
- Trends in the development of gas consumption in Czechoslovakia up to 1990;
- Natural gas and LPG utilization in the 1990s in Hungary.

Lastly, in the paper entitled "Use of gas for olefin production in Western Europe" the possibilities of using ethane and the higher hydrocarbons contained in natural gas for olefin production in Western Europe are considered.

I shall begin by briefly recapitulating the eight excellent papers on topic B.

I shall then complete this submission by further comments on the prospective trends of gas demand in the principal ECE countries, particularly in the light of recent reports prepared by various international organizations.

The contribution of natural gas to energy-saving in industry

The target which France has set itself is to reduce industrial energy consumption in 1985 to 90 Mtpe from the forecast of 106 Mtpe made before the energy crisis.

This would be a saving of 16 Mtpe, or 15 per cent of the original estimate of consumption in the industrial sector.

The target entails an investment of 15-20 billion francs, or 940-1240 francs per tpe saved annually. With a view to achieving this target, the authorities and energy distributors are informing industrialists of energy-saving problems and offering them their assistance in their particular spheres of competence.

Through the medium of the *Agence pour les Economies d'Energie*, the authorities have taken the following measures with regard to the operating conditions and efficiency standards of the equipment and the financial aid offered to industrialists:

- mandatory periodic visits of inspection by a qualified expert to all establishments containing thermal equipment with a capacity of more than 1160 kW;
- the establishment of new minimum standards for boiler efficiency and new boiler-room operating standards;
- a variety of financial incentives for investment in energy-saving, particularly by means of grants, which could be as much as 50 per cent of the investment in the case of pilot projects and 25 per cent in the case of current operations.

The power distributors for their part are ready to provide industrialists with technical assistance in the following respects:

- the training of operating staff;
- the monitoring and improvement of the operation of existing equipment;
- the modification of existing plant to increase its efficiency from the point of view of power consumption, particularly through the installation of recuperators;
- the choice of new equipment, especially when the introduction of new gas-industry techniques can substantially increase productivity while simultaneously reducing specific consumption.

To illustrate the different possibilities, the authors describe the characteristics of eighteen installations in which notable improvements in efficiency and productivity have been made. In more than half the cases studied, specific consumption has been reduced by between 15 and 60 per cent, while the increase in productivity often reaches 10 to 20 per cent. Lastly, related to the annual saving of energy, investment is generally in the 5-10 centimes/kWh range. Particularly impressive results have been obtained through the use of JET burners for the heating of ceramic kilns and copper-alloy smelting kilns.

Apart from the points it makes on energy saving, the paper submitted by Potié, Fournier, Leblanc and Millard also contains an interesting analysis of past and foreseeable medium-term trends for gas consumption in the principal industrial sectors.

Natural-gas savings in the household and tertiary sector in France

The aim of the authorities is how to bring total energy consumption in the household and tertiary sector in 1985 to a level 22 per cent, or 18 Mtpe, below that of the forecasts made before the energy crisis.

The paper describes the measures taken by Gaz de France to attain this objective, singling out four areas of action:

- provision of information for users and fitters;
- the application of proven energy-saving techniques;
- the introduction of new or still relatively unknown energy-saving techniques;
- the use of solar and geothermal energy together with gas.

Provision of Information for Users and Fitters

Advice on the proper use of gas has been given to all subscribers in messages sent out with the bills, while booklets providing fuller information have also been sent to new subscribers and to those who are changing their installations. Special booklets on the use of gas for heating are sent only to subscribers who are installing gas heating and to the fitters themselves. Following these information campaigns, specific consumption of gas by individual space-heating installations dropped about 15 per cent between 1973 and 1975.

The Application of Proven Energy-saving Techniques

Through the trade, efforts have been mainly concerned with:

- heating by individual boilers, which will remain at least 20 per cent more economical than collective heating;
- heating by individual gas radiators; it is believed that this highly flexible kind of heating can be 20 per cent more economical than the conventional kind of individual space heating;
- collective heating with improved regulation and possible heat recovery.

New or Little-known Energy-saving Techniques

The installations proposed are as follows:

- simultaneous improvement of insulation and regulation so that up to 20 per cent or more can be saved in some existing buildings;
- the use of collective heating as the basis, supplemented by individual heating, so as to recover the maximum amount of free heat while limiting heat transfers to apartments;
- the use of condensation boilers, which save between 15 and 20 per cent by comparison with the ordinary type of boiler;
- heating of the fresh air blown into buildings by means of the heat recovered from the stale air extracted; a saving of 20 to 30 per cent can be made;
- the heating of buildings by heat pumps with gas-driven engines and with heat from the exhaust gas and cooling water; it is estimated that a saving of 40 to 50 per cent can be made;
- the dual production of heat and electricity, which is about 30 per cent more economical than separate production; this type of installation would meet the safety requirements of certain public services, hospitals and computer centres.

The Application of Solar and Geothermal Energy together with Gas

Solar energy. The author describes six pilot operations which have been carried out or are under way in the southern part of the country. For individual-house heating and domestic hot-water supply, a saving of about 45 per cent was achieved. For only the heating of an office building, it is hoped to make a saving of 30 per cent.

Geothermal energy. The possibility of combining gas with geothermal energy depends on a number of parameters, and complex studies are needed to make such an operation pay. The author reports on two operations involving 2000 and 6000 dwellings in which geothermal energy covered 90 and 70 per cent of the requirements respectively.

Predicted Consequences of Measures on Gas Conservation and Efficient Utilization in Various Industries

The share of natural gas in the USSR fuel-energy balance, which was 18 per cent in 1970, is expected to be 24 per cent in 1980 and may increase to 26 per cent* in the future, according to forecasts over the longer term.

The consumption of natural gas for technological purposes is expected to increase very rapidly, so that it may be twice as high in 1976-1980 as in 1970-1975.

The use of secondary sources of energy led to a saving of 15.1 Mtce in 1975, and it is anticipated that 30 Mtce can be economized in 1980.

The use of external recuperators in furnaces would help to reduce natural-gas consumption by 30 per cent. In pusher-type furnaces with internal recuperation, energy consumption has been cut from 2 to 0.7 kWh/kg, and heating time from 6 hours to 1 hour, while furnace efficiency has been increased by 35 per cent.

Energy consumption in the USSR steel-smelting industry has been reduced by 39 kgce/t over 7 years, and further savings are expected with more intensive production of converter steel. Coke consumption per ton of cast iron was reduced by 14.3 per cent between 1967 and 1975 through the use of gas, oxygen and preheating.

Apart from the results obtained in the USSR itself, the report makes interesting references to the experience of other countries, such as the United States and the Federal Republic of Germany, particularly, as regards energy saving in industry.

The author concludes by listing a number of measures to economize on gas consumption in industry. The implementation of these measures and of others designed to improve the structure of the energy balance by using the different energy resources more rationally should enable the USSR to save more than 150 Mtce by 1980.

Sectoral Approach to the Development of Gas Consumption in France up to 1990

This paper analyses the prospective development of gas consumption in the residential, tertiary and industrial sectors and in thermal power stations. The trend of development is related to the target set by the Seventh Plan: to increase the share of gas in total primary energy requirements to 15 per cent by 1990, i.e. 42-44 Mtpe.

The Residential Sector

The proportion of gas in the energy consumption of the residential sector, which rose from 10 per cent in 1970 to 16 per cent in 1977, is expected to reach 20 per cent in 1985 and 22 per cent in 1990. The reasons for this decline in the rate of growth may be found, not only in the virtual saturation of the market for existing housing after the slowing down anticipated in the period 1985-1990, but also in the scattered nature of the new housing built, which consists for the most part of individual houses whose connection to the system requires heavy investment and in the rapid expansion of electric heating. Sales of gas in this sector, which were 80×10^9 kWh (288 000 TJ) in 1977, are expected to increase at the annual rate of 6×10^9 to 7×10^9 kWh (21 600 to 25 200 TJ).

---

*Although the author does not say so explicitly, these percentages seem to refer only to natural-gas production, the figures at the consumption level being about 2.5 points higher. The gap is mainly due to the volume of net oil exports, which is expected to decrease in future.

This progression corresponds to the installation of gas-fired central heating in 300 000 dwellings per year divided more or less equally between new buildings (35 per cent of all new dwellings) and the existing housing stock (slightly over 10 per cent of the present stock of badly heated dwellings already connected for gas).

The Tertiary Sector

The share of gas in the energy consumption of the tertiary sector is expected to develop in the same way as in the household sector, from 5.3 per cent in 1970 to 12 per cent in 1977, 17 per cent in 1985 and 18.5 per cent in 1990. Sales in this sector rose to 25×10^9 kWh in 1977 (90 000 TJ). The annual progression envisaged is of the order of 3×10^9 kWh (10 800 TJ), with the increase originating equally in the new tertiary sector and in the already existing one.

The Industrial Sector (excluding thermal power stations)

The share of gas in total consumption is expected to develop as follows: 8 per cent in 1970, 15 per cent in 1977 and 21 per cent in 1990. Sales in 1977 reached 104×10^9 kWh (374 400 TJ). The annual progression is expected to average 10×10^9 kWh (36 000 TJ) over the period 1977-1990.

In 1976 sales to the steel-making and metallurgical industries represented nearly 40 per cent of total sales of gas to industry, while the chemical industry took nearly 20 per cent. These branches are expected to drop a few points in favour of the engineering industries by 1990.

As to the price of the gas sold to industry, the author states that the price level for interruptible supplies should be lower than that of firm supplies and that the differential will vary according to the terms of interruptibility.

Thermal Power Stations

The decline in sales of gas to thermal power stations, which began in 1975, is expected to continue in the future. In that year, sales were 22×10^9 kWh (79 000 MJ).

General Balance

The author compares estimated consumption with gas supplies, and concludes that the contracts entered into will suffice to cover 85 per cent of requirements in 1985, whatever the outcome of the current negotiations.

To meet the seasonal variations in gas demand for space heating, the useful capacity of underground reservoirs will have to be considerably increased as requirements expand. Moreover, to cushion the impact of an unforeseen drop in supplies, sales to heavy industry must be under contracts with a relatively large interruptible element.

The author concludes that if the price of oil rises during the 1980s, the establishment of new gas-supply lines will be facilitated, and the result may be a shift in the balance of world energy supplies in favour of gas.

The Place of Gas in France's Energy Supplies

The paper explains the reasons for, and the conditions for the success of, French medium-term gas policy, and the situation as regards its continuation up to 1990-2000.

After defining the main lines of French energy policy between 1975 and 1985, which at the international level entails greater diversification of supplies and closer co-operation between the producing and consuming countries, the author refers to the limited contribution made by gas to the French energy balance in 1960-1973 (3.5 per cent in the former year and 8 per cent in the latter).

The oil crisis of 1973 led the Government to set itself the ambitious target of increasing the share of gas in primary energy resources to 16 per cent by 1985. There are several reasons which justify the decision to rely increasingly on gas, such as the possibilities of energy saving, the rebalancing of foreign trade and security of supply through stronger links with the exporting countries.

There are, however, several prerequisites for the success of this development policy:

balanced use of the sources of supply — national resources and those obtained from the different exporting countries;
judicious utilization, deriving maximum benefit from the specific advantages of gas; and
lastly, the need to make arrangements (interruptible supply contracts, underground reservoirs) for dealing with a break in supplies and/or daily and seasonal variations in demand.

The possibility of meeting the target for 1985 must be judged from the dual viewpoint of the supply and use of gas. On the supply side, the contracts already signed cover more than 85 per cent of the target of 37 Mtpe set for 1985. As to use, an extrapolation of the trend of demand over the last few years points directly to the target, and the chances of achieving it may therefore be considered good.

As far as the 1990 horizon is concerned, two special features of the French energy situation may adversely affect a gas-development policy, however beneficial it may appear to be:

first, increased reliance on nuclear energy should make it possible to provide industries and households with electricity at prices that are relatively independent of the international hydrocarbons market;
second, on the supply side, there will be a gradual decline in 1985-1990 in two sources which now provide the bulk of France's gas supply, namely, Lacq and the Netherlands.

And in fact, in view of the resources at present assured and those that would be needed for the major projects under consideration, the Energy Commission has thought it reasonable to maintain for 1990 the 15 to 16 per cent that was set for the share of gas in primary energy in 1985.

By the year 2000, Europe's supplies will largely depend on the stage of development reached by the natural gas liquefaction chains and the long-distance gas pipeline networks. If the economically accessible resources reach too low a level, sources of energy to supplement or replace natural gas could be drawn on, such as liquefied petroleum gases, the gas obtained from coal gasification and the hydrogen produced by electrolysis on the basis of the nuclear electric power available at off-peak hours.

Trends in the Development of Gas Consumption in Czechoslovakia up to 1990

The very rapid development of the Czechoslovak gas industry in the years following the Second World War was based on the maximum use of available coke-oven gas and the construction of plants for the gasification of brown coal under pressure. The natural gas from local deposits, which had not played a significant part before 1955, became the main source of supply from 1956 until 1967, when its place was taken by natural gas from the Soviet Union. Co-operation with the Soviet Union has influenced the subsequent development of the gas industry in Czechoslovakia, mainly through the transmission of natural gas in transit to the Western European countries and the German Democratic Republic, for this entailed the construction of a high-capacity pipeline which provided a means of supplying a considerable number of consumers in new areas.

Natural gas at present represents 8 per cent of primary fuel resources and more than 75 per cent of the gas distributed. Its share in the fuel balance will rise to 10 or 11 per cent by 1980 (86 per cent of the gas distributed) and to 14 or 15 per cent by 1990 (90 per cent of the gas distributed). Total gas resources should thus rise from 55 TWh (198 000 TJ) in 1975 to 110 TWh (396 000 TJ) in 1980 and 169-180 TWh (608 000-648 000 TJ) by 1990.

In 1975 total consumption of gas by industry and thermal power stations was 30 TWh (108 000 TJ) and is expected to rise to 70 TWh (252 000 TJ) by 1980 and 105-109 TWh (377 000-393 000 TJ) by 1990. The share of the chemical industry in this total is expected to undergo a rapid reduction from 28 per cent in 1975 to 14 per cent in 1980. The corresponding proportions for steelmaking will be 13 and 17 per cent and for thermal power stations 7.5 and 8 per cent. In the period between 1980 and 1990, the growth rate of consumption is expected to be higher than the over-all average in the chemical, steel, engineering and food industries, and in agriculture.

In the residential sector, the number of dwellings using gas is expected to increase from 1.8 million in 1978 (37 per cent of the total) to 2.1 million in 1980 (41 per cent of the total) and to 2.6-2.7 million in 1990 (48.5-50.5 per cent of the total). The corresponding sales of gas will be 8.4 TWh (30 200 TJ), 11.4 TWh (41 000 TJ) and 21.2-26.7 TWh (76 300-96 100 TJ). The upward trend anticipated for 1980 to 1990 is at its most rapid in the services sector, viz. 8.6 TWh (31 000 TJ) in 1975, 17.7 TWh (63 700 TJ) in 1980 and 39.5-45.3 TWh (142 000-163 000 TJ) in 1990.

Between 1975 and 1990 the share of industry in total sales is expected to fall from 65 per cent to 60-63 per cent in favour of services, which will account for 24-25 per cent in 1990 as against 19 per cent in 1975, while the share of the residential sector will remain stable at about 15 per cent. Changes in sales structure influence the pattern of the send-out chart over the course of the year. Send-out charts of monthly consumption in particular branches of industry or sectors are used for making forecasts, and the monthly charts are supplemented by weekly and daily charts, which are also used for purposes of forecasting and current operations.

Underground reservoirs are used in Czechoslovakia to keep the exploitation of the different sources of gas as regular as possible, and they now hold enough to cover 20-25 per cent of daily consumption. In 1990 their contribution may be raised to 28-30 per cent, an additional 8-10 per cent being provided by the installations to be built for liquefaction and the storage of LNG. Two other means of regularizing the send-out are to interrupt supplies of gas to consumers with access to another source of fuel and to deliver gas in the summer, for the drying of fodder for instance. Big consumers are also asked to confine to the winter heavy maintenance work on thermal equipment involving a reduction in gas consumption.

In preparing forecasts, preference is given to the use of gas in the categories of demand which are of most value to the economy or the development of certain state sectors. The main method of controlling increases in the amount used by the big industrial consumers is to give them a basic fuel quota. In the non-production sector, the development of gas supplies is regulated by the programme for supplying new towns and laying mains, etc.

Natural Gas and LPG Utilization in the 1990s in Hungary

The different government bodies concerned have forecast that total energy consumption in Hungary will rise from 257 Pcal (25.7 Mtpe or 299 TWh or 1 076 000 TJ) in 1975 to 440 Pcal (44.0 Mtpe or 511 TWh or 1 842 000 TJ) in 1990.

Despite the efforts being made to develop national resources, the share of imports in total consumption will rise from 46 per cent in 1975 to 65-67 per cent towards 1990. An endeavour will be made, however, to replace imports of crude oil by other, less costly, resources such as natural gas. The share of gas in the total balance would thus rise from 5.9 Mtpe out of 25.7 Mtpe, or 23 per cent, in 1975, to 13 Mtpe out of 44 Mtpe, or 29.5 per cent, in 1990. This expansion should be stimulated, both in Hungary and in the rest of the world, by the upward trend foreseeable for the share of gas in future hydrocarbon discoveries, particularly at depths of more than 3000 metres.

The structure of gas consumption will be consonant with the basic objectives, which are to encourage a moderate increase in household consumption, to satisfy industrial requirements through the use of gas as a primary source of energy and to employ natural gas in technological processes.

Production and Use of Natural Gas after 1975

The author predicts that output from the deposits that are now being worked will decline gradually up to 1985 and then more rapidly up to 1990. New discoveries will, however, compensate for this decline, and national production is expected to remain virtually constant at 6 Mtpe/year. The additional amount required to bring it up to 13 Mtpe by 1990 will be imported by gas pipeline alone or by tanker and then pipeline (contractual arrangements have already been made for 4 Mtpe). The shares of the major consumer sectors in total gas consumption in 1975 and 1990 are as follows:

	1975 (%)	1990 (%)
Thermal power stations and miscellaneous	50.3	41.2
Gas used as a raw material	13.5	19.5
Industrial technologies	25.7	25.5
Household and communal requirements	10.5	13.8

The composition of natural gas varies according to its source. Gas with a high content of inert matter (60-65 per cent) will be used in power stations, while gas without sulphur will be reserved for nitrate plants. Increases in household requirements are mainly anticipated in areas where transport and distribution networks already exist, provided that investment on extending the networks per 1000 m^3 of gas consumed annually are kept within certain limits. A ceiling has also been set for investment on connecting up furnaces scheduled for conversion from fuel oil to gas.

In order to cope with the seasonal and daily variations in demand, a number of underground reservoirs are shortly to be made available.

In the interests of international co-operation, Hungary will be able to offer large-scale storage possibilities when its gas deposits, which initially held reserves of 100 billion m^3, are exhausted.

Comparative Prices of Energy Resources and Expenditure

After describing the way in which gas demand will evolve over the long term, the author indicates the comparative prices of various fuels at the source of supply and for the end user. The domestic price of energy is maintained below the international price level for political and social reasons, and the author considers that gas produced locally will continue to be the cheapest source of energy in the future. Final consumer prices should be controlled so as to encourage energy saving.

Liquefied Petroleum Gas, Propane and Butane

Consumers of bottled gas are expected to increase in number from 1 620 000 in 1975 out of a total of 3 900 000 households to 2 100 000 in 1980 and 2 600 000 in 1990.

Production of propane and butane from natural gas will remain close to its present figure of 180 000 t/y up to 1990, but the development of heavy hydrocarbon refining and cracking may well double refinery stocks (93 000 t in 1975 and 210 000 t in 1990).

In 1990 sales of bottled gas to households and bulk sales to industry will be in approximately the same proportions as at present, i.e. 85 per cent and 15 per cent respectively.

Use of Natural Gas for Olefin Production in Western Europe

World production of ethane, butane and propane from natural gas, whether the gas is dry or oil-associated, is expected to increase very significantly in the future, and new outlets should be sought for these products.

Up to now, olefin production in Europe has been based on the steam-cracking of naphtha. While olefin requirements, however, are growing at the rate of 5 to 6 per cent, the supply of naphthas is limited by the stagnation in the volume of crudes refined and the rising demand for motor fuels.

The light feedstocks such as ethane, butane and propane could be used to supplement naphtha in existing installations in proportions of up to 20 per cent of the latter's capacity or as a substitute for it in new installations especially designed for this kind of feedstock.

The existing installations, which should be able to satisfy the demand for olefins up to 1988, are an important potential outlet for the heavy hydrocarbons extracted from natural gas: they could take 4 Mt/year if half the European steam-cracking units used a light feedstock for 20 per cent of their capacity. As to the cost of heavy fuel oil, high-grade petrol and naphtha per ton of product, which is 400, 680 and 630 French francs respectively, the use of light feedstocks as a substitute for naphtha would not affect the cost of the olefins produced provided that the cost of the feedstock was: 456 French francs for ethane, 569 French francs for propane, and 581 and 598 French francs for butane and n-butane.

Of the corresponding figures for new plants, viz. 657 French francs for ethane, 644 for propane and 591 and 632 for butane and n-butane, those for ethane and propane are definitely more advantageous.

TREND OF GAS DEMAND IN THE REGION OF THE ECONOMIC COMMISSION FOR EUROPE: CONCLUSIONS

1. *The Share of Gas in Total Energy Consumption*

The papers which have just been summed up show that the share of gas in primary energy consumption in 1990 is likely to be 15 to 16 per cent in France, 14 to 15 per cent in Czechoslovakia and over 29 per cent in Hungary.

In the case of the European Economic Community as a whole (the Nine), it is estimated in a recent report,* that gas may account for 17.2 to 18.4 per cent in 1985 as compared with 16.4 per cent in 1976. The individual forecasts for member countries are as follows:

19.5 per cent in Belgium	as against 19 per cent in 1976
12.0 per cent in Denmark	as against 0 per cent in 1976
17.4 per cent in the Federal Republic of Germany	as against 14.5 per cent in 1976
13.4 per cent in France**	as against 9.4 per cent in 1976
8.8 per cent in Ireland	as against 0 per cent in 1976
16.4 per cent in Italy	as against 15.5 per cent in 1976
7.7 per cent in Luxembourg	as against 8.8 per cent in 1976
30.3 per cent in the Netherlands	as against 44.5 per cent in 1976
17-23 per cent in the United Kingdom	as against 16.7 per cent in 1976

With respect to the United States, the paper submitted under topic A by the American Gas Association reckons that if field price controls were removed, the share of natural gas, excluding condensates, might rise to 26.9 per cent by 1985, with 25.3 per cent from domestic sources (27.3 per cent in 1976). Under the OECD hypothesis of a partial relaxation of price controls towards 1980, however, the share of gas in 1985 would be only 21.1 per cent, and the International Energy Agency forecast is even more modest — 18.1 per cent (see Table 1). In the United States as in most of the Western European countries, the share of gas is basically limited by the supply obtainable from domestic or foreign sources.

In the USSR the share of gas might move five points above its present level of 24-25 per cent, and in the other countries of Eastern Europe as a whole the present figure of 15 per cent might rise to 19-20 per cent by 1985.† Consumption in the USSR does not appear to be limited by the level of production, which is rising rapidly; and one of the policy problems there is the division of supplies between its own needs and those of the CMEA importing countries and the rest of Europe.

---

*See the report dated July 1977 on the achievement of the Community's energy policy objectives.

**According to official statistics in France, the figure is 15.5 per cent. The difference of 2 points is mainly due to the recent downward revision of estimates of primary-energy consumption in 1985.

†This group comprises Bulgaria, Czechoslovakia, the German Democratic Republic, Hungary, Poland and Romania.

TABLE 1

Share of Gas in Primary Energy Consumption in the ECE Region

Area and number of countries	1976 Mtpe	1976 %	1980 Mtpe	1980 %	1985 Mtpe	1985 %	1990 Mtpe	1990 %
ECE (9)	(UN 1978)		(OECD 1977)		(OECD 1977)		(IEA 1978)	
Combined with solar energy	207		216		216		254	
Oil, including bunkers	542		560		665		645	
Natural gas	156	16.4	203	19.0	241	18.9	247	
Primary electric energy, etc.	47		88		154		248	
Mtpe	953		1067		1276		1394	
Total: EJ	39.9		44.6		53.4		58.3	
FREE TRADE AREA (6)			(estimate)		(estimate)		(estimate)	
Combined with solar energy	18							
Oil, excluding bunkers	79							
Natural gas	6	4.1	6	3.6	10	5.1	12	5.4
Primary electric energy, etc.	43							
Mtpe	146		166		195		221	
Total: EJ	6.1		6.9		8.2		9.2	
CMEA (6)			(estimate)		(estimate)		(estimate)	
Combined with solar energy	238							
Oil, excluding bunkers	87							
Natural gas	59	15.1	70	15.2	100	17.9	130	19.7
Primary electric energy, etc.	6							
Mtpe	390		460		560		660	
Total: EJ	16.3		19.2		23.4		27.6	
USSR (1)			(estimate)		(estimate)		(estimate)	
Combined with solar energy	370							
Oil	379							
Natural gas	255	24.5	350	27.3	480	29.5	610	30.0
Primary electric energy, etc.	37							
Mtpe	1041		1280		1630		2030	
Total: EJ	43.6		53.6		68.2		84.9	
CANADA(1)			(OECD 1977)		(OECD 1977)		(IEA 1978)	
Combined with solar energy	16		20.3		24.1		32	
Oil, excluding bunkers	91		111.0		119.5		121	
Natural gas	35	23.1	46.7	19.1	53.8	19.3	61	
Primary electric energy, etc.	62		65.8		81.7		117	
Mtpe	204		243.8		279.1		331	
Total: EJ	8.5		10.2		11.7		13.8	
UNITED STATES (1)	(UN 1978)		(OECD 1977)		(OECD-AGA)		(IEA 1978)	
Combined with solar energy	350		376		493-428		657	
Oil + LNG, including bunkers	808		971		1031-994	(21.1	1153	
Natural gas	472	27.0	449	22.6	498-678	to	438	16.6
Primary electric energy, etc.	120		194		338-418	26.9)	390	
Mtpe	1750		1990		2360-2518		2638	
Total: EJ	73.2		83.3		98.7-105.4		110.4	

TABLE 1 (*contd.*)

Area and number of countries	1976		1980		1985		1990	
	Mtpe	%	Mtpe	%	Mtpe	%	Mtpe	%
OTHERS (5)			(estimate		(estimate)		(estimate)	
Combined with solar energy	36.3							
Oil, excluding bunkers	81.9							
Natural gas	3.3	1.9	6.3	3.8	11	5.1	20	7.2
Primary electric energy, etc.	13.1							
Mtpe	134.6		167		216		277	
Total: EJ	5.6		7.0		9.0		11.6	
TOTAL (29)								
Combined with solar energy	1253							
Oil	2033							
Natural gas	987	21.4	1131	21.0	1383-1563 (21.2 to 23.4)		1518	20.1
Primary electric energy, etc.	329							
Mtpe	4602		5374		6516-6674		7551	
Total: EJ	192.5		224.8		272.6-279.2		315.9	

Sources: Report by the Secretary-General of OECD, *World Energy Outlook*, 1977.
"IEA reviews of energy programmes", June 1978.
AGA, "The American point of view on the future of the gas industry".

Published forecasts up to 1990 and 2000 are few and far between. In the case of the European Community, some scenarios prepared by Community bodies envisage that in 1990 the share of gas in total requirements will be the same as in 1985, i.e. 18 per cent. The International Energy Agency in its report of June 1978 envisages a slight contraction in the share of gas between 1985 and 1990, from 19.2 per cent to 17.8 per cent (see Table 1). These forecasts are modest, in view of the volume of gas in the hydrocarbon deposits discovered, which rose from about 10 per cent before 1970 to over 70 per cent during the period 1972-1977, when 25 giant oilfields representing about 8 Gt, and 33 giant fields of gas totalling 18 500 Gm^3 were found.*

By the end of the century, the AGA estimates that consumption in the United States may rise to 33 x 10^{12} ft^3. Of this, 4.6 per cent would be imported (1.6 per cent from Mexico, Canada and three other countries in the form of LNG), which might represent 26.5 per cent of total primary energy consumption taken as equivalent to 134 000 TJ (3200 Mtpe).

The figures given above for the estimated share of gas in 1985 have been reproduced in synoptic Table 1, and supplemented by estimates for the six countries in the Free Trade Area (Austria, Finland, Norway, Portugal, Sweden and Switzerland) and the other countries in the ECE region as a group (Yugoslavia, Greece, Spain, Turkey and Malta). When added together, the results provide a basis for calculating a possible trend in primary energy consumption and in the share of natural gas in such consumption for the 29 countries of the ECE region taken as a whole, on the assumption that the share of gas will remain at its present level of 20 per cent and that OECD's estimate of the level of consumption in the United States in 1985 will prove correct. According to the AGA figures, the share of gas in the ECE region will improve by more than two points to 23.4 per cent.

---

*"Giant fields contain less than 1% of world fields but 75% of reserves", *Oil and Gas Journal*, 7 March 1977, pp. 164-169.

2. *Pattern and Evolution of Gas Consumption by Sector of Use*

If we examine the pattern of natural gas consumption by major consumer sectors, it will be found that the weight of a given sector can vary widely from country to country. In 1976, for instance, the share of "Households and other consumers" was more than 45 per cent in France, the Netherlands, the United States and the United Kingdom as against less than 30 per cent for the other countries in general. Thermal power stations accounted for less than 15 per cent in Czechoslovakia, France, Italy, Poland and the United Kingdom, but for over 50 per cent in the German Democratic Republic and Romania. The heaviest proportion of industrial sales is to be found in Poland (77 per cent) and the smallest in the Netherlands (25 per cent), followed by the United States (29 per cent).

The policy of rationalizing the use of energy should eventually narrow the structural differences in consumption between the industrialized countries. Divergencies will nevertheless remain owing to such factors as climate, rate of urban development and volume of national energy resources and means of financing, which may vary considerably from one country to another.

It should also be noted that the possible rate of penetration of the market by gas varies according to sector of use, and that the share of gas in energy consumption can more easily be expanded in the thermal-power-station or large-scale-industry sectors, for example, than of the household sector. On the whole, it will be found that the share of gas in over-all requirements rarely increases by more than one point per year (rate recorded in the United States from 1956 to 1960, when the share of gas rose from 23.4 to 28.4 per cent), but this rate has been far exceeded in certain small countries with abundant resources, such as the Netherlands.

Importance of Long-term Forecasting in View of the Length of Time Needed to Bring New Sources of Gas on Stream

Long-term forecasts of gas consumption are, of course, less uncertain in countries with large reserves, where the problems to be solved are mainly in the areas of technology, programming, co-ordination of the work of exploiting the deposits and developing a transport and distribution network.

In the case of countries that are heavily dependent on outside sources of supply, the different import possibilities must be weighed up before long-term forecasts are made. This inventory should be promptly followed by the opening of negotiations, which usually take 1 to 2 years. The phase of project study lasts for 2 or 3 years after a contractual agreement has been signed. The actual construction of the installations usually takes 3 to 4 years for ordinary operations. In all, 6 to 7 years will elapse between the start of the negotiations and the first deliveries of gas.

Large-scale projects which present special difficulties, however, such as the installation of underwater gas pipelines at great depths or production and transport in the polar regions, are liable to take even longer and may require as much as 8 to 10 years.

In these circumstances, the preparation of forecasts beyond 1985 is of great importance, and it is to be hoped that to the few long-term forecasting studies up to the 1990-2000 horizon that have so far been made many more will be added in view of the different possible scenarios for transmission by land or sea which are rendered feasible by the relative abundance — recognized by most experts — of natural-gas resources. From the economic standpoint, such transmission should be facilitated by the strengthening and extension of international co-operation and by the existence in Europe and the United States of transport and distribution networks, and of markets, that have reached an advanced stage of development.

The Contribution of Natural Gas to Energy-saving in Industry

N. Potié, P. Leblanc, G. Fournier and P. Millard

France

SUMMARY

Since 1974 France has made a determined effort to save energy. Of the 1985 target for energy-savings of 45 Mtpe, 16 Mtpe is scheduled to come from the industrial sector. This deliberate policy on the part of the public authorities, which involves very considerable investment, is being implemented under the auspices of the *Agence pour les économies d'énergie*.

The public authorities have assigned an important place in the satisfaction of industrial fuel requirements in 1985 to natural gas, which will cover 35 per cent against 20 per cent as at present.

In the interests of the rational management of energy consumption, future gas supplies to industry will have to be directed towards particular uses. However, the most immediate growth in gas sales will be for less economic uses because of the importance of the market for "steam" and "large furnaces", particularly since the gas industry must establish a schedule of contracts for low-priority supplies.

Companies distributing power in the form of NG or LPG have performed a very considerable amount of research in recent years into more advanced technologies, and this has made it possible to increase the productive capacity of thermal installations and reduce individual consumption.

New heating equipment has made it possible to improve transfers of heat and combustion yields. Such equipment includes GefGN variable pulse burners or conical-type burners with adjustable sections, on cement kilns, and flat-flame burners on steel-reheating furnaces.

Jet burners are being installed increasingly in the preheating zones of ceramic kilns.

The recovery of energy in all its forms has also been undertaken.

Lastly, the commissioning of total energy installations, most of them very large, has made it possible to offer users inexpensive electric power by the regeneration of hot gases for drying and the production of hot water or steam in post-combustion boilers.

Industrialists cannot disregard current developments and, prior to any new installation, they must look for the technology which yields the best results, both in productivity and in energy consumption.

Specific examples show very encouraging positive industrial results in respect of:

energy saving;
savings in raw materials;
improvements in working conditions;
protection of the environment.

La Contribution du Gaz Naturel aux Economies d'Energie dans l'Industrie

N. Potié, P. Leblanc, G. Fournier et P. Millard

France

RESUME

La France a entrepris depuis 1974 un effort considérable pour économiser l'energie. Sur l'objectif d'économie d'énergie de 45 Mtep en 1985, 16 Mtep devraient provenir du secteur industriel. Cette politique volontariste des Pouvoirs Publics, qui implique un effort d'investissements très important, est mise en oeuvre sous l'impulsion de l'Agence pour les Economies d'Energie.

Les Pouvoirs Publics ont assigné au gaz naturel une place importante dans la couverture des besoins en combustibles de l'industrie en 1985, soit 35 % contre 20 % actuellement.

Dans l'intérêt d'une gestion rationnelle des consommations d'énergie, les placements du gaz dans l'industrie devront à l'avenir être orientés vers des usages spécifiques. Toutefois, les développements les plus immédiats dans les ventes de gaz passeront par des usages moins valorisants en raison de l'importance du marché existant au niveau des usages "vapeur" et "grands fours", d'autant plus que l'industrie gazière doit constituer un portefeuille de contrat de fournitures effaçables.

Les sociétés de distribution d'énergie en GN ou en GPL ont fait de très gros efforts ces dernières années dans la recherche de technologies plus élaborées, permettant des augmentations de la capacité de production des installations thermiques et de diminuer les consommations spécifiques.

De nouveaux équipements de chauffe ont permis d'améliorer les transferts de chaleur et le rendement de combustion. On trouve ainsi des brûleurs à impulsion variable du GEFGN, ou du type conique à section réglable, sur les fours à ciment, et des brûleurs à flamme plate sur des fours de réchauffage d'acier.

De plus en plus, on installe sur les fours céramique des brûleurs Jet dans les zones de préchauffage.

Les récupérations de l'énergie sous toutes ses formes ont également été mises en oeuvre.

Enfin, la mise en place d'installations d'énergie totale, de plus ou moins grande importance, a permis d'obtenir pour les utilisateurs, de l'énergie électrique à bon compte par récupération des gaz chauds pour des opérations de séchage, de production d'eau chaude ou de vapeur dans des chaudières à post-combustion.

Les industriels ne peuvent ignorer l'évolution en cours et, avant toute installation nouvelle, ils se doivent de rechercher la technologie la plus performante, aussi bien sur le plan de la productivité que sur celui de la consommation d'énergie.

Les quelques exemples d'exploitation font apparaître des résultats industriels positifs très encourageants allant dans le sens des:

- . économies d'énergie,
- . économies de matières,
- . amélioration des conditions de travail,
- . sauvegarde de l'environnement.

ACTION DES POUVOIRS PUBLICS EN MATIERE D'ECONOMIES D'ENERGIE DANS L'INDUSTRIE

Particulièrement vulnérable en raison de la faiblesse de ses ressources énergétiques nationales, la France a entrepris depuis 1974 un effort considérable afin de réaliser une économie de 45 millions de tonnes d'équivalent pétrole (tep) en 1985, soit une réduction de 16 % de la consommation totale d'énergie prévue avant la crise à cet horizon (285 Mtep).

Ce chiffre est du même ordre de grandeur que la production d'énergie nucléaire attendu à la même époque (55 Mtep).

Pour l'industrie (hors secteur de la production d'énergie), qui utilise 35 % de l'énergie consommée, l'objectif est une économie de 16 Mtep en 1985. Cela implique un effort important d'investissement, dont le coût total, cumulé jusqu'en 1985, est estimé à 15 à 20 milliards de francs.

Une telle politique est mise en oeuvre sous l'impulsion de l'Agence pour les Economies d'Energie. Cette dernière, créée en Novembre 1974, est un établissement public à caractère industriel et commercial, placé sous la tutelle du Ministre chargé de l'énergie.

Sa mission est la suivante:

- informer et sensibiliser le public sur les économies d'énergie;
- animer la politique des Pouvoirs Publics dans ce domaine, c'est-à-dire proposer, mettre en oeuvre et suivre l'application des mesures incitatives ou réglementaires;
- promouvoir des techniques nouvelles pour économiser l'énergie et valoriser des sources d'énergie encore inexploitées.

Dans l'industrie, ce triple souci d'informer, de réglementer et d'inciter a conduit aux principales actions suivantes:

(a) Obligation instituée en 1974, pour tout établissement comportant au moins un équipement thermique d'une puissance supérieure à 3000 thermies/heure, d'un examen critique approfondi par un expert agréé des installations utilisant de l'énergie. Cette expertise, qui a concerné environ 4000 établissements, a été élargie en 1977 aux établissements comportant au moins un équipement thermique consommant plus de 1000 thermies/heure.

(b) En matière de réglementation, divers textes ont fixé des normes de construction et d'exploitation: rendements minima des chaudières en fonction des catégories de puissance, conditions d'exploitation et de construction des chaufferies, isolation thermique et normes d'équipement et de fonctionnement des installations de conditionnement d'air dans les bâtiments industriels, etc.

(c) Incitations financières aux investissements d'économie d'énergie: en raison de l'insuffisance aux yeux des Pouvoirs Publics des investissements réalisés à ce titre dans l'industrie aux cours des années qui ont suivi la crise, ces incitations ont été progressivement élargies, et prennent actuellement les formes suivantes:

- subventions allant de 15 à 25 % du coût des investissements correspondants, pour autant que le coût d'investissement rapporté à l'énergie économisée soit compris entre certaines limites;
- désencadrement du crédit, octroi de prêts à taux bonifié et procédure de financement par crédit-bail;
- amortissement accéléré pour les équipements correspondants;
- avantages particuliers accordés aux entreprises prenant des engagements précis d'économies d'énergie, dans le cadre d'accords sectoriels passés entre les syndicats professionnels et les Pouvoirs Publics; fin 1977, de tels accords avaient été souscrits par 37 secteurs professionnels correspondant à plus de 80 % de la consommation d'énergie dans l'industrie.

Afin d'inciter à l'innovation:

. la Délégation Générale à la Recherche Scientifique et Technique (DGRST) apporte son concours financier à des actions de recherche et développement en ce domaine;
. l'Agence pour les Economies d'Energie subventionne par ailleurs jusqu'à 50 % le coût d'"opérations de démonstration", correspondant à la première réalisation industrielle en matière de procédés ou d'équipements nouveaux susceptibles d'apporter des économies d'énergie appréciables au plan national, l'entreprise concernée s'engageant à ce que l'efficience en soit mesurée par un organisme agréé et à faire visiter l'installation aux industriels intéressés.

EVOLUTION PASSEE ET PREVISIBLE A MOYEN TERME DE L'UTILISATION DU GAZ DANS LES PRINCIPAUX SECTEURS INDUSTRIELS

Le gaz naturel couvrira en 1985 plus de 22 % de l'énergie totale consommée par l'industrie française, soit environ 17 Mtep. Rapportée aux besoins en combustibles (charbon, produits pétroliers, gaz) la part du gaz naturel consommée dans ce secteur passe de 20 % en 1976 à 35 % en 1985.

Cette place privilégiée inscrite dans le Plan a été associée à la volonté des Pouvoirs Publics d'économiser l'énergie.

Le gaz naturel est utilisé dans toutes les branches de l'industrie et participe à la plupart des opérations thermiques. Dans l'industrie chimique il devient aussi matière première de synthèse.

La place du gaz dans le marché industriel de l'énergie peut être analysée sous tres aspects:

(a) Une approche géographique de ce marché met en relief des variations importantes entre les parts respectives des différents combustibles au niveau des régions, avec des valeurs fortes pour le gaz naturel dans le Sud Ouest ou le Nord, zones historiques de pénétration du gaz naturel (gaz de Lacq, gaz de Groningue).
(b) Une approche sectorielle met également en évidence des écarts de pénétration importants. En 1976 le gaz naturel a couvert:

. 48,6 % de l'énergie consommée dans la "lère transformation des métaux",
. 17,2 % dans le secteur "plâtre, chaux, ciments",
. 5,9 % dans le secteur "textiles".

Le Fig. 1 traduit l'évolution des taux de pénétration des combustibles dans quelques branches industrielles: au déplacement important des consommations énergétiques du charbon vers les fuels pendant les années 1960, suit une période de pénétration du gaz naturel spécialement dans l'industrie des métaux (mécanique) et l'industrie de la céramique et du verre.

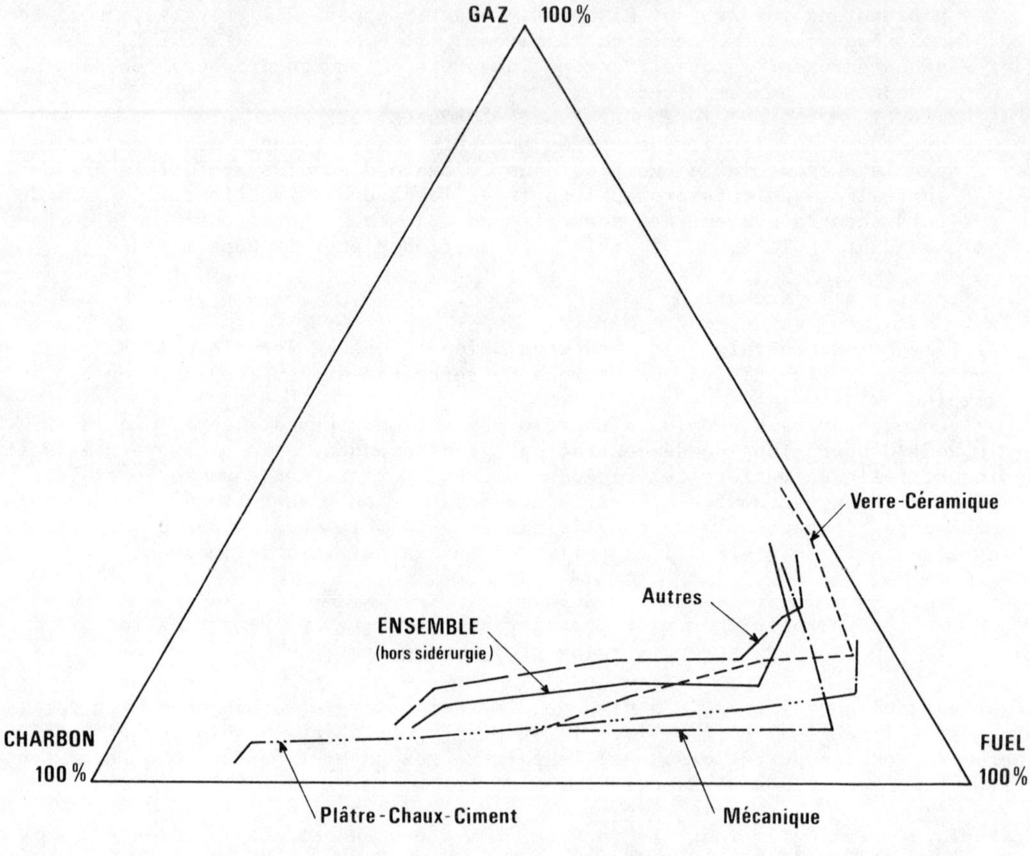

Fig. 1. Evolution de la penetration du gaz dans quelques branches industrielles entre 1958 et 1976.

Ce graphique montre également deux types d'évolution: le premier illustré par les secteurs "mécaniques" ou "verre céramique" où l'on note une tendance très affirmée pour le gaz et le deuxième illustré par le secteur "plâtre, chaux, ciment" et "autres" où la tendance est beaucoup plus douce. Ces évolutions sont caractéristiques de deux grandes catégories d'équipement, les premières en général plus spécifiques des combustibles gazeux, les secondes quasiment polycombustibles.

(c) Une approche par type d'usage met en lumière des aspects propres au gaz naturel. La classification des consommations par type d'usage fait ressortir pour le gaz naturel une nette prépondérance pour les usages four + séchoirs, usages pour lesquels les caractéristiques spécifiques du gaz améliorent considérablement les performances des matériels (chauffage direct de l'air, brûleurs spéciaux. . .).

Cette prépondérance est particulièrement mise en valeur dans la comparaison faite avec les fuels lourds pour lesquels les importances relatives des usages sont inversées. Les éléments chiffrés relatifs à 1976 confirment cette analyse (Tableau 1).

TABLEAU 1

	Gaz naturel	Fuel lourd BTS	Fuel lourd HTS
Four - séchoir	50 %	40 %	30 %
Vapeur	30 %	60 %	70 %
Synthèse	20 %	-	-
Part de chaque combustible dans le marché industriel	20 %	9 %	36 %

In Fig. 2 ci-après traduit l'évolution des poids respectifs des différents secteurs industriels dans la consommation de gaz naturel; le regroupement par secteur, permet en première approximation d'estimer la part relative des tres grands usages.

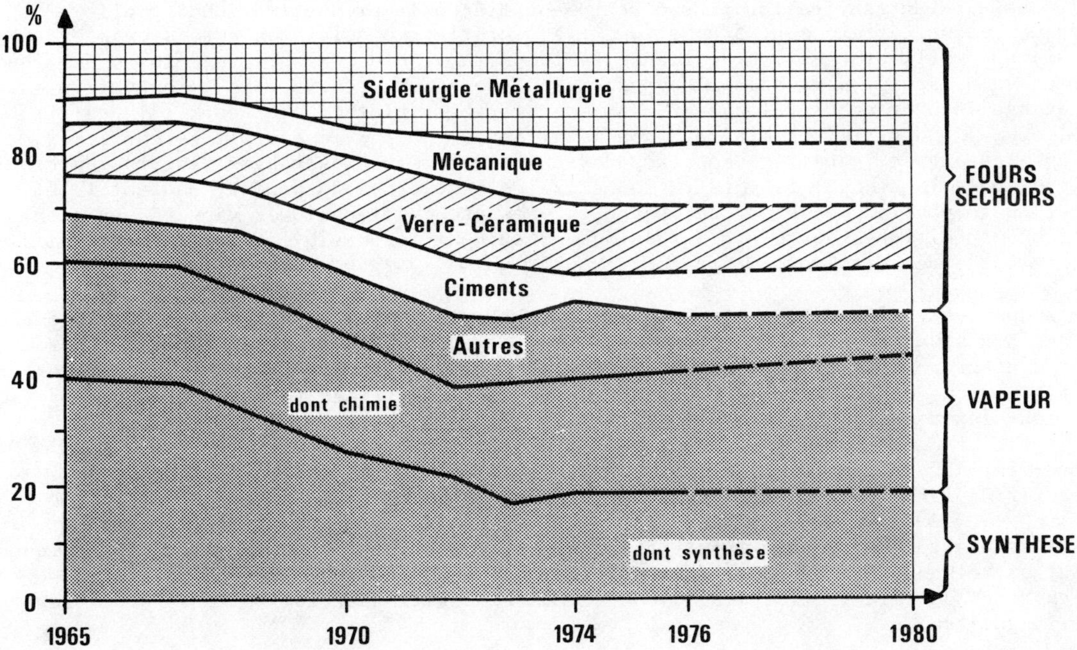

Fig. 2. Part relative des differents secteurs industriels dans la consommation de gaz naturel.

Parmi les usages industriels, une autre distinction s'impose: les usages de fabrication et les usages de confort (chauffage des locaux, renouvellement d'air).

Les premiers sont directement liés à l'évolution de la production industrielle, les seconds augmentent avec la recherche de meilleures conditions de travail dans l'entreprise.

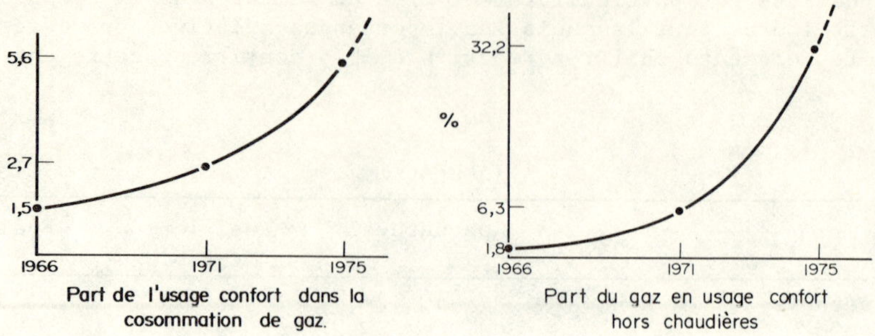

Fig. 3.

Les besoins de confort dans les entreprises ne peuvent que croître, mais l'évolution des consommations correspondantes se fera de façon plus douce, en intégrant les nouvelles dispositions réglementaires concernant l'isolation des bâtiments, la régulation des matériels, en tenant compte d'une meilleure utilisation des chaleurs "gratuites" dans les ateliers. Techniquement l'évolution se fera vraisemblablement vers des équipements spécifiques permettant une utilisation divisée plus souple, plus performante (générateur d'air chaud, chauffage par rayonnement, aérothermes, générateur de ventilation tempérée).

Les besoins de fabrication plus directement liés à la production industrielle augmenteront certainement à un rythme très inférieur à celui des années passées, mais une meilleure gestion de l'énergie conduira vers des usages gaz plus spécifiques, plus valorisants, plus économes en énergie. Un effort résolu vers l'innovation est nécessaire pour assurer l'avenir. A moyen terme la place du gaz dans l'industrie des métaux, l'industrie des minéraux non métalliques, la chimie devrait se confirmer, dans les autres industries sa place devra s'améliorer. Les usages vapeurs resteront importants longtemps, la pénétration dans les industries du papier, du textile et de l'alimentation passe par cet usage, le passage des techniques vapeurs aux techniques spécifiques se fera progressivement suivant la sensibilité et la perméabilité au progrès des différentes branches.

L'industrie gazière a constitué un portefeuille de ventes effaçables en vue de faire face aux aléas techniques, voire politiques d'approvisionnement en gaz de la France et d'assurer la continuité des fournitures "fermes" aux consommateurs domestiques prioritaires et aux industriels pour qui le fait d'utiliser le gaz est impératif de par les techniques employées.

Ces ventes représenteront environ 160 milliards de MJ (45 milliards de kWh) à partir de 1980, soit le 1/3 des ventes industrielles à cette échéance.

Par nécessité, ces ventes interruptibles correspondront à des usages non spécifiques du gaz tels que grands fours (cimenteries) ou vapeur et concerneront un nombre restreint de clients (1 % des 15.000 établissements industriels clients de l'industrie gazière).

L'importance de cet usage "obligé" du gaz nécessitera également des efforts notables en matière d'économie d'énergie.

ACTIONS CONDUITES EN MATIERE D'ECONOMIES D'ENERGIE

Les possibilités offertes aux industriels pour économiser l'énergie sont assez larges. Mais avant toute décision, ils font une analyse financière qui tient compte de l'investissement à envisager et de la durée de l'amortissement, mais également du pourcentage représenté par le coût de l'énergie dans le prix de revient du produit traité. Les augmentations de prix des combustibles depuis la crise ont sensiblement accru ce pourcentage.

L'industriel sera particulièrement motivé si des améliorations technologiques permettent d'augmenter la productivité de son installation thermique.
Les distributeurs d'énergie se préoccupent d'informer leurs clients sur le développement des techniques gazières et l'amélioration de la productivité des équipements thermiques qui en résulte.

Leur action porte principalement sur:

- le rappel des principes connus, mais pas toujours appliqués, tels que le contrôle sérieux des conditions de la combustion dans une enceinte thermique et la limitation d'entrée d'air parasite;
- la formation du personnel d'exploitation, par des journées d'information ou colloques traitant de problèmes concrets, ou par des rencontres dans le secteur d'industrie de l'utilisateur;
- la mise en place de technologies plus élaborées allant vers un approfondissement des phénomènes physico-chimiques se développant au cours du chauffage du produit;
- l'utilisation, chaque fois que c'est possible, de l'énergie potentielle représentée par la pression du gaz livré à la clientèle industrielle qui peut être utilisée, soit pour induire le comburant aux brûleurs, soit pour produire de l'électricité à l'aide d'une turbine de détente, soit pour produire des frigories, soit enfin pour pulvériser un combustible liquide ou du charbon pulvérisé;
- la récupération de la chaleur sensible contenue dans les produits de combustion et, éventuellement quand on en a l'utilisation, de la chaleur latente de ceux-ci;
- l'application de techniques gazières nouvelles permettant des augmentations importantes de la productivité et, en corollaire, une diminution de la consommation spécifique.

Les exemples ci-après matérialisent ces différentes possibilités.

Economies au Niveau de la Combustion

(a) Amélioration de la combustion — Four tunnel de cuisson d'émail

La cuisson de l'émail sur porcelaine associe un profil thermique à un phénomène physico-chimique. Elle s'effectue en trois phases:

- . combustion avec léger excès d'air, jusqu'à 1000-1100° environ;
- . combustion avec défaut d'air "dite réductrice", avec un rapport CO/CO_2 de l'ordre de 0,4/0,5 %, en vue de réduire les oxydes ferriques jusqu'à une température de 1300-1380°;
- . combustion finale stoechiométrique ou avec léger excès d'air jusqu'en fin de cuisson.

En fait, ces conditions de combustion sur des fours tunnels ou sur des fours cellules ne sont pas toujours respectées et il est souvent constaté, en particulier pendant la phase de combustion réductrice, des rapports CO/CO_2 beaucoup plus importants que la normale, atteignant et même dépassant 1.

L'explication bien souvent avancée pour justifier cet excès de CO réside dans la crainte d'entrée d'air parasite dans le four pendant cette période délicate de réduction des oxydes ferriques.

Une étroite collaboration entre les conseillers techniques du distributeur de gaz et les utilisateurs a permis de réduire les consommations spécifiques d'un four tunnel en portant l'action sur:

- le contrôle en continu pendant les essais des produits de combustion, en ramenant le rapport CO/CO_2 à 0,5 % dans la zone de réduction;
- la mise en place d'une régulation de pression à l'intérieur du four, permettant de régler celle-ci de telle façon qu'elle se situe à plus 1 ou 2/10e de mm de CE au niveau de la sole des wagonnets;
- la mise en place de nouveaux brûleurs garantissant un bon mélange combustible-comburant aux différentes allures de chauffage et d'atmosphère.

La productivité du four a été augmentée de 20 % par augmentation de la vitesse d'avancement des wagonnets (une meilleure isothermie du four a permis de diminuer le temps de séjour des produits à cuire dans le four: 1 wagonnet toutes les 60 minutes au lieu de 75 minutes).

La consommation spécifique a été réduite de 15 % en ramenant celle-ci à 24,66 MJ (6,85 kWh) par kilo de produits cuits. L'investissement, de l'ordre de 45 000 F en 1975, a été amorti d'après le client en moins d'un an.

L'économie d'énergie a été de 154 000 m^3 (n)/an.

(b) Utilisation du chauffage par rayonnement infrarouge (panneaux radiants, type Solaronics-Vannecke)

La cuisson d'émaux sur carreaux en céramique à une température de 1000° C donne des résultats très intéressants sur le coût d'exploitation, à comparer à celui obtenu avec l'énergie électrique pour une même opération.

Puissance installée au gaz: 505 à 523 kW pour la cuisson superficielle des émaux.

Production: 23 à 40 m^2 jour de carreaux, avec un temps de cuisson variant de 9 à 28 minutes selon la nature de l'émail (en moyenne 20 à 22 minutes).

Le prix de revient du chauffage au propane se situe entre 0,80 F et 2,74 F le m^2 (en moyenne, il est de 1,80 F à 2 F).

Ces résultats sont à comparer à ceux obtenus sur une installation chauffée électriquement (chauffage en armoire, c'est-à-dire mise en température de la masse céramique).

La production moyenne est de 20 m^2 en 12 à 12 h 30 et le prix de revient du chauffage se situe entre 5 et 6 F le m^2.

Le client a constaté, avec l'utilisation de panneaux radiants, une réduction des frais d'exploitation de 30 % et une diminution de la casse estimée à 10 %.

(c) Les panneaux infrarouges peuvent également être utilisés dans d'autres secteurs, par exemple dans le traitement du soufre

Une installation permet la vaporisation et la surchauffe jusqu'à 500/550°C (poussée à 800°C pour une autre fabrication).

A noter que le procédé classique en cornue donne un rendement de l'ordre de 18 à 22 % sur PCI avec du fuel lourd.

L'utilisation de panneaux radiants pour chauffer les cornues a fait gagner 20 % sur le rendement (PCI), ce qui porte celui'ci de 22 à 26 %.

Sur fours tubulaires, le rendement obtenu atteint 64,5 % PCI, même dans des conditions défavorables (dans l'exemple présenté, la charge était réduite de 15 %).

Ce résultat a été obtenu sans récupération poussée sur les fumées. En fait, celles-ci peuvent être utilisées pour la fusion du soufre solide.

D'autre part, il a été enregistré une réduction des pertes par reliquéfaction du soufre dans les chambres de condensation, alors que l'utilisation des cornues occasionne des pertes pouvant atteindre 20 %. L'investissement d'un four tubulaire est pratiquement le double de celui nécessaire pour le chauffage de cornues mais, compte tenu de la nécessité de remplacer pratiquement les cornues tous les ans et de l'amélioration du rendement, on peut considérer que l'amortissement s'effectue en moins de deux ans.

Economies au Niveau du Combustible

(a) Récupération et combustion de gaz résiduaires

Les industriels ont toujours cherché à récupérer les sous-produits combustibles de leur fabrication, afin de limiter leurs achats d'énergie primaire. Les gaz de hauts fourneaux, de cokerie, de raffinerie, les gaz sous-produits de certaines synthèses chimiques sont devenus des combustibles d'autant plus nobles que leur constance était mieux assurée.

La récupération des gaz de convertisseurs d'aciérie est en cours. Les liqueurs noires de production de cellulose et les écorces assurent environ 70 % des besoins thermiques de ces usines.

Tous ces produits font l'objet de la plus grande attention des responsables de l'énergie.

C'est ainsi que des gaz résiduaires à très bas pouvoir calorifique viennent de faire l'objet d'une récupération particulièrement intéressante à la sortie des réacteurs de la Compagnie Carbon-Black à Ambes (33); 54 000 m^3(n)/h de ces gaz à 250°C ayant un PCI de 1,780 MJ/m^3(n) (0,494 kWh/m^3(n)) sont envoyés par deux injecteurs tangentiels et brûlés dans une flamme en rotation dans un foyer d'incinération briqueté, où le temps de passage doit être supérieur à deux secondes. L'air de combustion est réchauffé à 120°C sur un échangeur de vapeur et à une température d'autant plus élevée que le PCI baisse jusqu'au minimum de 1,464 MJ/m^3(n) (0,406 kWh/m^3(n)).

Les gaz chauds à 1020°C alimentent un générateur de vapeur classique produisant 32,5 t/h de vapeur à 22 bars et 315°C avec un rendement global de 37,5 %.

Une partie seulement de ces gaz, les plus riches, étaient brûlés autrefois avec un rendement de 33 % malgré leur enrichissement avec 350 m^3(n)/h de gaz naturel. La suppression de cet enrichissement et la production supplémentaire de vapeur de l'ordre de 10 t/h représentent une économie globale d'environ 330 000 GJ/an (91,8 GWh/an). L'investissement nécessaire a été de 6,5 MF HT. Il correspond à 1,96 c/MJ (7,14 c/kWh) récupéré.

(b) Récupération de l'énergie de détente du gaz

Dans le cadre des économies d'énergie au stade initial de sa préparation, autrement dit au niveau du combustible avant son utilisation, la récupération de l'énergie de détente des gaz a été un problème régulièrement posé. Si la détente de la vapeur est classique, en particulier dans les systèmes chaleur-force, comme celle des gaz combustibles en turbine ou moteur à gaz, la récupération de l'énergie de pression des gaz naturels est peu fréquente avant combustion.

Depuis 1959, trois turbines CEM assurent, chacune, la détente de 33 000 m^3(n)/h de gaz naturel sur trois tranches de 125 MW de la centrale EdF d'Artix (64).

Conditions d'entrée du gaz	42 bars	- 300°C
Conditions de sortie du gaz	1,5 bar	- 150°C
Production	2600 kWh par tranche	

Le réchauffage du gaz avec de la vapeur de première surchauffe porte la consommation à 7,785 MJ/kWh (1,860 kWh/kWh) produits, soit les 2/3 de la consommation spécifique de la tranche, ce qui revient à dire que la marche avec turbine de prédétente améliore le rendement global de 0,8 %.

Plus récemment, en 1976, deux turbines HP et BP ont été montées sur le circuit de gaz d'alimentation d'une unité de synthèse d'acétylène à l'usine Rhone-Poulenc à Pardies (64), avec récupération de l'énergie de détente sous forme électrique d'une part, et frigorifique d'autre part, par l'intermédiaire de deux échangeurs à circulation méthodique dans les conditions de fonctionnement suivantes (Tableau 2).

TABLEAU 2

Débit de gaz 25 000 m^3(n)/h	Pression (bars abs.)	T(°C)
Entrée turbine HP	28,00	+ 20
Sortie turbine HP	9,50	- 36
Entrée échangeur 1	9,45	- 35
Sortie échangeur 1	9,25	+ 5
Entrée turbine BP	9,20	+ 5
Sortie turbine BP	3,70	- 40
Entrée échangeur 2	3,65	- 40
Sortie échangeur 2	3,40	+ 5

L'énergie électrique récupérée est de 900 kWh et l'énergie frigorifique de 3411 MJ (947,5 kWh) pour une température d'admission du gaz de 20°C. La réfrigération de la N-méthylpyrolidone améliore son pouvoir absorbant vis-à-vis de l'acétylène. Son débit peut être diminué, ainsi que le débit de vapeur nécessaire à sa régénération. On économise ainsi en moyenne 2,38 t/h.

L'économie globale pour la production d'électricité et de vapeur correspond à 3300 t/an de fuel pour un investissement voisin de 2,39 c/MJ (8,60 c/kWh) récupéré.

L'intérêt de cette opération méritait d'être noté malgré la conjonction assez rare de débits importants à détendre avec la récupération sur place de l'énergie frigorifique produite. Le réchauffage du gaz pour la seule production de kWh peut rendre quant à lui, l'opération un peu moins avantageuse.

(c) Récupération de chaleur associée à la neutralisation d'effluents gazeux

Parmi les opérations de traitements de surface, la cuisson de peinture en particulier est responsable, en dégageant des produits parfois toxiques et souvent nauséabonds, d'importantes nuisances pour l'environnement.

Certains industriels, notamment ceux implantés dans les agglomérations, se sont préoccupés du traitement de leurs effluents avant leur rejet à l'atmosphère. Les différents procédés de neutralisation par le gaz naturel se sont avérés être des moyens efficaces de lutte contre ce type de pollution, mais consomment d'importantes quantités d'énergie.

Pour diminuer les coûts d'exploitation de ces procédés, la mise en place de récupérateurs s'imposait.

Ainsi chez Ziegler à Onnaing ou encore chez Citroen à Aulnay, on a conçu (grâce à Heurtey et Caliqua chez le premier, Air Industrie chez le second), des installations intégrées "propres" comportant des ensembles: étuves de cuisson, incinérateur d'effluents, récupérateur de chaleur.

Les résultats obtenus chez Ziegler sur une ligne de traitement de surface en continu de bandes de tôles d'acier font apparaître, l'installation étant à plein régime, une puissance récupérée d'environ 9500 kW. Cette énergie thermique récupérée est, d'une part recyclée dans la chaîne de fabrication (au niveau de l'étuve elle-même et des bains de dégraissage), d'autre part utilisée pour le chauffage des locaux.

Parallèlement, la puissance nécessaire à l'incinération est de 7700 kW. La différence correspond à la récupération de chaleurs "gratuites": celle contenue dans les fumées avant traitement et celle dégagée par la combustion des solvants "pollueurs".

Chez Citroen à Aulnay-sous-Bois, la puissance récupérée à plein régime est d'environ 11 300 kW pour une puissance dépensée par les incinérateurs de 8200 kW.

Economies au Niveau des Installations

(a) Etanchéité et isolation

Dans le cadre des économies d'énergie au niveau de l'utilisation dans les foyers, une attention particulière a été portée, ces dernières années, à l'étanchéité des fours.

Ainsi, dans l'industrie céramique, de nombreux fours tunnels avaient, malgré des équipements de chauffe excellents, des capacités de production bien souvent restreintes par suite d'entrées considérables d'air parasite provenant des joints de sable, d'une mauvaise étanchéité entre les wagonnets et, parfois, par des fissurations importantes.

La réfection des maçonneries, l'alignement et l'approfondissement des joints de sable, l'addition d'une couche d'isolant fibreux entre le béton réfractaire et le béton de vermiculite sur les 84 wagons d'un four de la Société Roques à Dieupentale (82), ont permis une augmentation de production de 22 % au-dessus de 7900 t/mois et un gain de 9,2 m^3(n) de gaz naturel par tonne produite, soit 17 % de la consommation antérieure pour une dépense de 1,45 c/MJ (5,90 c/kWh) récupéré.

Aux Etablissements Gelis à Colomiers (31) on a réalisé sur un des fours le remplacement, dans la zone de chauffage, des réfractaires isolants composant la face interne de la voûte et des murs.

En voûte, de la vermiculite en vrac a été disposée sur 20 cm d'épaisseur sur une dalle en béton réfractaire. Le revêtement intérieur des murs a été recomposé avec des panneaux de 50 mm d'isolant fibreux Diatisol protégé par le mur en béton réfractaire.

Cette opération principale, liée à quelques aménagements secondaires, a conduit à une augmentation de production de 15 % au-dessus de 7500 t/mois avec une baisse de consommation de 4 m^3 de gaz par tonne produite, soit 8 % de la consommation antérieure.

Dans ce même esprit, pour obtenir une bonne étanchéité des fours tunnels, plusieurs de ceux-ci ont été réalisés en France depuis cinq ans selon une technique brevetée par la Société Hartmann de Belfort.

Ils sont formés d'un casing métallique autoporteur enfermant de l'extérieur vers l'intérieur des couches successives de laine de verre, de Kieselguhr-amiante, de béton de vermiculite et de béton réfractaire.

Ces fours permettent de diminuer les consommations spécifiques à la tonne cuite de 25 % environ par rapport aux fours anciens auxquels ils peuvent être comparés et, autant qu'on puisse en juger, à récupération constante. Les conditions de celle-ci sont, en effet, profondément modifiées car, en limitant les passages d'air parasite, on tend à ramener le four à son usage spécifique. Il cesse dans une large mesure d'être utilisé en générateur de gaz chauds pour les opérations de séchage. Les pertes par les parois, à la cheminée et dans les circuits de récupération deviennent raisonnables.

(b) Récupération de la chaleur sensible contenue dans les produits de la combustion des combustibles gazeux

La récupération de l'énergie contenue dans les produits de combustion s'effectue en général à l'aide de récupérateurs, de formes et de conceptions différentes, adaptés aux niveaux des températures et en fonction de la géométrie des installations thermiques.

S'il est relativement facile de concevoir la mise en place d'un récupérateur sur une installation neuve, il n'en est pas de même sur un four ancien dont les structures n'ont pas été prévues en conséquence et cela entraîne un investissement très important qui, bien souvent, met en cause l'intérêt d'une telle opération.

Il est apparu qu'une autre forme de récupération pouvait être envisagée. Il s'agit d'assurer le préchauffage des charges.

Ce procédé est facile et peu onéreux à réaliser pour préchauffer des pièces à des températures importantes avant leur introduction dans le four.

Cette technologie est appliquée, entre autres, pour le chauffage de lopins d'acier dans l'industrie de la forge. Les principales caractéristiques de ces fours sont les suivantes:

- . type poussant à plusieurs files;
- . récupération de la chaleur contenue dans les fumées du four pour le préchauffage des pièces;
- . chauffage dans les fours par brûleurs radiants à gaz.

Les produits de la combustion circulent dans des tubes à contrecourant des lopins qui se préchauffent en avançant.

Quand les lopins pénètrent dans le four proprement dit, ils ont été portés à une température d'environ 650°C. Le four avec ses brûleurs radiants amène la température aux environs de 1200°C.

Ce principe de chauffage effectué sur des lopins de sections et de longueurs différentes a permis d'atteindre des rendements supérieurs à 50 % pour une température mesurée à coeur des lopins de 1200°C.

Il est bon de rappeler, pour mémoire, qu'un four de forge classique sans récupérateur a un rendement de 15 à 20 %.

Si l'on veut faire une comparaison de rendement et de prix de revient sur ce type de four en prenant pour bases:

. pour chauffer un kilo d'acier à 1200°C, il faut environ 0,83 MJ (0,232 kWh);
. pour fournir un kWh, il faut 10,44 MJ (2,90 kWh) en énergie primaire nous aurons;

- Four à combustible classique sans récupérateur. Rendement moyen 15 % soit par kg d'acier

$$\frac{0,83 \times 100}{15} = 5,53 \text{ MJ } (1,546 \text{ kWh}).$$

- Four à combustible classique avec récupérateur. Rendement moyen 18 % soit par kg d'acier

$$\frac{0,83 \times 100}{18} = 4,61 \text{ MJ } (1,288 \text{ kWh}).$$

- Four à induction. Rendement moyen 70 % soit par kg d'acier

$$\frac{0,83 \times 100}{70} = 1,18 \text{ MJ } (0,331 \text{ kWh}),$$

ramené en énergie primaire: $1,18 \times 10,44 = 12,31$ MJ (0,959 kWh).

- Four avec principe de préchauffage de la charge. Rendement moyen 50 % soit par kg d'acier

$$\frac{0,83 \times 100}{50} = 1,66 \text{ MJ } (0,464 \text{ kWh}).$$

Si l'on estime que l'énergie du gaz naturel en kWh vaut en moyenne 0,043 F et qu'un kWh d'énergie électrique vaut en moyenne 0,120 F le prix du chauffage à 1200°C d'un kg de métal s'établit aux valeurs suivantes:

Four à combustible classique, sans récupérateur	$0,043 \times 1,546 = 0,0665$ F.
Four à combustible classique, avec récupérateur	$0,043 \times 1,288 = 0,0555$ F.
Four à induction	$0,120 \times 0,331 = 0,0397$ F.
Four avec préchauffage de la charge	$0,043 \times 0,464 = 0,0200$ F.

(c) Techniques nouvelles et perspectives

La chauffe ultra-rapide en four cellule de cuisson céramique. Le Centre de Recherches du Gaz de France s'est plus particulièrement intéressé aux cuissons ultra-rapides de produits céramiques en four intermittent. Les recherches ont abouti à la mise au point d'un four à cloche de 3/4 m^3 de capacité utile. Les réfractaires tapissant la cellule du four sont des réfractaires fibreux à faible inertie thermique et insensibles aux chocs thermiques. Le four est doté de huit bruleurs Jet métalliques de 140 kW de puissance unitaire.

Les essais de ce four et la détermination de quelques cycles de cuisson se sont effectués avec la collaboration de la Société Française de Céramique. En cuisson de grand feu de porcelaine (1380°), on a obtenu les résultats suivants:

- la durée du traitement est passée de 24 à 3 heures;

- la consommation spécifique est ramenée de 51,12 MJ/kg (14,2 kWh/kg) a 17,28 MJ/kg (4,80 kWh/kg). En cuisson de poterie émaillée (1000°), la durée de traitement est passée de 24 à 3 h 30, celle du dégourdi de porcelaine à 960° de 12 à 1 h 15, celle du décor de porcelaine à 1250° de 75 à 25 minutes etc. De même que pour la cuisson de grand feu de porcelaine, tous les cycles de cuisson font ressortir d'importants gains d'énergie.

Ces résultats de laboratoires reçoivent actuellement leur consécration industrielle: un four d'une capacité d'un m^3 utile est actuellement en construction aux Establissements Raynaud, porcelainiers à Limoges, Dupeux en est le constructeur.

Utilisation de brûleurs Jet pour le chauffage de fours creuset de fusion d'alliage cuivreux. Le chauffage des fours creuset de fusion d'alliage cuivreux était effectué jusqu'à ces dernières années à l'aide de combustible liquide presque exclusivement. Une technique récente d'utilisation du gaz naturel a ouvert une nouvelle voie particulièrement économique pour la fusion de ces alliages.

La Fonderie L. Martel à Saint-Omer a été l'une des premières fonderies de France à mettre cette technique en application. Cela a consisté à adapter sur les fours creuset de 400 points* equipés de réhausse, un brûleur Jet à grande vitesse, d'une puissance de 520 kW.

Le préchauffage des charges dans la réhausse et la vitesse des produits de combustion ont amélioré le rendement thermiqué des fours ainsi que le montrent les résultats suivants (Tableau 3).

TABLEAU 3

Type d'alliage	Température de coulée, °C	Temps de fusion		Consommation spécifique 2e fusion MJ/t		Economie
		Fuel	Gaz	Fuel	Gaz	
Monel NU 29	1400	3 h	2 h	18 655	8151	56 %
Cuproalu UA 30	1300	1 h 50	1 h 30	9958	6134	35 %
Bronze phosphoreux UERP	1200	1 h 10	1 h	5210	4033	23 %

Le séchage en direct au gaz des céréales. Le gaz naturel peut être utilisé sans difficulté, contrairement au fuel oil, pour effectuer du séchage en direct de céréales. Son emploi a donc permis de réaliser, par rapport aux installations classiques avec échangeur, des économies de combustible d'environ 20 %. Les consommations spécifiques rapportées au poids d'eau évaporée ont été observées sur toute une saison de séchage:

- à la Cooperative Dauphinoise à Lyon sur un séchoir Vim sans échangeur: 1,06 kWh/kg;
- à l'Union des Cooperatives Sud Seine & Marnaises sur des séchoirs Cominor convertis au gaz naturel ayant conservé leurs échangeurs, mais où les produits de combustion du gaz sont réintroduits dans les séchoirs: 1,10 kWh/kg;
- sur les meilleurs installations classiques avec échangeur: 1,33 kWh/kg.

Ces résultats sont obtenus:

---

*Point = volume occupé par un kilo de bronze fondu.

. dans le cas de conversion d'installations existantes, pour un coût dont le temps de récupération est variable, mais en général compris entre un et trois ans;
. dans le cas d'installations neuves, avec une différence d'investissement le plus souvent à l'avantage de l'installation gaz.

Dans un avenir proche, le gaz amènera encore de nouvelles économies:

Une expérience, aisément généralisable, effectuée sur un séchoir Cominor à Pont-sur-Yonne a permis d'abaisser la consommation spécifique à 3,20 MJ (0,89 kWh) par kg d'eau évaporée, tout en multipliant la productivité du séchoir par un coefficient compris entre 1,5 et 2.

CONCLUSION

Ces novations impliquent un investissement sensiblement plus important que celui des techniques classiques, mais l'augmentation de la capacité de production qu'elles permettent d'obtenir va les imposer obligatoirement.

D'autre part, les exemples cités dans ce texte font également apparaître que faire des économies d'énergie et augmenter la productivité des installations thermiques a pour corollaire une diminution des prix de revient. Nous assistons donc à une convergence allant dans l'intérêt de l'industriel.

Enfin, un industriel ne peut ignorer cette évolution car, très rapidement, il se trouvera dépassé par le concurrent plus dynamique qui n'hésitera pas à s'engager sur ces nouvelles bases.

Il semble également que, de plus en plus, le personnel d'exploitation tient à bénéficier de meilleures conditions de travail (limitation du rayonnement des fours) et disposer au moins de deux jours de congés par semaine.

On s'oriente donc de plus en plus vers des installations thermiques automatisées au maximum, et de fonctionnement intermittent permettant des montées en températures rapides et des refroidissements aussi rapides.

La technologie gazière a ainsi su s'adapter aux nouvelles réalités énergétiques. Les ingénieurs des diverses sociétés gazières ont participé largement à cette novation et, en collaboration avec les constructeurs, transmettent le message aux industriels.

D'autre part, l'utilisation d'air butané ou d'air propané comme gaz substituables au gaz naturel permet, dès maintenant, aux industriels de faire face à certains contrats pouvant inclure une clause d'interruptibilité momentanée ou saisonnière.

A long terme, l'utilisation d'hydrogène ou de gaz naturel hydrogéné, les possibilités offertes par la gazéification du charbon et les espérances de nouveaux gisements de gaz naturel permettent de considérer l'avenir sans pessimisme, à condition d'utiliser cette énergie de la façon la plus rationnelle.

N.B. Les noms de certains constructeurs ayant mis au point des installations permettant de réaliser des économies d'énergie ne constituent pas une sélection. Il est bien évident que d'autres constructeurs spécialisés dans les mêmes secteurs d'industrie sont susceptibles d'obtenir des résultats similaires.

Sectoral Approach to the Development of Gas Consumption in France up to 1990

A. Bouleuc

France

This paper analyses the growth of gas demand forecast in the residential, commercial and industrial markets. These prospects rely on the expected growth of the markets, and on the competitive standing of the different types of energy. They take into account such factors as future availability of gas, security margins needed for matching demand and supply, constraints due to the adjustment over time of a growing demand to a growing supply.

The prospects result in a share of gas in the total primary energy consumption of 1990 reaching about 15%.

GAS SALES ON THE RESIDENTIAL MARKET

Present Situation of Gas on this Market

TABLE 1

(10^9 kWh)	1970	1975	1977
Total energy consumption of the sector	377	440	500
including: heat uses	333	380	429
Gas sales[a]	36.8	64.8	79.8
gas/total energy consumption (%)	9.8	14.7	16.0
gas/heat consumption (%)	11.0	17.1	18.6

[a]Out of which sales for collective central heatings represent (in 10^9 kWh): 1970: 2.7
1975: 12.9
1977: 18.0

There are three traditional *uses* of gas in this sector: cooking, water heating and space heating. The structure of gas sales by use for 1977 is given below:

Cooking	10.0%
Water heating	19.9%
Space heating	70.1%
	100.0%

The two first types of uses have penetrated the market for a long time: 7 300 000 of dwellings are equipped with gas cooking appliances, and 5 000 000 with hot-water-producing appliances. Central heating, more recently developed, equips 2 700 000 dwellings: 900 000 individual central-heating equipments in individual houses, 1 million individual central heating equipments in flats, 800 000 collective central heating equipments in flats. In addition, there are about 800 000 gas heating separate appliances.

Gas Market in the Residential Sector

Flats account for 47% of the new dwellings completed in 1977, whereas individual housing units account for 53%. This is the first year where the share of individual houses exceeds 50% of the new dwellings: the ratio was only 37% in 1970 and it has grown quickly during the past years. One estimates that 91% of new dwellings are equipped with a central heating equipment.

The gas-supplied area covers about 63% of the dwellings completed. The extension of the network in the future is supposed to maintain this level.

In 1977, gas central heating equipments have been installed in 27% of the new dwellings.

For the next years, the objective is to raise this ratio up to 35%.

In existing dwellings there may be either a changeover to gas or the acquisition of gas central heating in dwellings which were not initially equipped. This last potential market may be estimated to 2 300 000 main dwellings in the gas supplied area, 1 400 000 of which being already connected to the distribution network.

So this market offers a large outlet to gas and, in 1977, more than 160 000 existing dwellings had been equipped with central heating installations.

Competitive Standing of Gas on the Domestic Market

On the French domestic market, gas has to compete essentially against electricity and gasoil.

Some factors play against the penetration of gas:

- a strong trend towards the scattering of new dwellings, which results in a growth of network investments;
- the fast development of electricity heating favoured by a low capital cost of the installation.

Other factors favour gas:

- the lost cost of transportation and the existence of huge storage facilities;
- the savings allowed, from a conservation point of view, by the use of gas;
- the continuity of the delivery of gas without any storage equipment for the consumer;
- its powerness and its flexibility from the point of view of the user, its cleanness (absence of sulphur, total combustion), the environmental acceptability of the network which is entirely hidden;
- its competitiveness on a cost basis for the user (the cost including capital, energy and maintenance costs);
- the development of new technologies: vent-hole appliances, condensation boilers, gas heat pumps.

The gas prices are, at this moment, a bit lower than the market prices, especially for supplies to collective heating installations. Once at the market level, gas price may grow at the same trend: in this hypothesis, there will still remain a premium value for gas.

In these conditions it seems to be quite realistic to aim at selling about 300 000 central heating installations per year.

Forecasts

The analysis of the domestic market shows that annual increases of 6000 to 7000 millions of kWh may be reasonably expected on this market. Given the overall growth of domestic energy consumption, the gas share is therefore expected to reach 20% in 1985 and 22% in 1990.

GAS SALES ON THE COMMERCIAL MARKET
(including the public sector)

Present Situation of Gas on this Market

TABLE 2

(10^9 kWh)	1970	1975	1977
Total energy consumption of the sector	165	189	209
including: heat uses	112	115	127
Gas sales	8.7	19.1	25.0
gas/total energy consumption (%)	5.3	10.1	12.0
gas/heat consumption (%)	7.7	16.6	19.7

Uses of Gas

One finds again in the commercial sector the types of utilizations already described (cooking, water heating, space heating), but with some nuances. Cooking corresponds here either to the utilizations of gas in restaurants, for catering or in the food business; water heating covers the utilizations of gas in laundries as well as in swimming pools. Gas is also used for steam raising (pressings). In the whole there are 240 000 customers in this sector with a wide variety of utilizations, space heating accounting for 62% of the sector consumption.

In 1976 the breakdown of gas sales by branches was the following:

Health	16.5%
Education	14.9%
Hotels, restaurants	12.8%
Sport, cultural establishments	6.7%
Offices	29.1%
Shops	20.0%
	100.0%

Gas Market in New Constructions

In the whole territory, new heated commercial buildings represent each year about 16.5 million m^2, 13 millions of which being located in the gas-supplied area: the corresponding needs of heat are estimated, respectively, 4500 million kWh and 3600 million kWh in the gas-supplied area.

Present forecasts show that the new needs are expected to stay at the same level for the next years.

In 1977 gas has held 29% of the market.

For the next years, Gaz de France aims to hold 35% of this market.

The Market of Gas in Existing Constructions

In existing buildings, changeovers from one fuel to another is the main outlet offered to gas. The total energy consumption and the ratios showing the penetration of gas have been given above (Table 2).

The breakdown of energy consumption by branches is as follows:

Health	14.8%
Education	20.1%
Hotels, restaurants	20.5%
Sport, cultural establishments	14.8%
Offices	21.8%
Shops	8.0%
	100.0%

In 1977 new installations using gas accounted for 1500 million kWh and this value is expected to stay at the same level in the next years.

Competitive Standing of Gas on the Commercial Market

The growth of gas on this market is conditioned by almost the same factors as those already analysed in the section forecasts earlier, with some peculiar features. The localization of commercial buildings has long been more concentrated, and they are therefore nearer the gas network than residential buildings. For example, around 80% of new commercial buildings are in the gas-supplied area; higher levels of consumption per customer are well fitted for the delivery of gas.

Forecasts

The above analysis of the market indicates that one may expect annual increases of gas sales of 3 000 000 kWh. These forecasts will lead to a share of 17% for gas in 1985 and of 18.5% in 1990.

GAS SALES IN THE INDUSTRIAL MARKET

Present Situation of Gas on this Market

TABLE 3

(10^9 kWh)	1970	1975	1977
Total energy consumption of the sector	617	642	688
including: heat uses	440	445	469
Gas sales	49.0	82.7	103.8
gas/total energy consumption (%)	7.9	12.9	15.1
gas/heat uses (%)	11.1	18.6	22.1

Uses of Gas

Gas is used in all the branches of industry with a great variety of utilizations and processes.

Gas is essentially sought in drying operations, cooking, grilling and incineration, smelting, water evaporation and heat treatments. Gas is also used for feedstock purposes in chemical synthesis.

Gas processes involve ovens, boilers, reactors and various other equipment.

Ovens may be described according to their type or their size. For example: rotary kilns for cement, tank furnaces for glass, reheating furnaces in metallurgy.

Boilers of all sizes are widespread whenever low-temperature heat is needed: it is often the case in the chemical industry, in the paper, textile, food and agricultural industries.

Reactors are used in the chemical industry for the production of ammonia, methanol and carbon black.

Other types of process involve gas:

controlled atmosphere generators for metal treatments;
gas burners and gas blowpipes for direct treatments (of glass, for example).

The breakdown of gas sales by type of process is given below:

Boilers	27%	
Ovens, dryers, stoves	47%	
Reactors	22%	(including gas used for feedstock in the chemical industry)
Others	4%	
	100%	

There are around 15 000 industrial customers of the gas industry.

In 1976 the breakdown of gas sales by branches of industry was the following:

Steel industry, metallurgy, electro metallurgy	18.6%

Mining and mechanical industry	13.0%
Glass, ceramics	12.0%
Chemical	38.5%
Textile, paper, food and agricultural industry	7.3%
Cement, plaster	7.8%
Others	2.8%
	100.0%

Gas Market in Industry

The French industrial sector (including the steel industry) has been using for several years as much energy as the residential and commercial sectors. The breakdown of this consumption by branch is quite different when one considers the whole energy consumption (including electricity) and the fuel consumption.

TABLE 4

	1975 Breakdown of energy consumption (fuels + electricity), %	1975 Breakdown of fuel consumption, %
Steel industry, metallurgy, electro metallurgy	33.6	34.6
Mining and mechanical industry	13.3	11.1
Glass, ceramics	5.8	6.8
Chemical	19.4	18.0
Textile, paper, food and agricultural industry	17.4	18.0
Cement, plaster	7.7	9.2
Others	2.8	2.3
	100.0	100.0

The energy needs of 1975 have been covered for:

18.8% by solid fuels,
28.2% by heavy fuel oil,
8.6% by light fuel oil, gasoil and LPG,
13.2% by gas supplied through a distribution network,
31.2% by electricity.

In the next years these industrial needs are expected to balance approximately the residential and commercial needs of energy.

Meanwhile, the structure of energy consumption by industrial branches is expected to change: by 1990 steel industry and metallurgical industry are due to lose some percents on their 33.6% present share while mechanical industry is due to increase its share.

Competitive Standing of Gas on the Industrial Market

On this market gas has to compete with oil products: naphtha, LPG, and above all gasoil and heavy fuel oil with either normal or low sulphur content.

Some factors play against the penetration of gas:

limitations of the gas supplied area,
cost of supplies with a bad load factor,
necessity to develop some sales on interruptible basis.

But some factors favour gas:

no storage facilities for the users;
no sulphur, no impurities, and environmental acceptability of combustion products;
controllability, and availability of high power;
specific processes such as radiant tubes, immersion systems, gas heat pumps.

As for price levels for industrial gas, one must consider separately firm sales and interruptible sales. For firm supplies of gas, the specific advantages of this fuel justify a premium value over heavy fuel oil with low sulphur content. At the beginning of 1978, the level of prices are not far (a little below) from this reference. As for interruptible sales, a discount must be made on the market price according to the type of use and hence of interruptibility.

Forecasts

The above analysis shows that on the average, between 1977 and 1990, annual increases of 10 million kWh may be forecast for gas sales. Gas will then control a share of 21% in the overall consumption of industry by 1990.

GAS SALES TO POWER STATIONS

Some of the French electric power stations have dual equipment with steam generating equipment fire either with gas or with another fuel (coal or heavy fuel oil).

These plants represent a peculiar non-final use of gas. They play a regulation role when supplies exceed demand for a limited period of time either seasonally in summer, or occasionally during the build-up period of a new supply. Thus, these plants contribute, in addition to interruptible sales to industry and to underground storage facilities, to match at any time demand and supply.

A few plants use gas with no dual equipment: Artix for instance, near the Lacq gasfield. Gas is supplied to these stations on a firm basis.

In the recent past, the amount of gas sales to power stations is given below:

in 1970 16 400 000 kWh,
in 1975 30 800 000 kWh,
in 1977 22 200 000 kWh.

For the future, gas sales will probably stay at the present level with however a falling trend due:

firstly to the allocation of gas rather to premium markets,
and then to the effect of the nuclear programmes on the primary fuel consumption of electric power stations.

GAS SALES IN THE FORECASTED ENERGY BALANCE TO 1990

Present Situation

Table 5 sums up the present situation and the past profile of gas sales in France.

TABLE 5

(10^9 kWh)	1970	1975	1977
Gas sales	110.6	197.5	230.8
Average annual growth rate on the last 5 years (%)	10.6	12.3	9.0
Breakdown of gas sales by sector:			
residential	33.5	32.8	34.6
commercial	7.9	9.7	10.8
industrial	43.6	42.0	45.0
power plants	15.0	15.5	9.6
	100.0	100.0	100.0

Forecasts

Sectorial considerations on the growth of gas sales to 1990 show that sales to all sectors (except to electric power plants) might reach 380 million kWh in 1985 and 470 million in 1990.

As noted above, the energy needs of the residential and commercial sector on the one hand and of the industrial sector on the other hand will be approximately equal.

In order to control an equal share of the different markets, gas sales to the residential and commercial sector have to be balanced with sales to the industrial sector.

French Energy Balance

These forecasts may be compared to the general orientations resulting from the last report of the Energy Commission of the National Plan which are summed up in this forecast of the primary energy balance of 1990 (Table 6).

TABLE 6

(10^6 t.o.e.)

Solid fuels	25 to 30	approx. 10%
Oil products	115 to 130	approx. 43%
Gas	42 to 44	approx. 15%
Hydro-electricity	14 to 15	approx. 5%
Nuclear electricity	70 to 75	approx. 25%
"New" types of energy	4 to 5	approx. 2%
Total	285	

Gas, in this forecast, would cover about 15% of the primary energy needs.

Such forecasts rely on security reasons for the energy supply of France, on savings achieved through the substitution of gas for oil products, on the rebalancing of external payments of France through exportations induced by gas contracts, on the rebalancing of the balance of trade between France and some countries, and also on environmental needs.

CONTROLLING THE DEMAND, MATCHING DEMAND AND SUPPLY

Given these orientations, the gas industry has then to define a policy:

- aiming at an optimal allocation of gas resources to the various markets with considerations of the market prices (mainly prices of oil products), of the characteristics of the demand (level of the demand, modulation), and of the development costs;
- and altogether meeting constraints which may be external constraints (potential markets, competitiveness of gas sales on the markets, all aspects already considered above), or also internal constraints.

A Demand Forecast Consistent with Supplies Forecast

On the whole, new supplies are negotiated so as to meet objectives of 37 Mtoe in 1985 and 42-44 Mtoe in 1990 set by the National Plan. All the negotiations are not yet over but about 85% of the 1985 needs are covered by the contracts still running in 1985. In addition one may recall, among other projects forecast in the eighties, the Segamo project which would give to France additional supplies of 10 to 20 Mtoe. Thus, viewed as a whole, the objectives described above may be considered as consistent with the supplies available in France.

One has then to specify what are the conditions for matching demand and supply. The characteristics of supplies are well known: rigidity of delivery, uncertainty of deliveries during the first years of a new supply, uncertainty on the availability of some LNG supplies . . . , and so we shall emphasize the demand characteristics and the consequences on the development policy on the different markets.

Demand Characteristics

A comparison between sales of the public distributions, which are made at 85% on the residential and commercial markets, and sales to customers located on the transmission system clearly shows two types of quite contrasting demand:

- demand to the public distributions is mostly a demand for heat. This demand for heat is quickly growing, at an average annual rate of +18.7% between 1970 and 1976 (the growth rate of the whole sendouts of Gaz de France being of +15.6% during the same period). Thus heat use accounts for 57% of the natural-gas sendouts, and, according to the objectives described above, they will reach 63% of the 1990 sendouts.

These uses result in a strong *seasonal modulation:* the sendouts of the public distributions in 1976/77 accounted for 65% during the five winter months from November to March, for 35% during the seven other months. The growth of heating use leads to the development of storage capacities filled with summer spare gas: useful storage capacities have grown from 745×10^6 m^3 (n) in 1970-71 to 2220×10^6 m^3 (n) in 1976-77 (i.e. at an average rate of growth of 20%).

These uses result also in a strong *sensitivity* of the sendouts of gas to changes of temperature: at a 50% level of risk on the whole winter period of 1976/77, the daily peakload would have reached 911×10^6 kWh (with heating uses being responsible for

80% of the load); at a 2% level of risk, the peak would have been increased from 15% to 20%.

The last characteristic of this demand is its atomicity: any correction on the residential and commercial markets takes a long time, any sharp action may carry out unexpected effects during a long period.

The characteristics of *industrial demand* are quite different: the sales are much less modulated, they depend on the activity of the branches, the number of establishments is much smaller, their consumption is much higher, they often may change over from gas to another fuel, thus enabling the contracting of sales on an interruptible basis or at short notice.

Consequences for the sales policy

These considerations result in two consequences for the sales policy:

the system must be flexible enough to control the growth of the demand, and to match permanently rigid and uncertain supplies to a stochastic demand; this flexibility is achieved by piloting the demand with industrial sales which have the least inertia;

the system must be secure enough to ensure with the lowest possible risk of failure, continuity of supply to firm customers; this is achieved by cutting off gas to interruptible customers who may turn to another fuel.

These reasons explain the weight given to industrial sales where interruptible customers may be found: the analysis of demand/supply balance shows that 46 million kWh of interruptible sales are needed in 1980. This potential is approximately at the level of the most vulnerable gas supply, and it enables the matching of the demand in critical situations, with the conjunction of severe weather conditions and of a failure in supply. Most of interruptible customers are industrial customers and industrial sales of gas must then be important enough to develop such interruptible contracts, generally linked with firm contracts. At the end of the contracts, there may still be a possibility of spot sales at short notice to those customers who are connected to the network.

The balanced allocation of sales to the industrial market and to the residential and commercial market is justified as a compromise between the best possible value of gas sold, and the costs of the storage facilities or of the interruptible sales which are necessary to meet peak demand. Such an allocation allows also a fair control of the demand and gives to the gas system its flexibility in spite of the rigidity of supplies.

CONCLUSION: FUTURE TRENDS IN ENERGY IN 1990

A sectorial approach of the growth of gas sales till 1990 shows that the growth of market makes possible to forecast the development of gas. Gas will then control 15% of the primary energy consumption in 1990 when its share is 11.5% in 1977. Such a growth is consistent with the security and the flexibility of the gas system which has to match demand and supply.

The past trends indicate a constant growth of gas share and, given the proximity of the year 1990, these trends will still weigh on the growth of the next years.

This means, of course, that these forecasts have been made without any hypothesis on a new crisis or on sharp modifications of the energy system. So these forecasts, in spite of their interest, are fragile.

The fact remains, however, that world gas consumption accounts for no more than 40% of oil consumption, when the ratio between reserves is nearer to 70%.

Natural Gas Savings in the Household and Tertiary Sector in France

M. Gendrin

France

SUMMARY

World energy resources are limited and their exploitation requires increasingly heavy investment. Recent growing awareness of these two facts has highlighted the need to try and save energy or to promote new forms of energy.

The gas industry in France has frequently experienced highly competitive conditions, so that problems of ensuring the most efficient use of gas energy have always been of concern to distributors and manufacturers of appliances. The events of 1973 have merely reinforced this concern and caused them to embark on new approaches.

In its energy-saving policy, Gaz de France is endeavouring to:

Encourage tenant-controlled gas heating: a statistical study has shown that this type of heating can lead to energy savings of 20-35% as compared with whole-house gas heating. It is also expected that substantial savings in consumption can be achieved through the use of gas radiator heating systems.

Contribute to the design and development of more economical systems by the *study of new techniques:*

- regulation and insulation techniques;
- basic and supplementary heating combinations;
- condenser boilers;
- dual flow with heat recuperation;
- absorption or gas-engine heat pumps;
- total energy systems.

Promote the development of *new forms of energy* by studying and proposing their use in association with gas.

These efforts have been concerned mainly with:

- solar energy;
- geothermal energy.

Supply users, consumer advisers and design offices with technical information on the use of gas in accordance with the regulations and in the most efficient way possible, through pamphlets, audio-visual material, films, lectures, personal visits, etc.

Inform domestic users about simple techniques for better use of their gas appliances. Unit consumption for space heating was reduced in this way between 1973 and 1975 by:

13% in flats;
15% in houses.

Contribute to the preparation of regulations on the use of energy through participation in specialist working groups set up by the administration.

These activities of Gaz de France to encourage energy saving constitute various aspects of an established and continuing policy to promote the most economical use of gas, with the following three objectives:

energy savings in accordance with the guidelines of the Energy Committee of the Plan;
a relative increase in the share of gas in the French energy market, in keeping with national energy policy;
fair prices reflecting the cost of gas on the world market.

Les Economies de Gaz Naturel dans le Secteur Residentiel et Tertiaire en France

M. Gendrin
France

L'industrie gazière en France n'a pas attendu les évènements de 1973 pour se pencher sur les problèmes des économies d'énergie; avant la découverte du gaz de Lacq (et après cette découverte, tant que le réseau de grand transport n'a pas été réalisé), le gaz de cokerie alors commercialisé était cher, sur un marché de libre concurrence. En conséquence, un effort tant technique que commercial était alors indispensable pour assurer la simple survie des entreprises gazières: il fallait proposer des solutions meilleures, donc plus économiques, que celles des concurrents. . . .

D'autre part, les ressources mondiales d'énergie primaire sont limitées et leur mise en oeuvre nécessite des investissements de plus en plus lourds. La prise de conscience de ces deux faits, pour récente qu'elle soit, n'en est pas moins réelle et alimente les réflexions de tous ceux qui, à quelque degré que ce soit, ont une responsabilité dans le développement économique de leur pays.

Depuis quatre ans, on voit ainsi se dessiner progressivement des nouvelles perspectives d'évolution de l'équilibre énergétique mondial, dans le cadre desquelles les Pouvoirs Publics français, particulièrement préoccupés par la faiblesse des ressources nationales, ont tenu à souligner les axes d'efforts qui doivent guider le comportement de chacun:

- l'énergie est une denrée rare et coûteuse: il est donc du devoir de tous de l'économiser et de l'utiliser avec la plus grande efficacité possible;
- tout doit être fait pour aboutir à une moindre dépendance et à une atténuation du coût de nos approvisionnements dont l'incidence sur l'équilibre de notre commerce extérieure restera importante pour de nombreuses années;
- la sécurité de ces approvisionnements exige une réelle diversification des sources.

La nouvelle situation ainsi créée n'a pas changé les structures de recherche et de fonctionnement de l'industrie gazière, tournées qu'elles étaient depuis toujours vers la recherche de meilleurs rendements, donc d'économie. Elle aura cependant contribué à accentuer ses efforts dans la recherche, d'une part des volumes définis par les Pouvoirs Publics correspondant à la part assignée au gaz dans la couverture des besoins en énergie primaire, d'autre part des techniques nouvelles permettant de placer ces volumes sur le marché. Enfin, Etablissement National et Service Public, Gaz de France se doit d'apporter son soutien actif aux utilisations du gaz économes en énergie.

Aussi le Gaz de France s'est-il fixé un certain nombre d'objectifs:

- *diffuser une information* (courrier service, brochure . . .) à l'intention des usagers domestiques, afin de leur expliquer ce qu'il convient de faire pour économiser l'énergie par l'emploi rationnel de leurs appareils à gaz;
- préconiser des *techniques "éprouvées" de chauffage au gaz* permettant de réaliser d'importantes économies de combustibles;
- contribuer à mettre au point et développer les produits les plus économes en énergie par l'*étude de techniques nouvelles;*
- favoriser la mise en oeuvre des *énergies nouvelles* en étudiant et proposant des solutions originales les associant au gaz;
- mettre à disposition des utilisateurs, prescripteurs, bureaux d'études, . . . les techniques gaz et les conditions de leur réalisation, au moyen de brochures, audiovisuels, films, conférences, visites, etc. . .;
- contribuer à la définition des textes réglementant l'utilisation de l'énergie en participant aux groupes de travail spécialisés mis en place par les Pouvoirs Publics.

INFORMATION DES USAGERS

Une action parallèle à celle de l'Agence pour les Economies d'Energie a été entreprise et menée à bien: des numéros du courrier service destinés aux abonnés ont traité à plusieurs reprises du bon usage du gaz.

Les consommations unitaires du chauffage individuel ont ainsi diminué de 1973 à 1975 de:

- 13 % en appartement,
- 15 % en maison individuelle

(la liste des brochures et les quantités diffusées sont indiquées en annexe).

TECHNIQUES EPROUVEES ECONOMES EN ENERGIE

Développement du Chauffage Individuel

Un effort commercial a été engagé en faveur du chauffage individuel au gaz. En effet, une étude statistique a démontré que celui-ci consomme de 20 à 35 % moins d'énergie que le collectif, le chiffre de 20 % devant rester un plancher, même dans l'avenir. Parmi les contributions importantes à cette orientation, notons:

- l'amélioration de la régulation du chauffage individuel avec l'apparition des robinets thermostatiques et de corps de chauffe mieux adaptés aux nouvelles caractéristiques des logements;
- la meilleure adéquation des installations de chauffage aux besoins thermiques des logements grâce à une meilleure connaissance des déperditions et des installations;
- la mise sur le marché des différents appareils à ventouse motorisée (chaudières et radiateurs) qui tout en présentant des rendements améliorés facilitent la pénétration du chauffage individuel (en particulier dans les logements existants dépourvus de conduit de cheminée en état de fonctionnement);
- l'analyse des facteurs de pertes en chauffage collectif, destinée à l'information des utilisateurs. Les pertes de canalisation peuvent varier de 10 à 4 % selon que les immeubles et les canalisations sont isolés ou ne le sont pas.

Promotion du Chauffage par Radiateur à Gaz Indépendant

L'apparition sur le marché de petits radiateurs à gaz d'une puissance inférieure à 2 kW, (certains appareils non encore homologués n'affichant que 0,6 kW de puissance) a permis de créer un produit gaz nouveau, le chauffage au gaz modulable.

En profitant de la modulation et d'un rendement d'exploitation amélioré, pour un chauffage de ce type bien utilisé, l'économie de combustible devrait atteindre 20 % par rapport à un chauffage individuel classique.

Améliorations en Chauffage Collectif

De nombreuses améliorations ont été réalisées:

- création de circuits de distribution, verticaux ou horizontaux plus faciles à réguler;
- technique de régulation par sonde extérieure et ralenti de nuit;
- régulations séparées par façades;
- utilisation de robinets thermostatiques et de corps de chauffe mieux adaptés aux pièces à chauffer;
- techniques de récupération de chaleur développées ci-après (Double flux, Base-appoint, Pompe à chaleur);
- recherches sur les améliorations de rendement des différents appareils à gaz.

TECHNIQUES NOUVELLES ECONOMISANT L'ENERGIE

Etude d'Amélioration de la Regulation et de l'Isolation en Immeuble Collectif

Plusieurs études de l'influence de la régulation et de l'isolation ont été entreprises et menées à bien:

- à Lille sur un bâtiment de 6 logements;
- à Charleville (étude du chauffage collectif et des pertes en ligne);
- avec l'Education Nationale, en ensemble de 10 opérations pilotes pour la régulation de bâtiments scolaires, zone par zone, classe par classe.

Des résultats de ces opérations, on retiendra que, dans l'existant, les économies qu'il est possible de réaliser par une bonne régulation dépassent 20 %.

Système Base + Appoint

Cette technique consiste à essayer de profiter des avantages du chauffage individuel (appoint) et du chauffage collectif (base). En effet, l'appoint permet de réguler finement la température intérieure et de récupérer les chaleurs gratuites au maximum, la base évitant, en partie au moins, les vols de chaleurs entre les appartements.

De plus, la température de base étant fixée vers 14°C, elle peut être sans inconvénients assurée par un chauffage par le sol, à basse température. Ceci est très intéressant pour les techniques nouvelles de production de chaleur économisant l'énergie qui sont pratiquement toutes à basse température, (par exemple: chaudières à condensation, pompe à chaleur, solaire, géothermie, etc. . . .).

Une étude très poussée a été réalisée sur ce type de chauffage par un bureau d'études; un certain nombre d'études ou d'opérations sont en cours:

en logements neufs

- Troyes	100 logements	Base à condensation
- Sarreguemines	70 logements	Base à condensation
- Marnes la Vallée	100 logements	Base à condensation
- Viroflay	50 logements	Base en pompe à chaleur gaz
- Bayeux	120 logements	Base en pompe à chaleur gaz
- Valence	150 logements	Base en pompe à chaleur gaz
- La Courtilleraie	6000 logements	Base en géothermie

en logements anciens ou tertiaires

- Siège du CNET HLM		Base à condensation
- Moulins	300 logements	Base à condensation
- Reims	résidence de personnes agées	Base à condensation
- Chateaudun	70 logements	Base à condensation
- Reims	résidence de personnes agées	Base en pompe à chaleur gaz

Les Générateurs à Condensation

Les combustibles gazeux contiennent une certaine proportion d'hydrogène qui, lors de la combustion, s'oxyde et se transforme en vapeur d'eau. Dans les matériels classiques les produits de combustion sont à des températures où cette eau est sous forme de vapeur, et on ne récupère donc pas dans l'appareil la chaleur latente de condensation de la vapeur d'eau. C'est pourquoi on exprime le rendement de ces appareils sur le pouvoir calorifique inférieur et non sur le pouvoir calorifique supérieur du combustible. La différence entre ces deux pouvoirs calorifiques représente justement cette chaleur de condensation de la vapeur d'eau.

Pour le gaz naturel, le rapport $\frac{PCI}{PCS}$ est de 0,9.

Pour travailler sur le pouvoir calorifique supérieur, il est nécessaire de refroidir les produits de combustion, de façon à provoquer dans l'appareil la condensation de tout ou partie de cette vapeur d'eau de la combustion. Il faut que, dans une zone de l'échangeur de l'appareil, il existe une température suffisamment basse, inférieure à ce qu'on appelle la température de rosée des produits de combustion, c'est-à-dire la température pour laquelle les produits de combustion se refroidissant sont saturés par la vapeur d'eau provenant de la combustion.

Pour les gaz naturels, et suivant les conditions de la combustion, cette température de rosée est de 55° environ. Plus on refroidira les produits de combustion, meilleur sera le rendement.

Les applications

Les générateurs à condensation existant sur le marché sont conçus pour des températures d'eau entrée-sortie de:

25°C et 45°C respectivement pour le cas de leur application au chauffage de l'eau des bassins des piscines;

35°C et 85°C respectivement pour le cas de leur application au chauffage de bâtiments et la production d'eau chaude sanitaire. Dans ce deuxième cas, la température moyenne des corps de chauffe (60°C au lieu de 80°C dans les systèmes classiques) nécessite de surdimensionnelr les corps ou systèmes de chauffe par rapport aux matériels classiques.

Cette contrainte est nécessaire pour obtenir des retours d'eau aux générateurs à la température la plus basse possible de sorte que le rendement soit le plus élevé possible; 97 % PCS à 25°C, 95 % PCS à 35°C contre 75 à 80 % PCS pour les matériels classiques.

L'intérêt à l'exploitation des générateurs à condensation conduit à envisager leur application non seulement comme par le passé en piscines (plus de 100 installations à ce jour en France) mais également en chauffage de bâtiments résidentiels et tertiaires, notamment dans la formule de chauffage "base + appoint". Dans ce type de chauffage, l'utilisation des générateurs "25-45" permet non seulement l'obtention d'un rendement exceptionnel, mais garantit de plus contre les inconvénients dénoncés à propos du chauffage par dalle: la température superficielle de la dalle n'excède pas 26°C.

Des exemples d'application ont été indiqués au paragraphe précédent.

Le Double Flux au Gaz et la Récupération de Chaleur

Les Pouvoirs Publics ont édicté un certain nombre de mesures règlementaires concernant l'isolation thermique des constructions et la régulation des installations de chauffage.

Ces mesures qui ont pour but de diminuer les besoins thermiques des bâtiments concernés donnent aux pertes par ventilation un poids relatif plus important que par le passé:

- au moins un tiers des déperditions totales dans le cas des logements;
- 50 % et parfois plus dans certains locaux tertiaires.

L'intérêt de la récupération de chaleur sur l'air extrait apparaît donc évident et réalisé économiquement dans un certain nombre de cas. Cette récupération se pratique dans les installations traitées en double-flux, c'est-à-dire combinant l'extraction et l'insufflation mécaniques de l'air, cas fréquent dans les immeubles tertiaires.

Dans les installations chauffées au gaz naturel les bilans thermiques peuvent encore être améliorés par la récupération des calories contenues dans les produits de combustion. Cette récupération est possible en raison de la pureté du gaz naturel. Certains constructeurs de chaudières ont tiré parti de cette qualité pour construire des générateurs spéciaux: les générateurs à condensation dans lequels la récupération de chaleur est intégrée.

Le double flux gaz

La technique du double flux associe l'extraction de l'air vicié et l'insufflation d'air neuf dans les logements.

Un échangeur de chaleur placé entre les deux flux permet de transférer à l'air insufflé une partie de la chaleur contenue dans l'air extrait.

La récupération de chaleur sur les produits de combustion peut se faire:

- soit séparément de la récupération sur l'air vicié;
- soit simultanément, par mélange des produits de combustion et de l'air vicié (cas de la VMC-Gaz en particulier).

De la solution retenue dépendra le choix du récupérateur.

Les récupérateurs à circulation de fluide secondaire

Ces récupérateurs sont constitués par des batteries de tubes à ailettes. Un circuit d'eau glycolée permet le transfert de chaleur entre les deux batteries.

Les récupérateurs à plaques

Dans ces récupérateurs, le transfert de chaleur se fait directement au travers d'une paroi.

Le corps de l'échangeur est constitué par un assemblage de plaques parallèles en métal, en verre ou en matière plastique. L'assemblage des plaques est conçu pour que les deux flux soient indépendants.

Les récupérateurs du type "caloduc"

Ces récupérateurs de chaleur sont constitués par des tubes à ailettes traversant les deux flux d'air et partiellement remplis d'un fluide frigorifique, en général du type "fréon".

Le fréon se vaporise dans la partie chaude du tube située dans l'air extrait, la vapeur s'évacue vers l'autre extrémité du tube où elle se condense cédent ainsi à l'air froid sa chaleur latente de vaporisation.

Le retour du liquide vers la partie chaude est assuré par gravité, capillarité, centrifugation ou par association de ces phénomènes.

Les récupérateurs rotatifs

Ces récupérateurs sont constitués par une roue poreuse tournant autour d'un axe parallèle aux deux flux d'air.

Le matériau de la roue (papier, amiante ou aluminium) accumule la chaleur et, passant d'un flux à l'autre, permet le transfert de chaleur de l'air vicié extrait vers l'air neuf.

Les économies de combustible prévues ou réalisées avec la technique du double flux sont de 20 à 30 % selon les réalisations.

Quelques réalisations ou projets:

Logement neuf

. à Laval, un immeuble de 28 logements (réalisé),
. à St-Etienne, un immeuble de 30 logements (réalisé).

Tertiaire
. Reims 2000, un ensemble de bureaux (réalisé),
. IOSS à Reims (en construction),
. Ecole vétérinaire de Nantes 30 000 m^2 (en projet).

Les Pompes à Chaleur à Gaz

On distingue plusieurs types de pompes à chaleur à gaz:

Les pompes à chaleur à moteur à gaz

Ce type de pompe se rapproche le plus des machines frigorifiques classiques. Son principe consiste à associer un moteur thermique alimenté en gaz naturel et un groupe frigorifique; d'autre part, il est prévu de récupérer sur le moteur une grande partie de la chaleur non utilisée.

Ainsi, compte tenu de cette récupération, l'énergie utilisable sur le moteur représente 80 % de l'énergie primaire dépensée; d'autre part, le coefficient de performance de la pompe était généralement compris, selon les solutions, entre 3 et 5, le coefficient de performance moyen annuel en énergie primaire de l'ensemble moteur + pompe sera compris entre 1,4 et 1,7 (mais peut dépasser 2 pour certains cas particuliers). Pour le chauffage des locaux, ceci revient à dire que la consommation de combustible serait environ divisée par deux.

Quelques projets ou réalisations

- Rouen: piscine patinoire 2 moteurs, réalisé en 1971;
- Viroflay: 46 logements, 1 moteur sur air extrait en projet;
- Bayeux: 120 logements, 3 moteurs sur air extrait décidé;
- La Coutilleraie: géothermie - chauffage de 6000 logements environ - décidé;
- Clermont Ferrand - Hôtel de Postes en projet;
- Paris - institut Pasteur - rénovation - en projet;
- Valence: 150 logements - 3 moteurs sur air extrait - en projet.

Les pompes à chaleur à absorption gaz

Ces pompes sont spécifiques gaz. Leur coefficient de performance en énergie primaire est d'environ 1,6. Une étude très poussée sur les possibilités de ces pompes et sur le marché a été réalisée par les soins du Gaz de France, afin de guider les recherches à entreprendre; de nombreuses applications sont possibles, mais compte tenu des appareils existants sur le marché, chaque cas doit être étudié de près.

Réalisations prévues

- Alès: en collaboration avec le Costic a été étudiée la mise en oeuvre d'une pompe à chaleur gaz à absorption fonctionnant par l'énergie solaire, avec un appoint gaz, pour le foyer restaurant de l'école des Mines d'Alès;
- Montbéliard: mise en place d'une pompe à absorption pour refroidir une salle d'ordinateur et utiliser ces calories à chauffer en partie le reste d'un nouveau centre EDF-GDF.

Les pompes à chaleur à éjection ou à compression avec turbines

Des conventions de recherche ont été passées avec des organismes extérieurs pour l'étude de ces solutions. De bons espoirs de réalisation sont fondés sur les pompes avec turbines en particulier pour les petites puissances (3 kW); ces solutions ne devraient cependant pas être disponibles avant 1984.

Les pompes à chaleur à absorption dans l'industrie

Une convention de recherche a été passée l'étude d'une pompe à chaleur à absorption capable de produire une température de 150°. Ces pompes permettraient d'utiliser dans l'industrie des sources de chaleur constituées par des rejets à 70°C s'intégrant ainsi directement dans certains process industriels (séchage, chauffage de bains, concentration, etc). Cette solution pourrait être disponible dès 1980.

Energie Totale

Le concept d'énergie totale recouvre, pour une installation industrielle, la production et l'utilisation d'énergie électrique, de chaleur ou de pression à partir de moteur thermique ou de turbines à gaz. Un document a été édité par l'ATG à ce sujet.

Les exemples de réalisation ou de projet sont très nombreux:

- Caisse Régionale du Crédit Agricole Mutuel de la Brie à Meaux,
- Caisse Régionale du Crédit Agricole Mutuel de Vienne à Poitiers,
- Caisse Régionale du Crédit Agricole Mutuel de Quimper,
- Centre Commercial Blois 2 de Villebarou,
- Centre Commercial Point Mil à Saint-Jean de la Ruelle,
- Unedic Cedida à la Chapelle Saint-Mesmin,
- Groupe "Assurances Générales de France" à Paris - projet,
- Hôpital Bichat à Paris - projet,
- Ensemble de bureaux avenue de la Grande Armée à Paris.

MISE EN OEUVRE DES ENERGIES NOUVELLES

Les Etudes poussées réalisées par l'industrie gazière sur l'association du gaz aux énergies nouvelles portent principalement sur deux domaines: le solaire et la géothermie.

L'Energie Solaire

L'énergie solaire, qui ne mérite ni le qualificatif de nouvelle, ni celui de gratuite, suscite un très vif intérêt, d'une part parce qu'elle est renouvellable, d'autre part parce qu'elle permet d'éviter, en partie, une trop grande dépendance de notre pays pour son approvisionnement énergétique.

Cependant pour les deux types d'application les plus courants, l'eau chaude sanitaire et le chauffage des locaux, une énergie d'appoint est nécessaire pour des raisons techniques et économiques. Cet appoint étant en général important, l'industri du gaz a pris l'initiative d'étudier les associations solaire-gaz, afin de fournir aux praticiens les indications nécessaires pour réaliser des mariages heureux et durables.

C'est ainsi que:

- une brochure très complète "Energie solaire et gaz" à été éditée en janvier 1978 par l'A.T.G.;
- une étude et un prototype de chauffage solaire par air chaud, associé au gaz, ont été réalisés et sont en cours d'essais à la Direction des Etudes et Techniques Nouvelles du Gaz de France;
- des études sont engagées sur des procédés originaux de stockage de longue durée de l'énergie solaire, en association avec des architectes et le C.E.A.;
- de nombreuses réalisations ont été encouragées et sont systématiquement suivies et analysées.

L'école des Mines d'Alès

(1) Chauffage des logements d'élèves (en service)

Il s'agit de chauffer 15 pavillons de 20 chambres, en chauffage collectif au gaz et eau chaude sanitaire solaire avec appoint gaz — 54 m^2 de capteurs sont installés par pavillon.

(2) Climatisation et chauffage du foyer restaurant (fin des travaux prévue en juin 1978)

Le foyer restaurant (550 m^2) sera chauffé en hiver et climatisé en été par une pompe à chaleur à absorption (à bromure de lithium) grâce à l'énergie fournie par les capteurs solaires ainsi que celle récupérée sur l'air extrait (en hiver).

L'ensemble de l'étude a été réalisé avec l'aide du Costic et les bilans d'exploitation seront faits par l'Ecole des Mines.

Résidence de la Palmeraie à Nyons (Drôme)

Deux immeubles d'habitation de 15 logements chacun seront chauffés par énergie solaire, l'appoint étaint fourni par une chaudière à gaz collective. L'ensemble sera mis en service vers le mois de juin 1978.

La maison de la Fédération Régionale du Bâtiment à Nice

L'étude d'un chauffage solaire associé à une pompe à chaleur électrique, l'appoint complémentaire était fourni par le gaz, a été réalisée par le Costic.

Les Maisons solaires de Toulouse Blagnac

Dix pavillons ont été réalisés:

- cinq avec chauffage et production d'eau chaude sanitaire par l'énergie solaire avec appoint gaz (30 m^2 de capteurs);
- deux à chauffage traditionnel au gaz;
- trois avec chauffage au gaz et production d'eau chaude sanitaire par l'énergie solaire avec appoint gaz (3,5 m^2 de capteurs).

Les résultats de cette expérimentation sont en cours de dépouillement, l'économie de combustible pour les cinq premières maisons a déjà pu être estimée à 45 %.

L'Hôtel des Impôts de Salon de Provence

L'Hôtel des impôts comprend un bâtiment en L à 3 niveaux. La surface totale est de l'ordre de 3900 m^2. L'énergie solaire et le gaz naturel sont les deux sources thermiques utilisées pour le chauffage des locaux. Le captage de l'énergie solaire est assuré par 310 m^2 de capteurs situés en terrasse, au-dessus des bureaux. La cuve de stockage solaire est constituée d'un réservoir de 20 000 litres.

Cette installation présente un caractère démonstratif de faisabilité et un caractère expérimental; ses résultats sont suivis par le Gaz de France à l'aide de divers appareils de mesure qui devront permettre d'établir un bilan précis de l'installation pour la saison de chauffe 1977-1978. Les résultats escomptés s'établissent à 31 % d'économie d'énergie.

La Géothermie et le Gaz

Depuis les évènements de 1973 qui ont marqué la fin du règne d'un pétrole bon marché, l'énergie géothermique a pris une place de choix parmi les énergies de substitution aux combustibles traditionnels.

Cependant en géothermie, il y a lieu de distinguer soigneusement les énergies de hautes températures (qui permettent la production directe d'électricité) des énergies basse température dont la seule utilisation possible concerne le chauffage des locaux. Dans ce dernier cas, il est toujours indispensable économiquement d'associer la géothermie à une énergie d'appoint, de manière à pouvoir rentabiliser les coûts d'investissement très importants de ce type de chauffage. Si le gaz, par ses qualités propres, puissance, propreté, souplesse, présente d'excellentes caractéristiques techniques pour cet usage, il faut cependant s'assurer dans chaque cas que les conditions économiques d'association du gaz à une opération géothermique correspondent bien à l'optimum pour chacun des partenaires. Seule une étude approfondie des possibilités offertes par la technique peut permettre de trouver une solution satisfaisante pour tous et permettant de réaliser les importantes économies d'énergie atten dues de ce mode de chauffage.

C'est dans ce cadre que l'utilisation du gaz en appoint a été proposée pour aider au développement de la géothermie.

Une étude de principe a été réalisée sur les possibilités d'associer le gaz et la géothermie; de cette étude, on pourra retenir qu'il n'y a pas de solution type, mais qu'au contraire, dans chaque cas, l'ingénieur d'études devra optimiser un ensemble de paramètres:

- *les paramètres de la géothermie:*
  . le débit artésien d'eau géothermale et ses possibilités d'accélération,
  . la pression artésienne pour la réinjection,
  . la température de l'eau géothermale,
  . la salinité de l'eau géothermale.
- *les paramètres des besoins en chauffage*
  . le nombre d'équivalents logements, ses déperditions moyennes, ses consommations moyennes,
  . le mode d'émission (radiateurs, convecteurs, sol, etc.),
  . la région climatique.
- *les paramètres d'équipement*
  . pompes d'extraction,
  . pompes de réinjection,
  . pompes à chaleur (fonctionnant au gaz),
  . production d'électricité (*par une installation d'énergie totale*) pour alimenter les pompes.
- *les paramètres tarifaires*
  . selon les solutions retenues, les tarifs proposés par le Gaz de France peuvent être différents pour permettre d'amortir les ouvrages d'amenée et de stockage du gaz naturel.
- *les paramètres financiers*
  . subvention de l'AEE ou de la DGRST,
  . emprunts et taux d'intérêt,
  . etc.

Les études faites pour des opérations dans la Région Parisienne ont montré qu'il est normalement nécessaire de prévoir l'installation en complément de l'énergie géothermique:

. de pompes à chaleur, à gaz,
. de moteurs à gaz pour fabriquer l'électricité nécessaire aux pompes d'extraction et de réinjection,
. de chaudières d'appoint à gaz.

D'autre part, chaque point d'extraction doit alimenter un ensemble de logements suffisament important (de l'ordre de 4000 logements pour un doublet de débit artésien de 100 m^3/h à 70°C).

Dans ces conditions, il est donc possible d'associer le gaz et la géothermie. Ces opérations restent cependant fort coûteuses pour les promoteurs, et leur rentabilité ne peut être déterminée qu'après de sérieuses études dans chaque cas.

Quelques réalisations:

Tarbes

Chauffage par énergie géothermique de 2000 équivalent logements (en particulier un grand hôpital). Les besoins globaux seraient de 21 millions de kWh, (la géothermie assurait 90% des besoins). Ce cas est particulier par le fait que l'eau géothermale de cette région n'a pas besoin d'être réinjecté (les investissements de forage sont ainsi divisés par deux).

Melun - Zac de la Courtilleraie - C'est une opération beaucoup plus importante:

- 130 millions de kWh environ dont 70 % fournis par l'énergie géothermale;
- 6000 logements environ à alimenter;
- une installation d'énergie totale avec 3 moteurs à gaz de 1000 kW sera installée sur le site pour alimenter les pompes géothermales et les pompes à chaleur;
- 3 pompes à chaleur électriques permettront de bien épuiser la chaleur de l'eau géothermale.

Cette installation très complexe sera gérée par un mini-ordinateur qui permettra en outre à la Direction des Etudes et Techniques Nouvelles du Gaz de France de dresser des bilans d'exploitation réels permettant de mettre en valeur cette réalisation remarquable à plus d'un titre.

CONCLUSION

En conclusion, les différentes actions entreprises par le Gaz de France pour les économies d'énergie traduisent les différents aspects d'une politique, en continuité avec celle du passé, qui encourage l'utilisation la plus économique du gaz tout en cherchant à satisfaire trois objectifs:

- un objectif d'économies d'énergie conforme aux orientations de la Commission de l'Energie du Plan;
- un objectif d'augmentation relative de la part du gaz dans le marché français de l'énergie, conforme à l'orientation de la politique énergétique nationale;
- un objectif de vérité des prix de vente, correspondant aux prix d'achat du gaz pratiqué sur le marché mondial.

DESCRIPTION ET ETAT DE DIFFUSION DE DOCUMENTS DANS LESQUELS FIGURENT DES CONSEILS DE BONNE UTILISATION DE L'ENERGIE

Brochure d'Informations de la Clientèle sur les Solutions Gaz et Contenant Divers Conseils ou Exposés de Solutions de Bonne Utilisation d'Energie.

Votre Chauffage Central au gaz

Brochure existant depuis 1971, remaniée sur des points de détail (branchements, tarifs, photos et schémas de matériels) au fil des rééditions.

A ce jour près de 500 000 exemplaires diffusés. La diffusion s'est faite par la Boite Postale 500, la brochure étant proposée lors des Campagnes Nationales Chauffage et, en 1975 et 1976, des Campagnes Economies d'Energie.

Votre Chauffage par radiateurs à gaz

Alors que la brochure "Votre Chauffage Central au gaz" demeurait bien adaptée aux Campagnes Economies d'Energie et que sa diffusion a pu être poursuivie, l'ancienne brochure "Chaleur facile par radiateur à gaz" n'était plus adaptée ni aux solutions récentes d'économie, ni aux matériels actuellement disponibles (appareils avec thermostat d'ambiance, appareils à circuit étanche, chauffage divisé par plusieurs radiateurs . . .). Elle a été remplacée en 1975 par "Votre Chauffage par radiateurs à gaz"; deux éditions totalisant 195 000 exemplaires, diffusion au 1er Février 1977, 156 000 exemplaires.

Le gaz dans la maison individuelle

Création 1975, tirage 50 000 exemplaires, diffusions 34 000.

Information Générale des abonnés

Le Gaz de France vous conseille

L'arrêté d'Octobre 1962 prescrit au distributeur de gaz d'informer les abonnés sur la sécurité d'utilisation.

Jusqu'en 1973, "Votre Guide" répondait à ce besoin (statistiques de diffusion non disponibles immédiatement).

En décembre 1973, "Votre Guide" a été remplacé par *"Le Gaz de France vous conseille"*, document plus complet, apportant une information qui va bien au delà des obligations de l'arrêté.

Les tirages succesifs depuis fin 1973 atteignent 1585 000 exemplaires dont *1385 000 diffusés* au 31 janvier 1977 suivant les modalités ci-après:

(a) Gaz de France vous conseille est diffusé par les Centres de Distribution aux nouveaux abonnés et à ceux faisant modifier leur installation. (Commandes des Centres, au 31 janvier 1977: 928 000 exemplaires).

(b) En outre, pour l'information des abonnés existants qui n'auraient pas eu la brochure ou qui l'auraient égarée, une opération de diffusion directe par voie postale a été lancée au cours de l'année 197 .

Deux cartes-réponses étaient jointes aux factures: l'une permettait de demander à la B.P. 500 la brochure "Le Gaz de France vous conseille", l'autre permettait de demander "Le livret de l'usager de l'électricité: pour connaître votre abonnement". Cette opération a permis la diffusion directe de 420 000 Gaz de France vous conseille

Documents de Bonne Utilisation d'Energie pour le Grand Public

Brochure "Manuel d'utilisation du chauffage au gaz"

Créée en 1971, elle doit être remise par les Centres à tous les abonnés faisant installer le chauffage au gaz. Elle a été proposée, en outre, directement aux utilisateurs et futurs utilisateurs lors des Campagnes Nationales "Chauffage Central" jusqu'en 1973 et "Economies d'Energie" (1974-75-76)* et diffusée par la B.P. 500.

Plusieurs tirages totalisent 300 000 exemplaires; la diffusion au 31 janvier 1977 atteint environ 260 000 exemplaires.

Dépliant isolation maisons individuelles neuves (orange 21X29,7)

Création 1975; trois tirages successifs totalisent 40 000 exemplaires; diffusion 35 000 environ.

Dépliant isolation habitations existantes (vert 21X29,7)

Création 1975; tirage 50 000 exemplaires, diffusion 40 000 environs.

Dépliant regulation du chauffage individuel au gaz (rouge, carré)

Création en 1975; plusieurs tirages totalisant 50 000 exemplaires; diffusion 49 000 exemplaires environ.

Dépliant avec le chauffage individuel au gaz, faites aussi des économies d'énergie (blanc)

Création 1975 pour Arts Ménagers et Foire de Paris (Conseils + appel de brochures).

Tirage 200 000 - Diffusion 127 000 -

Dépliant un simple geste suffit - fonctionnement automatique des chaudières (dépliant régulation simplifié pour foires et salons)

Création 1976 - tirage 27 500 exemplaires - diffusion totale.

Documents à l'Attention des Exploitants, Questionnaires etc. . . de Chaufferi Collectives

Dépliant Economisez l'énergie en contrôlant le rendement de combustion de votre chaudière (bleu 21X29,7, 3 volets)

Crée en 1974 - 4 éditions successives totalisant 43 000 exemplaire entièrement diffusé, rééditions à l'étude.

Brochure Le Gaz de France, l'Association Confort Régulation et le SNEC vous conseillent (bleue 21X29,7)

Créée en 1975, 3 tirages successifs totalisant 26 000 exemplaires environ; diffusés au 31 janvier 1977: 25 000 exemplaires.

Brochure Amélioration des installations existantes (suite de la brochure précédente) (bleue 21X29,7)

Créée en 1976 - 2 tirages totalisant 21 000 exemplaires - diffusion 6000 exemplaires.

---

*Pas de campagne publicitaire en 1977.

Documentation à l'Usage des Installateurs

Dépliant réglage des brûleurs (pour brûleurs de cuisinières): 2 tirages totalisant 40 000 exemplaires.

Ce document n'est pas destiné aux particuliers qui n'ont pas à toucher au réglage de leurs appareils, mais aux installateurs et à l'enseignement ménager - diffusion 35 000 exemplaires.

Fiches de maintenance d'appareils à gaz

Fiches de mise en route, d'entretien et de dépannage des chaudières murales - 10 fiches publiées depuis 1975, 2 en cours de réalisation, d'autres en projet - Tirage actuel 16 500 exemplaires de chaque fiche.

Diffusion notamment aux cellules gaz CET et FPA.

Guide pratique des installations de gaz et règles d'or d'installation du radiateur à gaz (pour mémoire ici, ce sont des règles d'installation et non d'utilisation des équipement)

Documentation "Monographies" sous Classeur

Isolation des maisons individuelles neuves chauffées au gaz

Juillet 1975 - Tirage 10 000 - diffusion au ler janvier 1977, 4650 exemplaires.

Régulation du chauffage individuel au gaz

Avril 1976 - Tirage 3000 - diffusion 3000 en réimpression.

Le gaz naturel au service de l'Industrie: chauffage et ventilation des grands locaux

Août 1976 - Tirage 3000 - diffusion 2510 au ler janvier 1977.

L'isolation thermique et le chauffage au gaz des immeubles collectifs neufs

Janvier 1978 - Tirage prévu 3000

Le chauffage des bains dans l'Industrie

Mars 1978 - Tirage prévu 1000

Document extérieur à la SDIG (rédigé avec le concours de Gaz de France)

- Le développement de l'énergie totale en France (brochure ATG) - 5000 exemplaires
- La récupération de la chaleur perdue dans l'industrie (document DETN).
- De nombreux articles écrits dans des revues scientifiques en particulier pour "Travaux Communaux".
- Le gaz et la réhabilitation de l'habitat existant 10 000 ex.
- Zones d'activité industrielle et gaz naturel.
- Le chauffage au gaz par rayonnement basse température.
- Récupération de chaleur sur les installations de gaz.
- Les maisons solaires de Blagnac.

Audiovisuels et Films

- L'isolation thermique et la régulation gaz.
- Récupération de chaleur en immeuble collectif neuf.
- L'Art du Feu (film destiné à aider les industriels à régler convenablement fours et brûleurs).

The Place of Gas in the Energy Supplies of France

J. P. Hauet

France

SUMMARY

For many years, the proportion of France's energy supplies accounted for by gas has remained considerably lower than in most other industrialized countries. As a result of the oil crisis of 1973-1974, an effort was launched to bring the share of gas in France's energy balance up to 15 per cent by 1985.

The question at present is whether this expansion could and should be continued beyond 1985. The problem must be considered in all its many aspects within the framework of an overall energy policy extending up until 1990.

In the case of France, particular account should be taken of the following factors:

- the exhaustion of national resources but the abundance of world resources;
- the advantages offered by gas from the point of view of diversification but the difficulties in securing new sources of supply;
- the French gas industry's possibilities of winning new markets but the difficulty of making headway in a highly competitive sector;
- the specific advantages of gas for users but also the competition from other forms of energy.

In these circumstances, a stabilization in relative terms at the level of 15-16% would seem to be a reasonably ambitious objective for gas up until 1990.

La Place du Gaz dans l'Approvisionnement energetique de la France

J. P. Hauet

France

La crise pétrolière des années 1973/74 a amené les Pouvoirs Publics à redéfinir les orientations générales de la politique énergétique française, de façon à limiter le coût de nos approvisionnements et à réduire leur vulnérabilité.

Les mesures prises dans le cadre du nouveau programme énergétique français à moyen terme (1975-85) peuvent être classées selon quatre axes principaux:

- limitation des consommations par une politique active d'économies d'énergie;
- mise en valeur des ressources nationales (hydraulique, charbon) et relance de la prospection des hydrocarbures;
- diversification accrue dans l'origine et dans la nature de nos approvisionnements importés;
- resserrement de la coopération internationale entre pays consommateurs et pays producteurs.

La politique gazière constitue un volet important de cet ensemble, qui doit permettre au gaz d'occuper en 1985 une part de 15 à 16 % dans notre approvisionnement énergétique.

Le présent document a pour objectif de rappeler les motifs qui ont présidé à l'adoption de cette politique et d'exposer les conditions dans lesquelles se présente sa poursuite jusqu'aux horizons 1990 ou 2000.

LE GAZ AVANT 1973

A la différence des Etats-Unis, qui depuis de très nombreuses années exploitent et transportent le gaz naturel de leurs gisements, la part du gaz, et tout particulièrement du gaz naturel, au sein du bilan énergétique français est demeurée jusqu'à une date récente relativement modeste. On peut citer, en particulier, les chiffres suivants: (Tableau 1).

En 1973, la part du gaz dans l'approvisionnement français n'atteignait pas 10 % alors qu'elle excèdait 15 % dans des pays de niveau de développement comparable. Etats-Unis, Royaume-Unis Italie.

C'est toutefois au cours de la période 1955-1970 que se sont progressivement trouvées rassemblées les conditions qui devaient permettre en France un rapide

TABLEAU 1
Evolution de la structure des bilans énergétiques

	1960			1973		
	Monde	Europe des Neuf	France	Monde	Europe des Neuf	France
Combustibles minéraux solides	52	69	55	28	24	17
Pétrole	31	27	31	47	59	67
Gaz naturel	15	2	3.5	19	13	8
Electricit é primaire (hydraulique ou n/cléaire)	2	2	10.5	6	4	8
Total	100	100	100	100	100	100

essor de l'industrie gazière sur lequel la nouvelle politique énergétique allait pouvoir le moment venu s'appuyer.

C'est notamment à cette époque que l'on construisit les installations de traitement et les premiers éléments du réseau de transport, permettant de mettre en valeur le gaz du gisement de Lacq et de l'acheminer vers les zones consommatrices.

C'est également vers 1965 que les techniques cryogéniques, mises au point grâce en particulier à des techniques françaises, permirent de liquéfier et de transporter par navires méthaniers le gaz naturel, ouvrant ainsi la voie à de nouvelles chaînes d'approvisionnement à partir de pays producteurs souvent éloignés.

Le taux moyen de croissance annuelle de la consommation de gaz enregistré en France entre 1960 et 1973 (+ 13,3 %) supérieur à celui des produits pétroliers (+ 11,9 %) pourtant considéré comme spectaculaire, atteste de la vigueur avec laquelle se sont rapidement mis en place les fondements de nos infrastructures gazières.

LA CRISE DE 1973/74 ET LES ORIENTATIONS À HORIZON 1985

Les Objectifs 1985

La crise pétrolière de l'hiver 1973/74 ayant clairement démontré la fragilité de notre approvisionnement, le Gouvernement a cherché à définir par chaque forme d'énergie de nouveaux objectifs qui soient en ligne avec la politique générale qu'il s'était fixé.

Concernant le gaz, les objectifs à moyen terme ont été particulièrement ambitieux:

17 Mtep, soit 10 % de l'approvisionnement énergétique en 1975

27 Mtep, soit 13 % de l'approvisionnement énergétique en 1980

37 Mtep, soit 16 % de l'approvisionnement énergétique en 1985

Le développement, au taux moyen de 8 % sur 1975-1985, doit être comparé au taux de croissance envisagé pour l'ensemble des besoins énergétiques sur la même période, soit 3,5 %.

Le rapprochement entre les deux chiffres marque nettement la volonté des Pouvoirs Publics de voir s'accentuer la pénétration du gaz dans tous les sectuers où son utilisation apparaît comme judicieuse, de façon à amener sa part de marché à un niveau comparable à celui constaté dans les pays dont l'activité économique est voisine de la nôtre.

Les Justifications de la Politique Suivie

Le recours accru au *gaz* répond à de multiples préoccupations:

- la première tient aux *économies d'énergie*, le gaz apportant à cet égard des solutions satisfaisantes.

Ainsi, dans le domaine de l'habitat collectif, le chauffage *individuel* au gaz permet de réaliser des économies d'énergie qui sont évaluées aux environs de 15 à 20 % par rapport aux solutions collectives. Pour la modernisation de l'habitat existant, le chauffage au gaz apparaît comme l'un des modes les mieux adaptés et les plus économiques.

Dans l'industrie, le gaz, par sa souplesse d'emploi, par la finesse des réglages qu'il permet, conduit également à une meilleure valorisation du potentiel énergétique du combustible utilisé se traduisant par une amélioration des rendements.

- Une seconde raison réside dans le rééquilibrage du *commerce extérieur* que procurent les importations de gaz. En effet, la mise en place de nouvelles chaînes d'approvisionnement en gaz réclame des installations beaucoup plus complexes que pour le pétrole et qui ne peuvent être fournies que par des pays industriellement développés.

Ces investissements, qu'il s'agisse d'usines de liquéfaction, de terminaux méthaniers ou de canalisations de transports, sont l'occasion pour les pays acheteurs de compenser par des fournitures de matériels ou des prestations de service une partie des sorties de devises afférentes aux achats de gaz proprement dit. Notre pays s'est d'ailleurs solidement installé dans certains de ces créneaux technologiques, notamment en matière de liquéfaction de gaz et de construction de navires méthaniers, et les travaux préparatoires au VIIe Plan ont montré que le flux des devises qui pouvait résulter de cette activité, était de nature à compenser à horizon 1980, au moins la moitié des dépenses d'achat de gaz à l'étranger.

- Une troisième raison concerne la *sécurité des approvisionnements*. L'expérience montre en effet que la passation des coitrats d'achats de gaz, crée, en raison de la spécificité des investissements qu'il impliquent, des liens bilatéraux plus forts que le simple achat de pétrole brut, ce qui va dans le sens d'une meilleure sécurité des approvisionnements.

Le développement du gaz permet également de parvenir à une diversification plus large dans l'origine de nos approvisionnements et de nouer ainsi des rapports commerciaux stables et durables avec des pays situés dans des zones, géographiques relativement diverses.

Il conviendrait enfin de signaler l'intérêt que présente le recours au gaz du point de vue de l'environnement et notamment de la tuite contre la poilution atmosphérique. Le gaz se compare à ce point de vue aux combustibles liquides à très basse teneur en soufre.

Les Conditions de Succès de la Politique Gazière

Bien que le recours au gaz présente en soi des avantages appréciables, la politique de développement gazier ne peut porter ses fruits que si un certain nombre de conditions essentielles se trouvent réunies:

1° - Il convient tout d'abord de parvenir à une *structure harmonieuse de nos approvisionnements* se traduisant par l'arrivée, au niveau de chacun des côtés de l'hexagone français, de flux gaziers d'une ampleur bien équilibrée permettant de satisfaire dans des conditions optimales les besoins du marché français. Dans cette optique, indépendamment des ressources tirées au gisement français de Lacq, des contrats ont été signés avec la Hollande, l'Algérie, les producteurs de Mer du Nord, l'URSS, l'Iran. De nouvelles negociations sont en phase exploratoire avec d'autres pays tels que le Nigéria. Des installations de réception de gaz liquéfié ont été aménagées au Havre, à Fos sur Mer près de Marseile et à Montoir de Bretagne près de St Nazaire en complément des arrivées par gazoducs du Nord et de l'Est de la France.

2° - Il est nécessaire également de parvenir à un *placement judicieux* des ressources en gaz permettant de valoriser au mieux les avantages spécifiques reconnus au gaz.

Dans le secteur du logement, le Gouvernement français s'est prononcé en faveur d'un équilibrage optimal entre les différentes formes d'énergie, dans lequel la part dévolue au gaz sur le marché des logements neufs devrait être de 35 % et égale à celle de l'électricité.

Dans l'industrie, les usages réputés peu valorisants tels que la combustion sous chaudières doivent être désormais, dans la mesure où ils ne concourrent pas à la constitution d'un portefeuille de clients interruptibles, découragés.

De même conformément aux directives communautaires, la combustion de gaz en centrales électriques doit être limitée au strict minimum.

3° - Malgré ses avantages par rapport au pétrole, le gaz, en tant qu'énergie importée, demeure une énergie vulnérable et il est nécessaire de prévoir les dispositifs permettant de *faire face à une interruption* plus ou moins prolongée et plus ou moins profonde des approvisionnements.

Les mesures prises par les Pouvoirs Publics français sont de deux ordres:

Constitution par les *vendeurs de gaz* d'un portefeuille de *clients interruptibles*, pouvant en cas de nécessité, convertir rapidement leurs installations au fuel-lourd ou au naphta.

Ce portefeuille, constitué de clients industriels et de centrales électriques, devrait atteindre 46,5 TWh en 1985 (40 gigathermies) et permettre d'assurer la continuité de l'approvisionnement de la clientèle ferme, dans le secteur domestique notamment en cas d'incident sur une chaîne d'approvisionnement.

Développement des *stockages souterrains* de gaz de façon à pouvoir répondre à la modulation journalière ou saisonnière des besoins et, le cas échéant, disposer de ressources d'appoint mobilisables en cas de difficultés sur certaines chaînes d'approvisionnement.

Les stockages actuellement disponibles en France ont une capacité utile de 32 TWh (27 gigathermies). L'objectif retenu pour 1985 consiste, par un effort accru de recherche et d'aménagement, à porter cette capacité à 73 TWh (63 gigathermies) représentant — en moyenne — deux mois de consommation.

La Réalisation des Objectifs 1985

En ce qui concerne les *approvisionnements*, les contrats actuellement souscrits permettent de satisfaire la demande à hauteur:

- de 100 % des objectifs retenus pour 1980 (27 Mtep);
- de 85 % à 90 % des objectifs retenus pour 1985 (37 Mtep).

Au niveau du placement du gaz, on constate que la progression de la demande de gaz demeuretrès soutenues et globalement bien en ligre avec les orientations retenues pour 1985.

TABLEAU 2
Evolution de la consommation de gaz
(en taux moyen ce croissance annuel)

	Réalisations	Prévisions	
	1973/77	1973-80	1980-85
Secteur industriel	+ 8,0 %	+ 8,6 %	+ 7,5 %
Secteur résidentiel et tertiaire	+ 9,9 %	+ 10,1 %	+ 6,8 %
Centrales électriques	- 4 %	- 2,4 %	- 2 %

On peut donc considérer à ce stade comme probable la réalisation des objectifs visant à assigner au gaz une part de 15 à 16 % de l'approvisionnement énergétique de la France en 1985.

Toutefois, l'expérience des trois dernières années a mis en évidence certaines difficultés dont la prise en compte est nécessaire pour apprécier correctement la place à réserver au gaz à plus long terme.

- Au niveau du *marché*, la concurrence dans le secteur du logement neuf, s'est révélée extrêmement vive, de la part de l'électricité essentiellement. Le gaz présente à ce point de vue le handicap d'exiger des investissements plus lourds au niveau de l'utilisateur, des extensions de réseau plus onéreuses et des normes de sécurité plus contraignantes. La part du marché occupé par le gaz dans le logement neuf n'atteint donc pas actuellement les 35 % envisagés et n'a pas excédé en 1977 27 %.

Globalement, le développement rapide du gaz dans l'habitat *existant* a permis de compenser le retard pris au niveau de l'habitat neuf, les mises en service de chauffages centraux au gaz ayant dans ce secteur dépassé en 1977 160 000 logements. Bien que le parc des logements susceptibles d'être modernisés soit important (4 millions de logements mal chauffés dont 2,3 situés en zone desservie en gaz), il est probable que, par rapport au rythme de pénétration actuel, un certain infléchissement sera constaté à moyen terme dont il conviendra de tenir compte.

- Au niveau de *l'approvisionnement*, plusieurs projets qui étaient considérés comme plausibles aux lendemains de la crise pétrolière sont aujourd'hui soit abandonnés, soit différés.

Parmi ceux-ci on peut citer:

- la mise en place d'un "grand collecteur" en Mer du Nord,
- le projet EUROGAZ à partir de l'Algérie,
- les projets d'approvisionnement en gaz liquéfié à partir de l'Iran.

Par ailleurs, certaines incertitudes sont apparues quant aux conditions de fonctionnement des usines de liquéfaction dont il n'existe encore qu'un petit nombre d'unités à travers le monde.

LES PERSPECTIVES 1990-2000

Malgré l'intérêt évident de la pour suite d'une politique de développement du gaz, compte tenu des avantages spécifiques de ce combustible et compte tenu de l'importance des ressources actuellement mises en évidence dans le monde, il convient de prendre en compte les considérations qui précèdent dans la définition des objectifs 1990.

Ceci est d'autant plus impératif, qu'au niveau français, deux facteurs nouveaux doivent intervenir de façon significative entre 1985-1990.

- En ce qui concerne *le marché de l'énergie*, le recours accru à l'énergie nucléaire engagé depuis 1974 par les Pouvoirs Publics à un rythme soutenu, devrait se traduire par la mise à disposition des utilisateurs industriels ou domestiques d'électricité d'origine essentiellement nucléaire, à des conditions de prix relativement indépendantes du marché international des hydrocarbures. En outre, les énergies nouvelles (solaire notamment) pourront prendre progressivement sur certains marchés, et notamment sur le marché du chauffage, une place non négligeable.

La concurrence entre formes d'énergie se trouvera ainsi renforcée. Il appartiendra à l'industrie gazière de s'y adapter et de proposer de nouvelles formules: pompes à chaleur, systèmes mixtes: énergies nouvelles gaz. Mais il paraît difficile d'extrapoler au-delà de 1985 les taux de croissance très élevés retenuspour les prochaines années et qui correspondent pour une part importante à un rattrapage du passé.

- Au niveau des *approvisionnements*, la période 1985-90 est marquée pour la France par le déclin progressif des deux ressources qui assurent actuellement l'essentiel de son approvisionnement: le gaz de Lacq et le gaz de Groningue (Hollande).

Les ressources ainsi disponibles pour l'horizon 1990 n'excèdent pas 25 Mtep et il apparaît dans ces conditions que la possibilité pour la France de s'attacher dans les prochaines années de nouvelles sources d'approvisionnement constituera, plus que l'état du marché, le facteur essential de détermination de la place du gaz dans notre bilan énergétique des années 1990.

Parmi les grands projets susceptibles d'aboutirfigure le projet SEGAMO de gazoduc transméditerranéen dont la part revenant à la France pourrait être de 10 à 20 Mtep. Des pourparlers sonten pours également avec le Nigéria, dans le cadre d'un consortium européen et des ressources nouvelles peuvent être espérées à partir de la Mer du Nord.

Mais la concurrence internationale sera vive du fait notamment du déficit occasionné en Europe par l'épuisement progressif du gisement de Groningue. Elle le sera d'autant plus que les besoins des consommateurs seront élevés et c'est dire l'importance que pourront avoir sur les approvisionnements gaziers de l'Europe, le succès ou l'échec des politiques de conservation de l'énergie et de mise en valeur des ressources nationales engagées dans les différents pays.

En définitive, il a paru raisonnable à la Commission de l'Energie du Plan de proposer au Gouvernement de retenir pour le gaz un objectif de *43 Mtep* en 1990, permettant de stabiliser au niveau de 16 % la part du gaz dans le bilan énergétique de la France.

Celui-ci évoluerait approximativement de la façon suivante: (Tableau 3).

Le placement de ce gaz se répartirait de façon équilibrée entre les deux grands secteurs (industrie et résidentiel et tertiaire) et la consommation de gaz évoluerait approximativement de a façon suivante: (Tableau 4).

Après une pénétration très rapide entre 1975 et 1985, le gaz reviendrait ainsi progressivement à un taux de développement plus voisin du taux moyen de croissance des besoins énergétiques français (3,25 % de 1985 à 1990).

TABLEAU 3
Evolution du bilan énergétique français

	1977	1980	1985	1990
Charbon	31,2	32	29	26
Gaz	20,5	27	37	43
Pétrole	104,9	110	101	111
Electricité primaire	20,8	30,5	61	86
Energie nouvelles	-	0,5	2	4
	178,4[a]	200	230	270

[a]Compte-tenu de 1 Mtep d'achat d'électricité à l'étranger.

TABLEAU 4
Evolution de la consommation de gaz de 1985 à 1990
(en taux moyen de croissance annuel)

Secteur industriel	+ 3,4 %
Secteur résidentiel et tertiaire	+ 3,7 %
Centrales électriques	- 5 %

A horizon 2000, il est clair que les problèmes d'approvisionnement prendront une importance encore plus grande du fait de l'amenuisement des réserves auxquelles l'Europe peut aujourd'hui faire appel. La politique à mener paraît dans ces conditions devoir s'inspirer des principes suivants:

- *Développement de la coopération internationale*, entre pays européens notamment. Le continent européen commence à être traversé par un réseau qui met dès maintenant chaque pays en liaisan étroite avec ses voisins. Cette interconnexion européenne rend possible une entraide permettant de faire face à des difficultés qui pourraient être à défaut difficilement surmontables. Elle justifie également une démarche concertée dans les actions commerciales et dès à présent le Gaz de France se trouve associé avec un certain nombre de partenaires: la République Fédérale Allemande, la Belgique, l'Italie, les Pays-Bas, l'Espagne, l'Autriche ou la Suisse.

Il put être noter que la France jouit du point de vue du transport maritime d'une place relativement privilégée en Europe, ce qui, dans les années à venir, pourrait la placer au centre de différents courants d'approvisionnement et justifier l'implantation sur son territoire d'équipements à vocation internationale.

- *Dévelopement des efforts de recherche technologique* pour parvenir à la mise sur lemarché de produits énergétiques complémentaires ou substituables au gaz naturel.

Le gaz de pétrole liquéfiés. Il est possible que des disponibilités accrues de butane ci de propane en provenance de gisements pétroliers apparaissent sur les marchés internationaux dans les prechaines années, venant compléter la production propre de raffineries. Ces GPL pourraient trouver de nouveaux marchés dans l'industrie chimique, dans la carburation automobile et dans le secteur domestique comme appoint aux énergies nouvelles.

La gazéification du charbon. Les Pouvoirs Publics français viennent de décider d'apporter un soutien financier public à deux projets auxquels se trouvent associésle Gaz de France et les Charbonnages de France.

. L'un d'hydrogazéification de la houille par voie nucléaire utilisant dans un réformeur de méthane la chaleur produite par un réacteur à haute température;
. L'autre de gazéification in situ en couches profondes et sous pression.

L'hydrogène. La fabrication d'hydrogène pour électrolyse apparaît à horizon de 10ou 20 ans comme un corollaire du recours accru à l'énergie nucléaire, permettant de tirer parti de la capacité de production d'électricité disponible en heures creuses.

Cet hydrogène pourrait dans un premier temps jouer un rôle important comme matière de base de l'industrie chimique, de la sidérurgie ou de l'hydro-désulfuration des fuels. Ultérieurement, si l'évolution des coûts le permet, une utilisation de l'hydrogène comme combustible n'est pas à exclure, le stockage et la distribution ne posant pas de problèmes insurmontables, compte tenu notamment des infrastructures gazières qui, à horizon 1990-2000, pourraient se trouver disponibles à cette fin.

Trends in the Development of Gas Consumption in Czechoslovakia up to 1990

J. Danhelka

Czechoslovakia

SUMMARY

Trends in the development of gas consumption in Czechoslovakia up to 1990 are evident from forecasts of the growth of natural-gas resources. Almost all the gas is to be obtained from imports.

Gas resources in 1980 will be twice those of 1975 and by 1990 they are expected to be 53-63% higher than in 1980.

The share of natural gas in primary resources will increase from the existing 8% to 10-11% in 1980 and 14-15% in 1990.

In the development of consumption up to 1980, consumption in the industrial category is to increase faster than in the household and service sectors.

Between 1980 and 1990, on the other hand, consumption is to increase faster in the non-productive sector, mainly in services. Gasification in the household sector will increase from 35.5% in 1975 to 41% in 1980 and to 48.5-50.4% in 1990, when specific gas consumption per consumer in the household sector is expected to reach 7000-8500 Mcal.

The change in the structure of consumption in favour of the non-productive sector, and the use of gas for heating in all categories, will lead to greater imbalance in the annual consumption diagram. According to forecasts, the ratio of a minimum-consumption day to a maximum-consumption day is to increase from 1:2 in 1975 to 1:2.8-3 in 1990 (town gas from 1:2.7 to 1:3.4-3.5).

Seasonal and peak capacity for meeting winter demand will therefore be increased continuously. It is expected that on a maximum-consumption winter day 62-65% of natural gas supplies will come from basic resources, 28-30% from gas-holders and 8-10% from peak stations using liquefied natural gas.

In addition to measures taken by suppliers, ways are being sought to improve the pattern of demand among consumers (combined systems, increased summer consumption, carrying over of repairs to the winter period).

The main method used for planned regulation of the development of demand in industry is the allocation of fuel bases for future gas consumers at the level of the

Ministry of Fuel and Energy. Other means of controlling the growth in demand for gas are price policies and regulation of the share used for heating.

The development of the Czechoslovak gas industry during the years following the World War II was very rapid. Its basic aim was the greatest possible use of the free coke-oven gas for consumption and the building of gasworks for the production of town gas by brown coal gasification under pressure. Up to 1955 the natural-gas supply was negligible.

The mentioned conception of sources development together with the build up of the gas pipelines network operating under the 2.5 MPa pressure enabled the area supply of consumers and gasification on a larger scale. Since 1956 the natural gas from new deposits and since 1967 the imported natural gas from the USSR have become significant supply sources.

During the following years the development of the Czechoslovak gas industry was influenced by the co-operation with the Soviet Union, especially by the transit of Soviet natural gas to Western European countries and to the German Democratic Republic; it was necessary to construct the transmission gas pipeline, the capacity of which largely surpassed the capacities of gas pipelines which were in operation at that time. Although this pipeline has initially been operating as the transmission pipeline, it has enabled the provision of our own distribution needs and the supply of gas to many consumers in new areas.

SOURCES DEVELOPMENT AND STRUCTURE ESTIMATE

Natural gas represents about 8% of primary fuel-energy sources in Czechoslovakia at present. Coal prevails in the balance and therefore large possibilities of using gas exist in all consumer sectors. Besides natural gas, the town gas produced mainly by brown coal gasification under pressure is also used in western parts of our state.

The rate of future consumption increase will be mainly influenced by attainable natural-gas sources. The importance of town gas after 1980 will continually decrease and by about 1990 it will be used for supplying only the areas in the vicinity of its production.

The forecast for development of natural-gas and town gas sources in comparison with the years 1970 and 1975 is given in Table 1.

TABLE 1

	1970	1975	Ratio 75/70	1980	Ratio 80/70	1990	Ratio 90/80
Total available sources (Tcal)	28 525	47 369	1.66	95 000	2.01	145 000-155 000	1.53-1.63
Sources structure natural-gas, % of total	68.2	74.8		85.7		about 95	
town gas, % of total	31.8	25.2		14.3		about 5	

The yearly rate of sources rising is about 4.5-5% in the period 1980-1990. It is expected (with regard to limited possibilities of our own production) that nearly the whole growth of natural-gas sources will be carried out by natural-gas imports.

According to the above-mentioned estimate the proportion of natural gas in the fuel-energy balance will attain 10-11% in 1980 and 14-15% in 1990. It is expected that this proportion of natural gas in primary sources will also last after 1990.

ESTIMATE OF GAS CONSUMPTION AND ITS USE IN INDIVIDUAL SECTORS

The volume of net supplies of natural gas and town gas for consumption in the years 1970-1976 is given in Table 2.

TABLE 2

	1970	1975	1976
Total Tcal	23 550	39 382	44 547
wherein			
natural gas, Tcal	15 241	28 427	33 425
town gas, Tcal	8 309	10 955	11 122

The main gas consumer is industry. The proportion of the non-production field is much lower although it has increased a little during the last few years; the comparison of the consumption in individual sectors in the period 1970-1976 is as in Table 3.

TABLE 3

	1970		1975		1976	
	Tcal	%	Tcal	%	Tcal	%
Industry	16 669	70.8	25 641	65.1	28 881	64.8
Domestic	3738	15.9	6307	16.0	7210	16.2
Commerce	3143	13.3	7434	18.9	8456	19.0

The possibilities for using gas are great later on, first of all for main consumers: chemical industry, iron and steel basic industry, mechanical engineering, the food industry and electric power generation.

In the years 1980 and 1990 the total gas consumption in the industry is expected to be as shown in Table 4.

TABLE 4

	1970 Tcal	1975 Tcal	Ratio 75/70	1980 Tcal	Ratio 80/70	1990 Tcal	Ratio 90/80
Industry total	16 669	25 641	1.54	60 400	2.36	90 000-94 000	1.49-1.56

Gas consumption for various branches of industry up until 1980 is shown in Table 5.

The forecast for the year 1990 is an increase of about 40% in comparison with the year 1980; the total gas consumption in the industry will probably reach 90 000-94 000 Tcal. The consumption structure will probably alter. It is expected that

TABLE 5

	1970 %	1975 %	Ratio 75/70	1980 %	Ratio 80/75
Chemical industry	31.7	27.7	1.34	14.4	1.23
wherein raw material	11.3	17.1	2.32	11.5	1.59
Power stations	7.6	7.5	1.53	8.2	2.57
Iron and steel basic industry	10.8	13.0	1.85	16.6	3.02
Non-ferrous metals basic industry	3.6	3.1	1.32	2.1	1.60
Fabricated metal products machinery and equipment	16.8	13.0	1.19	12.8	2.32
Food industry	6.4	7.9	1.87	4.6	1.38
Other industry and building industry total	23.1	27.8	1.85	41.3	3.5

the consumption development rate in the chemical industry, iron and steel basic industry, fabricated metal products machinery and equipment, food industry and agriculture will be higher than the total average rate. In individual branches priorities will be directed towards technological and direct consumption while the consumption for conversion and steam production will be limited.

The development of gas consumption in households (domestic market) from 1970 to 1976 is given in Table 6.

TABLE 6

	1970	1975	1976
Total domestic consumption (Tcal)	3738	6307	7210
Total number of consumers in households ($\times 10^3$)	1387	1729	1802
Special consumption, Mcal/consumer	2695	3648	4001
% of households gasification	30.2	35.5	36.7

It is expected that the rate of gas introduced to households in the next years will remain about the same. In the year 1980 the number of consumers will be 2 080 000, i.e. 41% gasification. The specific consumption will probably be 4700 Mcal/consumer, thus making the total consumption 9800 Tcal.

With regard to hitherto existing experiences the rate of consumers development will probably not change up to 1985. After 1985 the rate will decrease a little; there will be more difficult economic conditions for gaining new consumers in areas with scattered buildings.

The specific consumption will be in principle influenced by heating, the growth of which will be regulated according to the building of peak capacities. It is expected that the total number of consumers in households in 1990 will reach 2.6-2.7 millions (48.5-50.4% gasification) and the specific consumption will reach 7000-8500 Mcal per consumer.

It is expected that the gas consumption in households in 1990 will be that shown in Table 7.

TABLE 7

Consumers number	2 600 000 - 2 700 000
Specific consumption	7000 - 8500
Total consumption of population	18 200 - 22 950
% of the total of expected sources	12.1 - 15.2
% of the total of expected net sale	13.4 - 16.9

Large possibilities of gas utilization are found in the commercial market. According to the growing significance of a third sector it is expected that the rate of development in this sector in the years 1980-1990 will be the greatest.

Thus, the total development of final consumption in comparison with the years 1970 and 1975 will probably be that shown in Table 8.

TABLE 8

	1970	1975	Ratio 75/70	1980 Tcal	Ratio 80/75	1990 Tcal	Ratio 90/80
Industry	16 669	25 641	1.54	60 400	2.36	90 000-94 000	1.49-1.56
%	70.8	65.1		70.7		60.1-63.1	
Domestic	3738	6307	1.69	9800	1.55	18 200-23 000	1.86-2.35
%	15.9	16.0		11.5		13.4-16.9	
Commercial	3143	7434	2.37	15 200	2.04	34 000-39 000	2.27-2.58
%	13.3	18.9		17.8		24.1-25.2	
Total	23 550	39 382	1.67	85 400	2.17	142 200-156 000	1.67-1.83

The increase in the domestic and commercial proportion in 1980 is only of transient character as the total sources are growing faster than they can be used in these sectors; it also connects with possible investment measures. It is expected that after 1980 the rate of consumption development in the non-production sectors will be higher than in industry; the total consumption structure in 1990 shows an increase in the commercial proportion in comparison with 1975 and the domestic proportion will be on the level of 1970-1975.

This expected development is favourable not only from the standpoint of the effectiveness of gas utilization but also from the point of view of ameliorating the environment.

SEASONAL COURSE OF CONSUMPTION

The evaluation of the consumption course for particular consumer categories and determination of the consumption-curve pattern for the whole year is very important for predicting the future consumption development. It enables one to foresee the influence of the structural consumption changes upon the total course of consumption and at the right time to determine technical or other measures for using gas supplies from the point of view of seasonal variation, especially in winter.

In Czechoslovakia, considerable attention is paid to the question of gas-consumption development exploration. Data is systematically collected and evaluated with the aim of determining and specifying more precisely a model set of annual consumption diagrams for characteristic consumption purposes; these diagrams are used for requirement estimates.

As an example some model diagrams are shown which were derived on the basis of evaluation over a period of years (natural gas) (see Table 9).

On the basis of diagrams of selected industrial branches and of other consumer groups the global model diagrams are elaborated. As an example the following model diagrams are shown (see Table 10).

The model diagrams are confronted with real results and their accuracy is being constantly augmented.

Besides diagrams, the yearly consumption weekly and daily diagrams for selected areas are included. The results are useful not only in the prediction of future trends and in the practical influence on the consumption growth in certain categories but also in practical supply control and the execution of short-term supply régimes (quarterly, monthly, weekly).

The conducted studies show that the yearly consumption diagram is getting more profound according to changes in structure up until 1990 in favour of the non-production sectors and a wider use of heating. According to present estimates, the development will be as shown in Table 9.

TECHNICAL MEANS TO MEET THE CONSUMPTION PEAKS

With the ratio of summer to winter consumption getting worse it seems advisable from the technical and economical points of view to reliably meet the gas consumption in the course of the whole year.

The basic tendency not only with natural gas but also with town gas is the equitable utilization of basic gas production sources, i.e. the natural-gas deposits and brown-coal gasworks. Staggering of supplies also appears feasible for imported natural gas. The consumption conditions indeed permit a certain lack of uniformity; practically, however, the supplier tends to attain a uniform gas export throughout the year.

Underground gas storage represents at present the basic means to equalize the yearly gas consumption pattern in Czechoslovakia. Underground gas storage meets 20-25% of the natural gas daily consumption in winter.

As far as the town gas is concerned one storage tank caters for the supply diagram equalization; it meets 10-13% of winter daily demands while the rest is being met by seasonal sources producing town gas by natural gas cracking and operating only in a winter or transient periods.

TABLE 9

Per cent of the Yearly Consumption

	1	2	3	4	5	6	7	8	9	10	11	12
Chemistry – technology	8.0	8.0	7.8	8.4	8.8	8.5	8.0	8.8	8.7	8.5	8.3	8.2
Chemistry – energy supply	11.5	9.8	10.1	8.5	7.6	6.6	6.5	6.5	7.1	8.5	9.0	9.3
Power stations, heating plants (yearly, consumption)	6.0	6.0	6.0	9.0	10.0	10.0	13.0	12.0	10.0	7.0	6.0	5.0
Ditto – summer consumption	2.0	2.0	2.0	5.0	10.0	20.0	20.0	20.0	10.0	5.0	2.0	2.0
Metallurgy	7.8	7.5	8.4	8.7	8.0	8.5	9.0	9.0	9.2	8.8	7.5	7.5
Mechanical engineering	9.0	8.8	8.8	9.3	8.8	8.3	6.5	6.5	9.0	9.2	8.0	7.7
Food industry	8.8	8.2	8.5	7.0	8.2	8.6	8.7	7.9	7.8	8.5	9.2	8.6
Building and other industry	9.4	8.7	9.4	8.4	7.5	7.4	6.8	7.2	7.8	8.8	9.3	9.0
Big consumptions of non-industrial character												
heating	16.4	13.6	12.1	8.8	3.1	2.3	2.0	1.9	2.6	8.9	12.8	15.5
other	8.8	8.3	9.1	8.7	7.7	7.5	7.4	7.3	8.1	8.6	9.5	9.5
Domestic and commercial												
heating	17.7	14.7	14.7	8.3	3.7	–	–	–	2.0	8.2	14.0	17.3
other	9.1	8.3	9.3	7.6	7.5	7.3	7.2	7.2	8.2	8.7	9.2	10.4
Total diagram												
heating	16.9	13.8	13.0	8.6	3.3	1.5	1.3	1.2	2.4	8.7	13.2	16.1
other	9.3	8.5	9.3	8.2	7.8	7.5	7.0	7.3	7.8	8.8	9.3	9.1

TABLE 10

	1975	1980	1990
Ratio of the yearly supply to the highest daily supply			
natural gas	278	260-270	225-240
town gas	230	210-220	190-200
Ratio of the lowest to the highest daily supply			
natural gas	1:2.0	1:2.2-2.5	1:2.8-3.0
town gas	1:2.7	1:3-3.2	1:3.4-3.5

The winter gas-consumption equalization by means of underground storage tanks also has its technical and economical problems, disregarding the fact that the construction of these storage tanks takes longer and that they can be sited only under certain geological conditions. Therefore we plan the construction of peak capacities for natural-gas liquefaction in the future and storing this liquefied gas up to the period of highest demand. As far as the town gas is concerned the peak demand will be met by increasing the seasonal sources capacity.

We expect that the above-mentioned measures enable the following layout of meeting the highest daily supply in the year 1990:

	%
Natural gas	
Basic supply sources (own exploitation, import)	62-65
Underground storage tanks	28-30
Peak liquefied natural gas sources	8-10
Town gas	
Basic supply sources (brown coal gasworks)	60-65
Seasonal sources (natural gas cracking)	20-25
Underground storage tank	10-15

Disregarding the measures taken by the gas supplier there exist also possibilities for gas consumers. A means for equalizing the total pattern is represented by two-fuel heating systems (coal-gas or heating oil-gas) which are utilized either according to a given plan (in winter the consumer uses another fuel) or in cases of a sudden growth of gas demand.

The augmentation of gas-supply reliability is favourably influenced by means of introducing gas consumptions which are typical for summer seasons (e.g. forage drying houses in agriculture). We are also trying to agree with important gas consumers in industry that larger repairs during which the gas consumption decreases will be transferred to the winter season.

As far as the town gas is concerned the best utilization of capacities is attained by concentrating complete overhauls of gasworks producing town gas by gasification under pressure — if they necessitate a considerable reduction or interruption of gas production — to the summer season; in this manner the highest capacity of production in winter is secured.

The use of natural-gas-cracking plants for equalizing the seasonal consumption variations proved in practice very successful. However, it is necessary to bear in mind that the demands in winter, which otherwise would be met by the town gas underground storage tank, are transferred to the natural-gas underground storage tank which must be accordingly designed.

METHODS OF INFLUENCING GAS CONSUMPTION

The prognosis of gas-consumption development prefers gas use in those categories where a high economic effectivity is attained. At the same time the priorities given by state authorities are taken into account regarding not only the needs of the whole society but also the development of individual regions of the state. It is further necessary to retain the balance between overall consumption and sources not only in yearly volumes but also during maximum consumption periods; thus the gas supply will be continuous.

The basic method of influencing gas consumption of large industrial customers and of other large customers is the method of assigning fuel allowances. On the basis of consumer application and of submitting information on the consumption purpose, on considered gas appliances and on the expected consumption development the Fuel and Energy Ministry — having received the appropriate recommendation from the Directorate General — issues the fuel allowance which is used in the design of new gas-consuming plants and later on for making contracts between the gas consumer and supplier.

The development of gas supplies in the non-production sectors is influenced by the programme of new towns gasification and of consumer connections. Special attention is devoted to the connection of new heating appliances for domestic and commerce to keep the growth of demand peaks within the supply possibilities of the gasworks.

Price politics in gas supplies represents a supplementary part of the general means of influencing gas use in the desired way. The prices of individual gas types are uniform for individual consumer categories all over the state.

The natural-gas prices for works are uniform; on principle the consumed m^3 is paid without regard to the branch or the art of using. An expectation is valid on the gas for chemical treatment which is 25% less expensive than for other big consumers. Enterprises with a consumption of less than 60 000 m^3/year pay besides the basic tariff a fixed monthly fee.

The tariff for household consumers is lower than for big consumers; either a simple tariff (m^3 supplied) or a composite tariff (fixed monthly fee and cheaper m^3 supplied) may be used.

Special agreements with consumers (e.g. summer consumption, the use of two-fuel heating systems, repairs in winter seasons) disregard the prices.

RÉSUMÉ

The tendencies of gas-consumption development in Czechoslovakia up to 1990 correspond with the estimate of the expected growth of natural-gas sources; we expect that practically all natural gas will be imported.

The capacity of gas sources in 1980 will be twice as high as in 1975, their further growth up to 1990 being 53-63% compared with 1980.

The share of natural gas in primary sources will grow from 8% (at present) to 10-11% in 1980 and to 14-15% in 1990.

It is expected that the growth of industrial consumption up to 1980 will be higher than the consumption growth in domestic and commercial sectors.

On the other hand, it is expected that in the years 1980-1990 the highest consumption growth will be in the non-production sectors, first of all in commerce. The household gasification will grow from 35.5% in 1975 to 41% in 1980 and to 48.5-50.4% in 1990; in this year the specific gas consumption will probably attain 7000-8500 Mcal/household consumer.

The change of consumption structure in favour of the non-production sectors and use of heating in all categories will cause the consumption diagram to be more profound. It is expected that the ratio of the lowest to the highest daily consumption will grow from 1:2 in 1975 to 1:2.8-1:3 in 1990 (for town gas from 1:2.7 to 1:3.4-1:3.5).

According to this situation the importance of seasonal and peak capacities for meeting winter demands will grow. We expect to meet 62-65% of the winter highest daily supply of natural gas by basic sources, 28-30% by underground storage tanks and 8-10% by liquefied natural-gas peak stations.

Besides the measures taken by gas suppliers there are also possibilities of improving the consumer consumption diagram by two-fuel systems, summer consumption, repairs in winter.

The basic method of influencing gas-consumption development in industry is the method of assigning fuel allowances to future gas consumers. A supplementary part of the general means of influencing gas consumption are price politics and control of the growth of heating.

Natural Gas and LPG Utilization in the 1990s in Hungary

A. Bán

Hungary

The long-term goals of natural gas (NG) utilization should be determined by the balance of complex energy sources available and the amount of consumption, with respect to optimum exploitation of the sources.

Actually it is not necessary to deal with the places of origin of the sources. Trends and price developments the world over indicate for the coming 30 years an increasing ratio of NG among the energy sources. In Hungary during the next 15 years continuous efforts will be made to explore more potential petroleum and NG reserves. The degree of exploration of the areas deeper than 3000 metres in the sedimentary basins of Hungary is still rather unsatisfactory. Experiences around the world and also the results of exploration in Hungary indicate that in these depths the discovery of prevailingly NG, gas condensate reservoirs and that of natural hydrocarbon pools of big gas cap and narrow oil rim can be expected.

With respect to examinations carried out by different governmental organizations, it is expected that the energy consumption of the country will increase by 4.1 to 4.3 per cent per year. The total energy consumption in 1990 will be nearly double that of 1975. The importance of electric energy utilization will increase within the overall energy consumption. In the decade 1980 to 1990 electric energy will make up 50% of the total energy growth. Accordingly the shape of energy supply will be determined to a greater extent by the energy base of the electric power stations to be installed. Energy consumption will increase from 257 Pcal per year in 1975 to 440 Pcal per year in 1990.

Local coal will be a source of energy for the power stations to be installed after 1980, later atomic power is envisaged.

The dependence of the Hungarian energy supply on imports will increase despite maximum utilization of her domestic sources. The ratio of imports will increase from a yearly 46% in 1975 to about 65 to 67% per year till 1990. Considering the fact that the price of crude oil is increasing faster than the price of NG, and further, that oil reserves are more restricted and that the bringing home of larger amounts of oil creates some transportation problems, it seems to be more appropriate with respect to reducing expenses to strive for the reduction of crude-oil consumption, replacing crude oil with other sources of energy, such as, for example, natural gas.

In the determination of the structure of consumption the basic aim is to increase the communal consumption of the population in a moderate ratio, and further, to satisfy the requirements of the factories utilizing NG as a basic raw material and to supply technological works with NG.

NATURAL GAS PRODUCTION AND UTILIZATION AFTER 1975

Domestic NG production is planned partly from existing natural-gas plants and partly from new plants to be installed. Table 1 shows the sources of exploitation.

TABLE 1

Item:	1980	1985	1990
	(billion m^3)		
From existing plants	5.4	4.8	1.5
From new reserves in connection with known reserves	0.4	0.7	1.7
From new reserves to be discovered by further exploration	-	-	2.3
From natural gases of high inert content	-	0.5	0.5
Total	5.3	6.0	6.0

The yield of actually known pools will decrease until 1985 in a smaller ratio and following to 1990, to a greater ratio. The difference between the actual production and reduced production should be made up by the yield of already known but not yet produced reserves and partly by the utilization of known NG reserves with high inert content, having a low calorific value thus suitable only for industrial use. The production foreseen for 1990 includes also some amount from the prognostic reserves indicated by geophysical methods as probable ones but not as yet explored by drillings.

Therefore keeping the actual yearly supply going, the domestic production is planned on a constant level. Table 2 shows the requirements and the sources. The missing amount should be imported either by transmission lines or in the form of liquid gas discharged by tankers in some foreign ports, forwarded to the inland also by pipe lines. The sources contain some natural hydrocarbon gases mixed with CO_2 as well. These should be delivered to the consumers after having enriched their CH content by certain methods. The enrichment can be carried out either by separation through cold energy, or by stripping, or by mixing with some CH gas.

During the following 10 years in addition to the investments necessary to keep the production level, some new gasfields of 2000 to 2500 million m^3 capacity per year should be established at every 5 years and provision should be made to process the NG produced and to connect the new fields with the existing gas-grid system. The bulk of NG should be produced later on by the Algyõ field. The yield of the older significant pools, such as Hajduszoboszló, Szank, and the gas pools in Bókés county, will become considerably decreased by 1985 providing excellent possibilities for underground storage in the depleted reservoirs.

The utilization of the NG with a high inert gas content is necessary to keep up actual production levels. At the same time it is also reasonable from an economic point of view. The amount of CO_2 and nitrogen may reach as much as 60 to 65%, yet the gas might be well utilized as fuel in thermal power stations, such as the Gagarin

TABLE 2

Natural Gas Sources
(amounts in Gm^3)

	1975	1990
Domestic production	5.3	6.0
Imported	0.6	7.0
Total	5.9	13.0
Division of imports:		
contracted	0.6	4.0
planned	-	3.0
Total	0.6	7.0

Division of Natural Gas Utilization
(amounts in Mm^3)

	1975	1990
Town gas manufacturing	190	-
Thermal and electric power production	2700	4660
Chemical raw material	780	2500
Industrial technologies	1490	3260
Population and communal supply	420	1760
Other (loss)	210	610
Total consumption	5900	13 000

or Tisza power plants. The above utilization is favourable also because it helps to save high-quality NG, or fuel oil, which can be utilized countrywide more efficiently.

In the coming years, due to increased imports, a slight rise of the sulphur content of the NG to be sold can be expected. It is intended to supply the fertilizer factories with sulphur-free gas of domestic production, i.e. the nitrogen plants of pét will be supplied from Algyõ and it is hoped for that the other two plants in northeast Hungary also might be provided for with sulphur-free NG from some pools to be discovered later. The expenses of NG grids being rather high, the extension of NG consumption by the population is foreseen mainly in areas with already existing transmission and distribution systems.

In 1985 to utilize 1000 m^3 NG some 3200 Florins should be invested without the transmission line. Due to greater efficiency in distribution these expenses should be reduced to 1200 Florins only by 1990.

During the coming 10-year planning period, some sixty-five more dwellings (towns and villages) should be attached to the NG grid as is shown by Table 3. Altogether some 200 localities will be provided for with piped NG. Due to purchasing difficulties (high price), quality requirements, and environmental pollution problems connected with other energy sources, it is expedient to continue the switching over of technological plants, industrial ovens, home-heating boilers from coal, fuel-oil and bunker-oil to NG utilization.

TABLE 3

Position of Piped Gas Supply
(Households provided for)

	1975	1990
With town gas	357 000	18 000
With NG	350 000	1 172 000
Altogether:	707 000	1 190 000
from which those also supplied for heating purposes	306 000	743 000
Position of Bottled Gas Supply		
	1975	1990
Number of consumers (1000 households)	1600	2600

While planning the expansion of regional NG utilization, the analysis of ancient data and the calculation of the components of gas consumption have been chosen as a working method. The analysis of data from the past, the so-called trend-extrapolation, is based on the assumption that the actual situation can also be projected for the future. Yet some corrections are necessary with respect to technical development and to changing requirements.

From the factors influencing gas consumption, the following should be considered according to some investigations carried out by the Hungarian Committee for Technical Development (OMFB):

The ovens actually utilizing fuel oil should be converted to utilize NG if the expenses of the switch-over are in the range of about 23 000 Florins per m^3 per hour.

The new apartments with individual or central-heating systems (including roof boilers of housing blocks) should be provided for with NG in the districts already supplied with NG, if the specific expenses of distributing network installation do not exceed 16 000 Florins per apartment.

In the densely populated areas of the big towns with existing piped gas supplies, the flats and communal consumers still utilizing coal or fuel oil should also be converted to NG utilization. The specific limit of expenses in this case also being 16 000 Florins per flat or per consumer.

All ovens in a district with an existing gas distribution network still utilizing fuel oil might be converted to utilize NG.

Based on optimalization examinations it seems to be appropriate to increase the utilization of NG in thermal power plants. The fuel requirements of the existing power plants and of those under construction should be satisfied. According to the above examinations the bulk of fuel requirements of these plants should be covered with NG.

In addition to the determination of yearly requirements, the satisfaction of seasonal and daily peak demands should be dealt with as well. Some related investigations have shown that in the case of consumers in the countryside the storage capacity of the pipe line is usually sufficient and the reduced pressure can be re-established during the periods of low consumption. In contrast, the capital having a more heat-degree-sensitive consumer circle must be provided for with sufficient storage capacity to satisfy daily peak demands. The storage space must be such as to permit a quick change of withdrawal, such as, for example, liquid-gas storage or storage of NG in aquifers, etc.

To satisfy seasonal peak demands underground storage should be established in the distributing grid. The situation is favourable in this respect since after 1980 some depleted gas pools will be available where underground storage can be established. In Hungary underground storage has already been in use since 1963 satisfying peak demands of local consumers in some counties.

The construction of a relatively high-capacity underground storage in the depleted reservoirs of Hajduszoboszló gas pool is under construction to receive a part of the NG imported from the Soviet Union beginning in 1979. The reservoir is only 100 km from the big gas-transmission line passing through Czechoslovakia to Western Europe, therefore it is suitable to adjust the parameters of the system and to increase its safety. In Hungary three underground stores will be installed by 1980, Kardoskut, Ederics and Hajduszoboszló, with an additional capacity of 9.5 million m^3 per day, providing 1 760 000 m^3 per hour withdrawal capacity at peak demands in 1980.

Up to 1990 an additional withdrawal capacity of 710 000 m^3 per hour should be installed.

The area distribution of big European consumers and NG sources raises the issue of establishing underground storage of high withdrawal capacity in the vicinity of the big consumers. In this respect Hungary is able to offer excellent facilities in the frame of international co-operation if her gas pools having some 100 billion (thousand million) m^3 original NG reserves become depleted.

The pools are not far — some 100-200 km only — from the international transmission lines. They are suitable to increase safety in operation of the system and to satisfy peak demands (Fig. 1).

The consideration of the geographical situation is recommended of the country lying centrally in Europe and providing with her plain surface the possibilities of easy construction for transit lines to be built in the future towards Western and southern Europe. A safe supply could be provided for by underground storage and multiple transit lines.

EXPENSES AND PRICE RATIOS OF ENERGY SOURCES AND THEIR EFFECT

Despite the sharp increase of crude oil and NG delivery prices the economic order of rank of the energy sources remained unchanged even since 1973. With respect to Hungary the related index numbers are as follows when considering coal at present as being 100%: local NG: 38%, local crude oil: 55%, Soviet crude oil: 99%, Soviet NG: 99%, the recently developed Eocene coal: 161%, capitalist crude oil: 297%.

The price of local energy sources is below the international price level. The relatively low domestic price of the energy sources reflects some economic and socio-political considerations.

It can be stated that the cheapest source of energy remains also for future perspectives the local NG. Therefore, domestic production possibilities should be exploited to the maximum extent. The solid foundation of exploitation and exploration should be secured. The price of imported NG will most probably become equal to the world market price of fuel oil.

To promote higher efficiency in utilizing the available NG, it should be distributed in an optimum way.

The price system of energy sources should be further developed in such a way as to promote increased economy in utilization. The latest consumer price changes do not promote a more powerful utilization of NG. If the price of coal in 1977 is

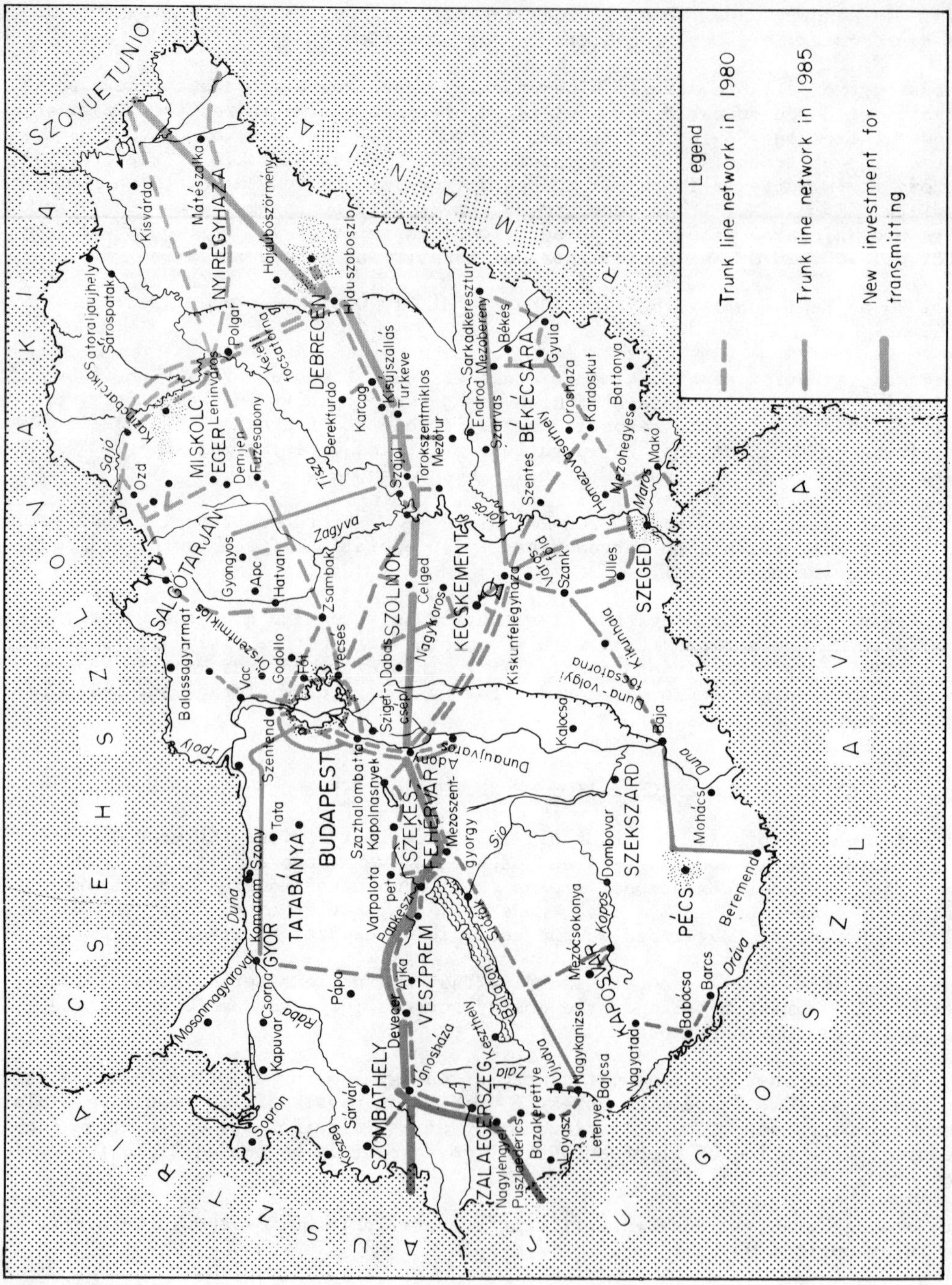

Fig. 1. Variation. Examination of transmitting by the construction of a new transit-pipe line.

considered as being 100%, then the price of NG is 174% and that of town gas 229%. At the same time the price of household fuel oil gives an index number of 209 irrespective of the fact that the purchasing price of crude oil underwent a more significant rise than that of NG. The price of bottled PB-gas is significantly different, showing an index number of 246% at the gas station and of 300% when delivered at home.

TABLE 4

Date about the Sources of LPG
(in thousand metric tons)

	1975	1990
Domestic production	213	410
out of which:		
from processing of NG	120	180
from crude-oil refining	93	210
Imported	28	0-60
Total	241	400-450

Data about PB-gas Utilization
(in thousand metric tons)

	1975	1990
Sold in bottles	212	350-380
Industrial and others	29	50-70
Total	241	400-450

The problems related to peak demands should be solved as well. Through price policy some pressure should be exerted to reduce consumption at the time of peak demand. In addition some more critical considerations should be made relating to the reality of contracted delivery obligations.

LIQUID PETROLEUM GASES (LPG)

The provision of LPG is one of the most successful services rendered in Hungary. In 1975 some 1 623 000 consumers have been listed not counting those utilizing tourist bottles only.

By 1980 about 2.1 million and by 1990 about 2.6 million consumers should be served according to current plans. The number of households in Hungary makes some 3.9 million in comparison.

The above numbers of consumers seem to be rather high. It is due to the fact that most of the dwellings are of the single-family-house type and a large part of the population is living on rather scattered individual farms. The rather low thermal density would require high investment expenses for the construction of distribution pipe systems and, consequently, the realized specific marketing expenses are lower than in the case of piped gas supply. It is also an advantage that the gas condensate content of domestic NG is rather high, making the separation of LPG and other components not only economic, but also necessary from the point of view of transportation.

The source of LPG supply is domestic production and imports. At the same time the export of seasonal excess amounts is appropriate.

Actually LPG is produced from every NG pool where its production is economic. There are two basic processing plants: one at Hajduszoboszló (constructed in 1964/65 with French co-operation) and the other one at Szeged-Algyõ (constructed in 1975/76 with Soviet assistance). As NG production increased in the latter years LPG production also increased. Today the domestic NG production has reached the peak of its possibilities, thus the amount of LPG obtained from NG remains also constant at the level of about 200 000 metric tons per year.

In the future hydrocarbon exploration will go down below 3000 m and experience shows that in greater depths the discovery of more gas condensate pools can be expected, containing a higher amount of LPG.

Another source of LPG production is the crude-oil refining industry. The amount of LPG produced in the refineries quickly increased in recent years. A further increase is planned in the coming 10-15 years. Crude-oil refining continues to increase, though in a more moderate rate. The heavy components left by primary distillation will be further refined by the destructive procedure, rendering another source of LPG production. Thus in 1990 a duplication of LPG production can be expected with respect to 1975 in the field of petroleum refining.

The preferred field of LPG utilization remains in the providing of the population with bottled gas, and it should be extended, in addition, not only to cooking purposes but also to water boilers and room heating. To serve completely a supply of individual family houses and communal consumers as well as small-scale industrial enterprises, which are not easily supplied by bottled gas, the supply in small-scale tanks should be introduced. The tanks are either mobile or fixed. The former can be exchanged. The latter can be filled by tank-cars. In the case of delivery in tanks, only propane should be supplied making the use of an evaporation instrument superfluous. The bulk delivery of LPG will increase marketing profits. At the same time the chance of accidents will be reduced by the elimination of frequent connection and disconnection of supplying and consuming equipment. The filling is carried out at the filling plants established in the vicinity of gas-processing plants and refineries. A larger amount of liquid gas is stored under pressure in Horton-spheres having a capacity of 250 to 1000 m^3 each.

The construction of new filling stations and the reconstruction in conjunction with extension of some old filling stations will be completed in Hungary by the beginning of the 1980s.

Thus the districts which can be supplied economically will be formed. In these districts the distances are such as can be covered by a bottle-transporting truck within the prescribed working time. At the same time containerized transportation shall be introduced and the ratio of home delivery should be increased. Home delivery is actually not very significant and it is restricted to the capital only. The exchange of empty bottles for full ones is being carried out actually at some 3000 exchange stations.

The increased ratio of home delivery will promote more comfort for the purchaser, will increase safety and reduce expenses.

So far, no sufficient storage capacity has been made available to balance differences between production and demand. The reservoirs of the processing plants and filling stations rendered a very restricted possibility only. By 1980 the construction of a cooled atmospheric reservoir of 30 000 m^3 capacity will be finished at Algyõ-Szeged. To promote safe bottling the filling stations should be provided for with pressurized Horton-spheres of 1000 m^3 capacity wherever it is possible.

I

Expected Consequences of Measures of Gas Conservation and Efficient Utilization in Various Industries

N. Fedorov

Union of Soviet Socialist Republics

Moreover, implementation of these measures would bring many processes to a higher technical level and, as a result, would considerably increase the output and improve the quality of various products with simultaneous reduction in fuel requirements. This will produce beneficial effects on the air basin of industrial centres.

Realization of measures of economic and efficient gas utilization in various industries should include:

- determination of optimum areas of gas utilization and its use chiefly for technological purposes and as feedstock for the chemical industry;
- revealing the routes to gas conservation;
- development of scientifically proved specific gas-consumption rates for concrete processes (plants);
- determination of lines along which the most promising methods of gas utilization and optimum designs of burners and combustion equipment should be developed;
- optimization and automation of thermal processes;
- development of new technological processes;
- up-dating of existing equipment and designing new methods and techniques of utilization of gas and secondary energy resources.

Implementation of the measures to improve the structure of fuel-energy balance and to use every fuel and energy resource in a more rational way is to save, in the USSR, over 150 MMt of fuel with a heating value of 7000 kcal/kg.

Over the last 10 years the world energy consumption has approximately increased by a factor of 1.5; gas-consumption growth rate proving even higher and is expected to be maintained by the year of 2000.

Seventy-eight per cent of the world gas supplies (natural and synthetic) are consumed in the Soviet Union and in the United States. Gas is mainly used in power generating plants (21 per cent), in industry, particularly in the chemical industry, metallurgy and steel-making (46 per cent) as well as for domestic and commercial purposes (33 per cent).

In 1970 the portion of natural gas in the Soviet fuel-energy balance was 18 per cent. In 1980 it is expected to amount to 24 per cent, while in the future it may reach 26 per cent.

In different parts of the globe (the COMECON countries, the USA, France, Japan, etc.) numerous programmes and laws regarding efficient and economic fuel utilization have been and are being developed.

These works are done along with the intensifying of the exploration of new natural gasfields and SNG production and are necessitated by the need to conserve fuel-energy resources and to utilize them in the most rational way. Consideration is given to standards setting a minimum permissible efficiency of fuel usage with due reference to scientifically proved rates of its consumption; methods for estimating fuel-utilization efficiency are being developed; the governmental control over gas distribution serves to reduce its uneconomic consumption to a minimum.

One of the ways of a better utilization of fuel-energy resources is to optimize the fuel-energy balance and to determine optimum areas for different fuels.

In 1976-1980 NG requirement for technological purposes is predicted to be twice as high as in 1970-1975.

In the Soviet Union, in the United States and some other countries, measures are being taken to reduce the specific volume of gas burned in boilers and to substitute it by other fuels.

Currently, natural gas and fuel oil are replaced by coal in cement-clinker kilns. In the United States 76 per cent of cement plants do operate on coal and by 1980 90 per cent of industrial plants will have been converted to coal.

The prospects for gas consumption in industry will greatly depend upon the costs of gas and alternate fuels.

New price systems are being developed and the present gas prices are being revised.

Natural gas can also be conserved due to advanced designs of heating equipment, new technological processes and methods for NG utilization, novel insulating materials; optimization and automation of technological processes, better recovery of the flue gas heat; wider utilization of secondary energy resources; introduction of "wasteless" technologies.

Among the most promising areas for NG utilization are the following: direct iron reduction; furnace burners providing intense heat transfer to the materials and articles being heated; combined steam-power plants for power generation, steam raising, and heating water for central heating systems; heat-pump systems providing energy conservation; fuel cells for direct power generation.

Nowadays there are known several projects of using certain fuels as reducing agents. For example, in Canada a plant is being designed with two types of fuel as a reducer: natural gas/coal and natural gas/fuel oil. In Peru seven companies are engaged in designing a plant with a solid-type reducer. In the next 10 years NG appears to enjoy wide usage for producing metals by direct metal ore reduction.

Gas-fired fluidized-bed furnaces help greatly reduce specific fuel consumption and improve thermal efficiency of metal heat-treating processes. Fluidized-bed furnaces are known to have been employed for wire annealing and tool-steel treatment. In this case the maintenance costs are 30-50 per cent lower than those with conventional furnaces and salt baths.

Considerable fuel economy arises from the utilization of secondary energy resources. Due to this factor, in the Soviet Union 15.1 MM tonnes of fuel with a heating value of 7000 kcal/kg were saved in 1975 and in 1980 conservation is to be 30 MM tonnes of such a fuel.

In connection with an ever-increasing scale of converter steel smelting, thought is given to the development and introduction of methods for utilizing converter gases formed during the converter purging with oxygen, which results in general fuel economy.

An application of recuperators would improve furnace efficiency up to 70 per cent, 20 per cent of the heat being lost with the flue gases and 10 per cent to the environment. Recuperators help reduce NG consumption in furnaces by 30-40 per cent, intensify furnace operation and reduce metal heating time by 30 per cent without affecting its quality. Thus, due to recuperators installed in annealing furnaces, air can be heated up to 530°C at a gas temperature of 1100°C and NG requirement can be reduced by 30 per cent.

Soviet experience shows that gas requirements for gas-fired furnaces equipped with recuperators are far lower.

Internal-recuperation furnaces, where flue gases are removed via furnace walls and roof fabricated of porous or multihole brick are finding commercial application.

In internal-recuperation furnaces the factor of convective heat emission on the furnace walls amounts to 230 W/m^2 instead of 30 W/m^2 in the conventional furnace case, the lining gets heated in a shorter time and leads to higher operating temperatures. All this contributes to better furnace efficiency, lower fuel consumption and reduced melting losses. For instance, the efficiency of a pusher-type furnace increases by 35 per cent; specific energy consumption goes from 2 to 0.7 kWh/kg; heating time reduces from 6 to 1 h.

The internal-recuperation concept is widely employed in the Comecon countries, thereby saving about 25-45 per cent of energy consumption. Fairly good results can be achieved by placing radiant burners in the furnace roof in combination with permeable furnace walls.

Substitution of brick linings for ceramic fibres in kilns saves about 20-25 per cent of gas fuel: one has an improved furnace configuration and, hence, a smaller internal surface area, a reduced number of burners, high-quality insulation, lower heat requirements for lining heat-up and lower heat losses.

During 7 years specific energy consumption in the USSR steel-smelting industry decreased by 39 kg of fuel/t with a heating value of 7000 kcal/kg, further decrease is expected with more extensive production of converter steel.

The lowest fuel (coke and fuel oil) consumption per tonne of produced steel is achieved in Japan and equals 490 kg/t, including 435 kg of coke (1973). Over a period of 1960-1973 the specific fuel consumption in metallurgy has decreased in Belgium by 24, in the Federal Republic of Germany and Great Britain by 18, in the USA by 15, in Japan by 18, in France by 9 and in Sweden by 2 per cent.

In the USSR blast-furnace industry, coke consumption per tonne of cast iron was reduced over a period from 1965 to 1975 by 14.3 per cent due to the utilization of gas, oxygen and preheated blast.

Limited supplies of NG make it essential, on the one hand, to intensify exploration of new NG fields and to improve its transportation systems, and on the other hand, to take measures to realize efficient gas utilization.

Implementation of these measures would bring many processes to a higher technical level and, as a result, would considerably increase the output and improve the quality of various products, and simultaneously reduce fuel requirements. This will produce beneficial effects on the air basin of industrial centres.

It follows from the above considerations that measures for economic and efficient gas utilization in various industries should include:

the determination of optimum areas of gas utilization and its use mainly for technological purposes and as feedstock for chemical industries;
revealing the routes to gas conservation;
the development of scientifically proved specific gas-consumption rates for concrete processes (plants);
the determination of lines along which the most promising methods of gas utilization and optimum designs of burners and combustion equipment should be developed;
the optimization and automation of thermal processes;
the development of new technological processes;
up-dating of existing equipment and designing new processes and equipment operating on NG and secondary energy resources, providing economic gas utilization and preventing air pollution.

Implementation of the measures taken to improve the structure of fuel-energy balance and to use every fuel and energy resource in a more rational way is to conserve in the USSR over 150 MMt of the fuel with a heating value of 7000 kcal/kg.

REFERENCES

1. Kopitov, V. F., *Energy Supplies and Conservation of Gas and Other Fuels*, VNIIEGAZPROM, Moscow, 1976.
2. *Revue gênéral de thermique*, Janvier 1977, tome XVI, no. 181, pp. 96-99.
3. *Gas Abstracts*, 1977, vol. 32, No. 12, p. 241.
4. Stirikhovich, M. A., *Energietechnik*, 1977, vol. 27, No. 3, pp. 97-103.
5. XXV Soviet Communist Party Congress documents, Moscow, 1976.
6. *AGA Monthly*, 1977, vol. 59, No. 8, pp. 5, 6, 26.
7. *Chem. Eng.*, 1977, vol. 84, No. 4, pp. 40, 42, 44.
8. *GWF (Gas Erdgas)*, 1976, August, vol. 117, No. 8, pp. 350-352.
9. *Chem. Eng.*, 1977, vol. 84, No. 5, pp. 128, 130, 132.
10. *Gas Wärme Int.*, vol. 26, No. 4, pp. 175-186.
11. *Prom. Energhetika*, 1977, No. 9, pp. 14-16.
12. *Moscow Energetic Institute Proceedings*, 1977, Issue 318, pp. 100-105.
13. *Gas Eng. and Manag.*, 1977, vol. 17, No. 6, pp. 180-194.
14. *Ind. Heat*, 1977, vol. 44, No. 5, pp. 36-37.
15. *Prom. Energhetika*, 1977, No. 9, pp. 9-10.
16. *Gas Wärme Int.*, 1977, vol. 26, No. 4, pp. 186-190.
17. *Amer. Ceram. Soc. Bull.*, 1977, vol. 56, No. 4, pp. 406-407.
18. Arakelov, V. E. *et al.*, *VNIIENERGOPROM Proceedings*, 1976, Issue 8, pp. 11-24.
19. *Iron and Steel Int.*, 1977, vol. 50, No. 2, pp. 87-88, 90-92.

Use of Natural Gas for Olefin Production in Western Europe

P.-H. Chaubernard

France

SUMMARY

Rapid advances in the working of deposits, particularly in the North Sea, are making substantial quantities of natural gas and related gases available on the Western European market. The petrochemical industry, greatly attracted by light feedstocks of high hydrogen content, has started to use these gases for the production of ammonia and methyl alcohol.

The report brings out the importance of ethane and liquefied petroleum gases as steam-cracking feedstocks in olefin production. It considers the consequences of replacing naphtha by these feedstocks on a steam-cracking unit's output of the principal intermediate products (ethylene, propylene, butadiene, aromatics). It then goes on to study how existing installations can handle the new feedstocks, and gives economic data enabling these feedstocks to be compared with naphtha for the purposes of existing and new installations.

Utilisation du Gaz Naturel pour la Production d'Olefines en Europe de l'Ouest

P.-H. Chaubernard

France

GENERALITES

Jusqu'aux années 1970, la demande de pétrole était aisément satisfaite par la ressource et les hydrocarbures à bon marché, utilisés dans tous les domaines, ont été à la base des progrès spectaculaires qui sont apparus au cours de ce 20ème siècle.

Aujourd'hui, la demande s'est accrue, la ressource, si elle est encore suffisante, ne pourra peut-être demain satisfaire la demande. On voit se dessiner deux tendances:

- La mise en exploitation de gisements d'hydrocarbures qui n'était pas rentable jusqu'ici. C'est ainsi que l'on assiste à la récupération d'importantes ressources de gaz naturel associé à la production du brut et dont la commercialisation a été jusqu'ici freinée par les coûts d'exploitation.
- La substitution aux hydrocarbures d'autres ressource, qui deviennent disponibles ou compétitives. C'est ainsi que dans le domaine de l'énergie, des substituts se mettent actuellement en place.

Ces deux tendances conduisent inévitablement au réexamen des possibilités d'usage du gaz naturel traditionnellement voué en Europe au domaine de l'énergie. Au moment où la pétrochimie cherche à ouvrir l'éventail de ses charges, le gaz naturel lui offre des molécules légères riches en hydrogène qui lui permettront de produire aisément de l'ammoniac, du méthanol, de l'hydrogène, de l'acide acétique et des oléfines.

De nombreuses installations fonctionnent déjà en Europe, pour la production d'ammoniac et de méthanol. Le domaine le moins développé est celui de la production d'oléfines qui exige des molécules comprenant au moins 2 atomes de carbone et c'est ce domaine que nous allons étudier.

PRODUCTION D'OLEFINES

Le vapocraquage ou craquage en présence de vapeur d'eau fournit la majeure partie des besoins en éthylène, propylène, butadiène et participe à la fourniture du benzène et d'autres aromatiques. Son développement qui a été très rapide dans les années 60 se poursuit actuellement au rythme de 5 à 6 % par an.

La matière première, traditionnelle en Europe pour le vapocraquage, est le naphta dont la production est actuellement freinée par la stagnation du traitement de pétrole brut, et la demande encore croissante de carburants pour l'automobile.

Le vapocraquage recherche donc des charges nouvelles et il peut se tourner soit vers des charges lourdes (gas oil, distillat sous vide) soit vers les charges légères (ethane, propane, butane). Ces dernières peuvent être fournies par le traitement du gaz naturel et des gaz associés.

Notre étude examinera deux aspects du vapocraquage des charges légères:

- le traitement dans des installations existantes;
- la construction d'installations nouvelles conçues spécialement pour ces charges.

Le premier aspect peut paraître le moins intéressant. Il n'en est rien, car le faible taux relatif d'expansion du débouché des oléfines joint à l'excédent de capacité actuel, font que la construction d'installations neuves avant 1988 sera le fait de circonstances exceptionnelles.

Les vapocraqueurs existants représentent une capacité importante, car si la moitié des vapocraqueurs Européens utilisait pour 20 % de leur capacité une charge légère, la quantité de charge nécessaire serait d'environ 4 millions de tonnes par an.

INCIDENCE DU TRAITEMENT DES CHARGES LEGERES SUR LA PRODUCTION

Le vapocraquage des charges légères conduit à des rendements différents de ceux obtenus avec le naphta.

Le tableau 1 met en évidence ces différences.

TABLEAU 1

Rendements de différentes charges

	Ethane	Propane	Butane	n Butane	Naphta
Gaz (% pds)	14	29	21,7	20,2	14,2
Ethylène	75,1	41	31,2	37,5	29,8
Propylène	3	16,4	20,9	18,4	16,8
Butadiène	1,8	2.9	3,0	3,3	4,8
Autres C4	1,4	1,3	12,4	10,8	6,0
Benzène	0,8	2,4	2,5	2,4	4,2
Essence	0,7	2,3	5,5	4,8	18,9
Résidu	0,2	2,2	0,3	0,1	2,8
Pertes	3,0	2,5	2,5	2,5	2,5
	100,0	100,0	100,0	100,0	100,0

Nota: Pour tous les rendements l'éthane est craqué jusqu'à extinction.
Le propylène est une coupe à 90 % de propylène.
Le butane craqué est constitué de 30 % d'isobutane et 70 % de n butane.
Le naphta est une coupe totale.

Le produit principal est l Ethylène, et les autres produits (propylène, butadiène, benzène, etc. . .) sont des coproduits. Pour 100 t d'éthylène, on a en tonnes =

Charge	Ethane	Propane	Butane	n Butane	Naphta
Ethylène	100	100	100	100	100
Propylène	4	40	67	49	56
Butadiène	2	7	10	9	16
Benzène	1	6	8	6	14

Ce tableau montre que l'éthane ne fournit pratiquement que de l'éthylène.

Pour les LPG, il y aura par rapport au naphta un déficit de propylène et surtout une production de butadiène réduite de moitié.

Comme actuellement, la sévérité du vapocraquage sera ajustée pour modifier le rapport ethylène sur propylène et rapprocher la production du débouché. Le déficit en propylène, s'il existe, pourra être comblé à partir d'autres sources comme le craquage catalytique qui se développe rapidement en Europe.

Le déficit en butadiène sera plus difficile à compenser.

Le vapocraquage de charges plus lourdes pourra y contribuer. Si la situation l'impose, le butadiène pourra être produit comme aux USA par déshydrogénation des butènes.

Le déficit en aromatiques contribuera à assainir le marché actuel.

EMPLOI DES CHARGES LEGERES DANS LES INSTALLATIONS EXISTANTES

Aspect Matériel

Les vapocraqueurs modernes utilisent une dizaine de fours qui peuvent être équipés pour recevoir des charges différentes. Ils sont capables, moyennant de petites modifications, de traiter correctement des LPG. Le fonctionnement du reste de l'installation ne sera pas fortement perturbé si les LPG constituent une fraction d'environ 20 % de la charge. Il pourra en résulter cependant dans certains cas, une légère diminution de la capacité de production d'éthylène.

Le traitement partiel du propane impose la purification du propylène produit dont la pureté est diminuée par suite de la dilution avec du propane non converti.

Le traitement de l'éthane issu du craquage du naphta dans des fours spéciaux est déjà une pratique courante dans les unités actuelles. Il pourrait se développer en traitant l'éthane en injection à 20 ou 30 % dans les fours de naphta.

Dans tous les cas, la suppression totale de la charge naphta n'est pas possible, mais une substitution partielle au niveau de 20 %, ne devrait pas poser de problèmes de fonctionnement dans la plupart des cas.

L'obstacle matériel le plus sérieux provient du transport et de la mise en place des charges légères. Les unités existantes se sont en général rapprochées des raffineries sources de naphta et rares sont les unités proches des lieux de production de LPG et qui pourront être ravitaillées par conduites. Dans la plupart des cas, il faudra installer d'importants stockages au départ et à l'arrivée pour permettre l'utilisation de moyens de transport économiques. Chaque cas est à étudier en particulier.

Dans la suite de l'étude, nous ne tiendrons pas compte des modifications à apporter aux installations ni des frais de mise en place des charges qui sont différentes dans chaque cas.

Aspect Economique

La comparaison des frais de traitement de différentes charges dans une installation existante ne fait intervenir ni les frais fixes ni les frais de capital. Seuls comptent les frais suivants =

(a) frais de combustible et d'utilités (vapeur, électricité etc. . .);
(b) frais de matières de charge;
(c) valorisation des coproduits obtenus.

Les frais de combustible et d'utilités ont été exprimés en combustible à 10000 th/t (C).

La valorisation des coproduits obtenus a été calculée en prenant des valeurs proches des prix de contrats européens actuels des différents produits. L'essence produite a été rattachée à la valeur du supercarburant européen (S).

Tous ces frais et rémunérations ont été exprimés pour 1t d'éthylène dans le tableau 2.

TABLEAU 2

Installations existantes, Frais en francs par tonne d'éthylene

	Ethane	Propane	Butane	n Butane	Naphta
Frais proportionnels					
Combustible et utilités	1,07 C	1,15 C	1,38 C	1,13 C	1, 3 C
Charge	2,11 E	2,44 P	3,20 B	2,66 B_o	3,35 N
Valeur des coproduits	- 104	- 585	- 977	- 762	- 1043
	- 0,258 C	- 1,01 C	- 1,24 C	- 0,936	- 0,865 C
	- 0,009 S	- 0,054 S	- 0,173 S	- 0,125 S	- 0,617 S
	- 104	- 585	- 977	-762	- 1043
Total frais proportionnels	0,812 C	0,14 C	0,14 C	0,194 C	0,435 C
	- 0,009 S	- 0,054 S	- 0,173 S	- 0,125 S	- 0,617 S
	1,33 E	2,44 P	3,20 B	2,66 B_o	3,35 N

C Valeur d'une tonne de combustible à 10 000 th/h.
E Valeur d'une tonne d'éthane.
P Valeur d'une tonne de propane.
B Valeur d'une tonne de butane (70 % de n C4 et 30 % de i C_4).
B_o Valeur d'une tonne de normal butane.
N Valeur d'une tonne de naphta.
S Valeur d'une tonne de supercarburant.

Pour obtenir la valeur équivalente de chaque charge par rapport au naphta, il suffit d'écrire que les frais pour l'obtention d'une tonne d'éthylène, sont les mêmes à partir de naphta et d'une autre charge.

C'est ce qui a été fait au tableau 3 où une application numérique donnant des valeurs au combustible, au supercarburant et au naphta, permet d'attribuer une valeur à chaque charge lénère.

Le n butane est la charge la plus attrayante avec une valeur un peu inférieure à celle du naphta.

Le propane n'est pas très intétessant et l'éthane dépasse à peine sa valeur combustible qui est de l'ordre de 450 F/t avec nos hypothèses.

EMPLOI DES CHARGES LEGERES DANS DES INSTALLATIONS NEUVES

Dans ce cas, l'installation est optimisée pour le traitement d'une charge particulière.

L'investissement en MF pour une unité de 450 Kt/an d'éthylène, est décomposé ci-dessous en 2 termes: l'installation proprement dite, les installations annexes (Infrastructure, chaudières, réfrigération, stockages, moyens d'expédition, etc. . .).

	Ethane	Propane	Butane	n Butane	Naphta
Installation	658	762	925	830	963
Annexes	450	450	544	544	544
Total	1108	1212	1469	1377	1507

Les frais fixes de personnel, d'entretien, les frais généraux et les charges de capital ont été évalués dans le tableau 4 où ils ont été ajoutés aux frais proportionnels pour obtenir les frais totaux.

Le tableau 5 permet comme plus haut de définir des valeurs d'équivalence par rapport au naphta pour les différentes charges légères.

Comme on pouvait s'y attendre, la comparaison des tableaux 3 et 5 montre que les charges légères ont une meilleure valorisation lorsque les installations de traitement sont spécialement étudiées pour elles.

On note en particulier que le n butane est équivalent au naphta ainsi que le propane.

CONCLUSION

Cette étude apporte des résultats préliminaires qu'il faut utiliser avec prudence. Les principaux sujets d'incertitude sont:

- les valeurs adoptées pour les produits chimiques et pétroliers qui doivent être adaptés dans le temps à chaque cas particulier;
- les rendements obtenus avec les différentes charges qui, dans notre étude représentent des moyennes, mais qui peuvent varier avec la qualité de la charge (naphta), la configuration des fours et surtout la sévérite de l'opération.

Pour une opération précise, il serait nécessaire de faire une étude particulière.

TABLEAU 3

Installations existantes. Valeur equivalente de chaque charge par rapport au naphta

	Différence P. R. au Naphta				Valeur F/T	Valeurs numériques			
	Ethane	Propane	Butane	n Butane		Ethane	Propane	Butane	n Butane
Frais	- 939	- 458	- 66	- 281		- 939	- 458	- 66	- 281
	- 0,377 C	+ 0,295 C	+ 0,295 C	+ 0,241 C	400	- 151	+ 118	+ 118	+ 96.5
	- 0,608 S	- 0,563 S	- 0,444 S	- 0,492 S	680	- 413	- 383	- 302	- 335
Proportionnels	+ 3,35 N	+ 3,35 N	+ 3,35 N	+ 3,35 N	630	+ 2110	+ 2110	+ 2110	+ 2110
	1,33 E	2,44 S	3,20 B	2,66 B_o		607	1387	1860	1590,5
	E	P	B	B_o		456	569	581	598

Mêmes conventions que pour le tableau 2.

TABLEAU 4

Installations neuves. Frais en francs par tonne d'éthylene

	Ethane	Propane	Butane	n Butane	Naphta
Total frais proportionnels	- 104	- 585	- 977	- 762	- 1043
	0,812 C	0,14 C	0,14 C	0,194 C	0,435 C
	- 0,009 S	- 0,054 S	- 0,173 S	- 0,125 S	- 0,617 S
	1,33 E	2,44 P	3,20 B	2,66 B_o	3,35 N
Frais fixes					
personnel	11	11	11	11	13
Entretien	78	88	106	98	110
Frais généraux	116	129	152	142	160
Charges de capital (20 %)	480	539	652	611	669
Total	685	767	921	862	952
Frais totaux	581	182	- 56	100	- 91
	0,812 C	0,14 C	0,14 C	0,194 C	0,435 C
	- 0,009 S	- 0,054 S	- 0,173 S	- 0,125 S	- 0,617 S
	1,33 E	2,44 P	3,20 B	2,66 B	3,35 N

Mêmes conventions qu'au tableau 2.

Les besoins du marché d'éthylène n'imposent pas dans un proche avenir la construction de nouvelles installations.

La construction d'une installation spécifique pour un produit ne peut s'envisager que lorsque l'approvisionnement de la charge est assuré avec certitude pour une période longue d'environ 15 ans.

Certaines installations de ce type sont prévues en liaison avec les nouvelles sources de Mer du Nord:

- Noretyl à Rafnes (Norvège), projet ESSO à Mossmoran (Ecosse), Unité ICI/BP à Wilton (UK) pour la dernière installation, il existe une flexibilité entre n butane et naphta, et il semble que cette solution soit possible, sans augmenter sensiblement l'investissement.

Le traitement de LPG dans des vapocraqueurs existants est une solution un peu moins intéressante mais plus réaliste. Les vapocraqueurs sont en général liés à une raffinerie pour leurs approvisionnements en naphta. Au cas où des LPG provenant des champs de brut et de gaz viendraient se substituer au débouché de la raffinerie, celle-ci disposerait alors de sa propre production qui pourrait régulièrement, sans grands frais, être livrée aux vapocraqueurs. Le seul critère important est le prix des LPG qui conditionne l'intérèt de l'opération.

Parmi les charges étudiées, la meilleure est le n butane qui craque facilement avec un rendement élevé en éthylène et sans réduire la capacité de l'installation.

Tableau 5

Installations neuves. Valeur equivalente de chaque charge par rapport au naphta

	Différence P. R. au Naphta				Valeur F/t	Valeurs numériques			
	Ethane	Propane	Butane	N butane		Ethane	Propane	Butane	n Butane
Frais	- 672	- 273	- 35	- 191		- 672	- 273	- 35	- 191
	- 377C C	+ 0,295 C	+ 0,295 C	+ 0,241 C	400	- 151	+ 118	+ 118	+ 96,5
	- 0,608 S	- 0,563 S	- 0,444 S	- 0,492 S	680	- 413	- 383	- 302	- 335
totaux	+ 3,35 N	+ 3,35 N	+ 3,35 N	+ 3,35 N	630	+ 2110	+ 2110	+ 2110	+ 2110
	1,33 E	2,44 P	3,20 B	2,66 B_o		874	1572	1891	1680,5
	E	P	B	B_o		657	644	591	632

Mêmes conventions que pour le tableau 2.

L'éthane, bien valorisé dans une installation spécialement conçue est crédité dans le vapocraquage d'une valeur équivalente à sa valeur combustible. Il présente par ailleurs l'inconvénient de ne produire que de l'éthylène.

Le propane demande une modification importante pour la purification du propylène.

Le butane qui est un mélange de normal butane et d'isobutane doit être séparé en ses constituants pour être bien valorisé. Le n butane pourrait ainsi être destiné au vapocraquage et l'isobutane à l'alkylation. La mise en place d'alkylation en Europe devrait permettre une large utilisation de ce schéma.

Heating and Air Conditioning using Solar Energy with an Auxiliary Gas-fired System for the Sophia Antipolis Experimental House

P. Lessieur and J. R. Fellous

France

SUMMARY

The Total Compagnie Française des Pétroles Group has for several years been conducting various studies and experiments on the use of solar energy in housing for:

production of hot water for household use;
heating;
refrigeration.

As the entire heating and cooling requirements of a building cannot, at the present stage, be satisfied in an economic manner by the use of solar energy, an auxiliary heating system using traditional sources of energy such as gas, fuel oil or electricity must be provided.

In addition to studies of solar energy installations proper, research has also been done in various areas which are closely related to the basic programme, namely, on:

the principles governing the integration of traditional forms of energy in a solar energy system and the various constraints inherent in each type of energy; and
the extent to which the specific problems of the auxiliary energy system influence the automatic regulation system, inevitable losses and the inflexibility of the system.

Use of a two-way heat pump system (heat pump with an absorption machine):

the integration of an auxiliary heating system in such solar energy systems must be carried out in a manner that satisfies the following requirements:

immediate pull down;
rapid adjustment of heat production to the actual load;
low-capacity operation without impairment of efficiency;
low consumption of electric power by the burners and draught fans;
limitation of inevitable losses;

ability to operate at low temperatures in winter and high temperatures in summer in relation to the enthalpy values necessary for the operation of the absorption machines for cooling or as heat pumps;

all of which have been identified as a result of studies based on the use of a gaseous fuel.

It seems that an auxiliary energy system using gas is highly suited to solar applications from the economic as well as the technical viewpoint.

The experiment now proceeding with the Sophia Antipolis experimental house will enable our research to be extended later on and provide valuable insight into the technology of heat pumps based on solar energy and gas.

Chauffage et Climatisation à l'Energie Solaire avec Appoint Gaz pour la Maison Expérimentale CLISOTOS à Sophia Antipolis (Nice)*

P. Lessieur et J. R. Fellous

France

La maison pilote CLISOTOS est la concrétisation des études conduites par le groupe TOTAL pour l'utilisation de l'énergie solaire dans l'habitat: chauffage, refroidissement et production de l'eau chaude sanitaire.

Le but de l'opération est non seulement la réalisation des systèmes de climatisation ayant comme source énergétique prioritaire l'énergie solaire, mais aussi l'analyse, en détail, des meilleures conditions d'utilisation de l'énergie d'appoint, le gaz dans ce cas particulier.

Pour atteindre cet objectif les études n'ont pas été limitées à l'installation des équipements pour la production du chaud et du froid, mais elles ont été élargies à tous les moyens actifs et passifs dont on peut disposer pour créer un confort thermique dans un bâtiment: structure, fenêtres, isolation, protection solaire, température des fluides etc. . . .

Il a été constaté, en effet, qu'un choix bien adapté de la structure et de la conception architecturale permettait de réduire sensiblement le coût global de l'ensemble.

La maison CLISOTOS a été donc conçue de façon à combiner harmonieusement les méthodes passives et actives de climatisation solaire, chaque paramètre étant optimisé en fonction des conditions d'utilisation et des données climatiques de la région.

La solution choisie devait satisfaire les exigences suivantes:

- utiliser à l'aval de la production d'énergie thermofrigorifique des équipements de grande série traditionnels;
- utiliser le système dans différents climats sans changer les principes de fonctionnement et la technologie des équipements;
- gérer convenablement l'énergie par un système de régulation à micro processeur avec possibilité d'optimisation pour chaque climat;
- utiliser des capteurs plans, avec un fluide à une température maximale de 90-95° C en été et 30-40° en hiver.

---

*CLISOTOS: CLImatisation SOlaire TOtal — Sofretès.

DESCRIPTION DE LA MAISON PILOTE CLISOTOS

Il s'agit d'une villa d'environ 200 m^2 utiles dont l'architecture générale est caractérisée par l'utilisation de deux éléments de forme triangulaire. Un triangle est utilisé comme habitation tandis que l'autre est utilisé comme bureau. Les figs. 1, 2 et 3 donnent une idée de la maison.

Fig. 1. La maison solaire avec le toit capteur.

Le système de captation de l'énergie solaire est intégré dans l'élément de toiture incliné à environ 20° qui constitue le toit du local technique. La maison est réalisée avec une structure lourde et très bien isolée:

K = 0,25 kcal/hm^2 ^oC pour les parois (isolation extérieure),
K = 1,8 kcal/hm^2 ^oC pour le vitrage.

La fig. 4 montre l'évolution de l'amortissement du flux et de la température en fonction de l'emplacement de l'isolation.

CLIMATISATION ET CHAUFFAGE

Les systèmes de climatisation et de chauffage sont indépendants pour chaque module triangulaire et de ce fait il est possible de conduire simultanément différentes expériences.

Le système de climatisation est réalisé par ventilo-convecteurs. L'air de renouvellement est traité centralement et chaque pièce a un thermostat pour la régulation de la température.

Le système de chauffage de base est constitué par des serpentins noyés dans la dalle

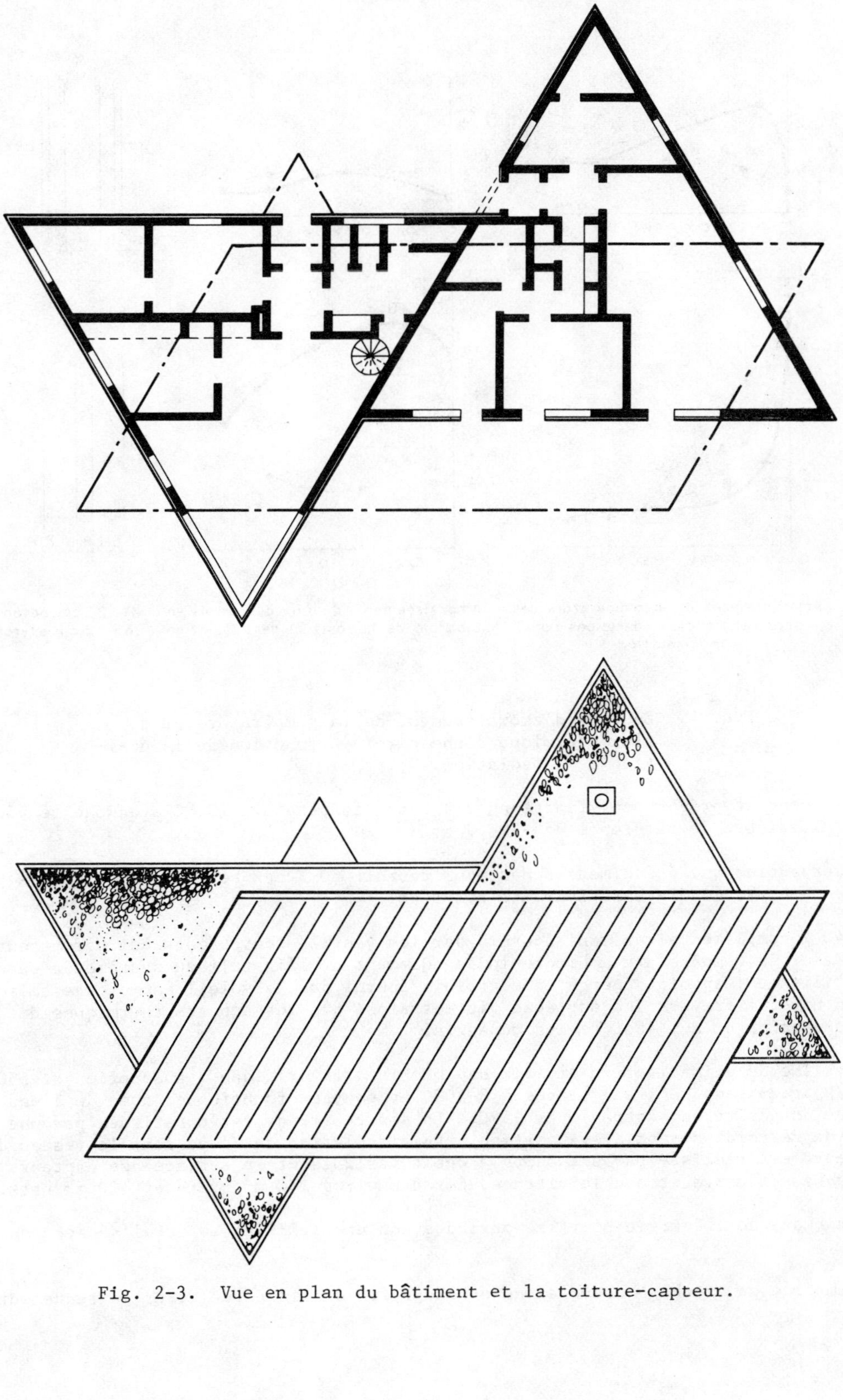

Fig. 2-3. Vue en plan du bâtiment et la toiture-capteur.

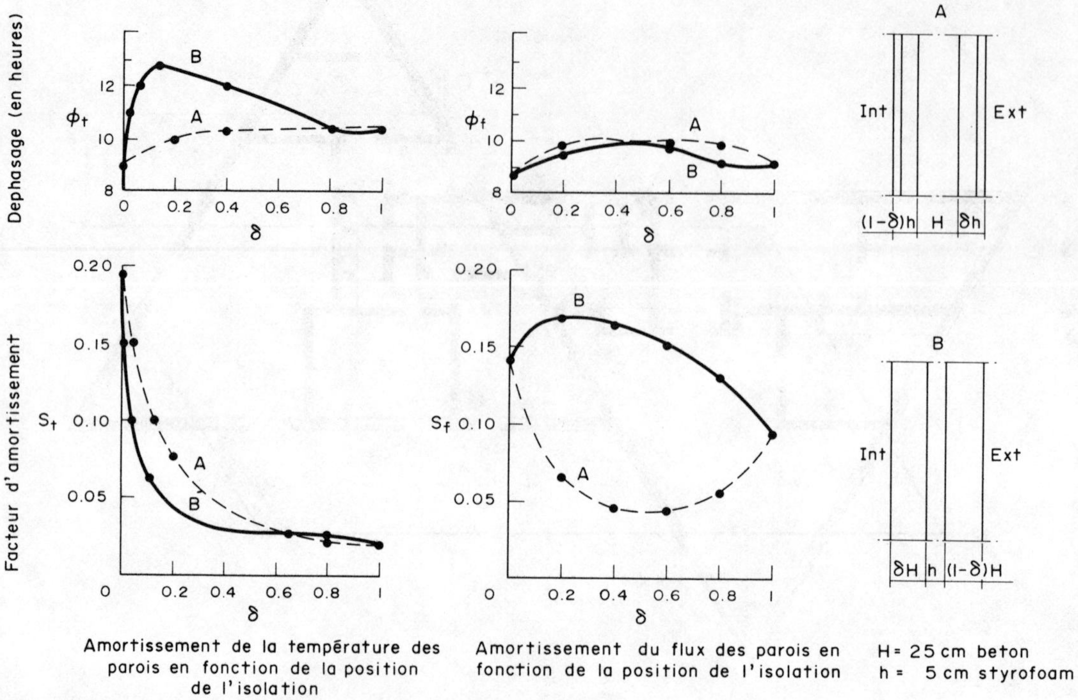

Fig. 4. Courbes d'amortissement de la température et du flux thermique d'une paroi en fonction de la position de l'isolation.

du plancher et prévus pour fonctionner à basse température (30°C avec une température extérieure de référence de -2°C).

Ces serpentins ont été dimensionnés pour fournir 60 % des besoins calorifiques pour la température minima extérieure de référence.

Les 40 % complémentaires sont fournis par les ventilo-convecteurs qui permettent de régler la température ambiante indépendamment de l'inertie du système de base, en utilisant l'appoint énergétique gratuit fourni par le soleil en moyenne saison et en hiver à travers les surfaces vitrées ainsi que les apports thermiques dus aux occupants et aux équipements ménagers.

La système de climatisation emploie une machine frigorifique à absorption de 5500 frig/h produisant de l'eau glacée à 8-10°C et pouvant fonctionner avec de l'eau chaude au générateur entre 70 et 95°C. La source froide est constituée par une tour de refroidissement à évaporation. La circulation de l'eau dans le réseau de capteurs est réalisée par une pompe à débit variable et la surface des capteurs, à couche sélective et double vitrage, est d'environ 140 m^2 bruts et 120 m^2 nets.

Le stockage de l'énergie pour les périodes non ensoleillées est réalisé avec de l'eau.

Il comporte un stockage chaud et un stockage froid afin d'améliorer le rendement de la machine frigorifique en évitant les pertes énergétiques résultant d'un nombre important d'arrêts et de redémarrages.

Le principe de fonctionnement du stockage chaud est défini dans les figs. 5, 6, 7, 8, 9.

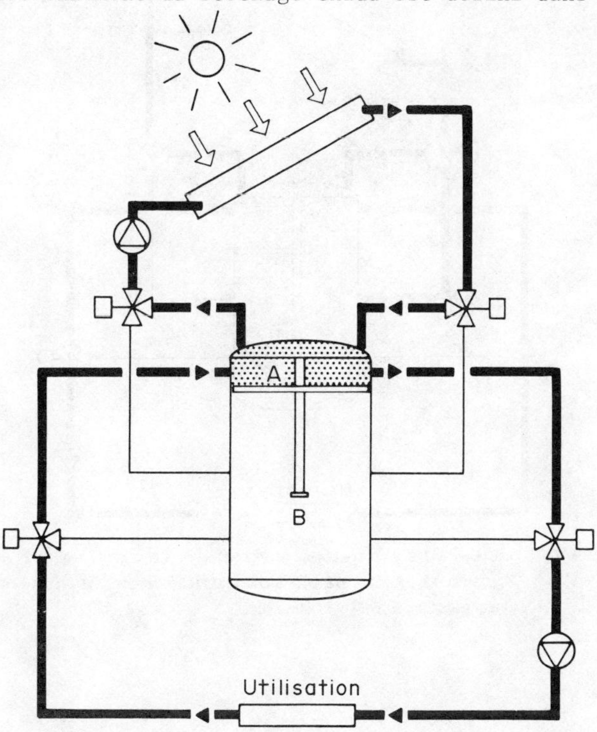

PHASE 1 : DEMARRAGE : L'ENERGIE SOLAIRE COLLECTEE EST EGALE OU TRES PEU SUPERIEURE AUX BESOINS DE L'UTILISATION. LE RESERVOIR "A" EST A SON VOLUME MINIMAL.

Figs. 5-6-7-8-9. Stockage de l'eau chaude système TOTAL-CFP pour différents cycles de fonctionnement.

Ce dispositif est nécessaire ici, car la climatisation fonctionne 24/24 h pour le module habitation et 16/24 h dans le module bureaux et en conséquence la variation du bilan thermofrigorifique est très importante et risque de pénaliser considérablement les performances du système.

L'indispensable appoint énergétique est réalisé par une chaudière à gaz dont le régime d'exploitation a été adapté aux problèmes spécifiques.

L'eau chaude sanitaire est obtenue par un système utilisant l'énergie solaire et l'éventuel appoint est réalisé par chauffage instantané au gaz (chaudière mixte).

La fig. 10 représente le schéma de principe de l'installation.

PRODUCTION DE CHALEUR

La production d'énergie thermique est réalisée par deux systèmes:

1. les capteurs solaires,
2. la chaudière d'appoint au gaz.

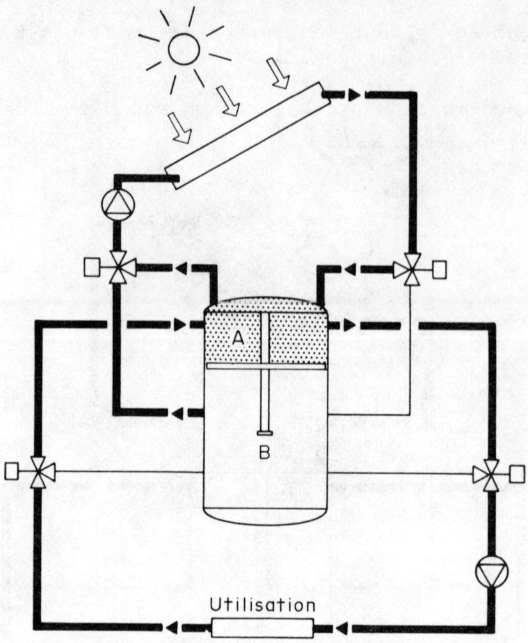

PHASE 2 : L'ENERGIE SOLAIRE COLLECTEE PAR LE CAPTEUR EST SUPERIEURE AUX BESOINS DE L'UTILISATION. LE VOLUME DU RESERVOIR "A" AUGMENTE.

Fig. 6.

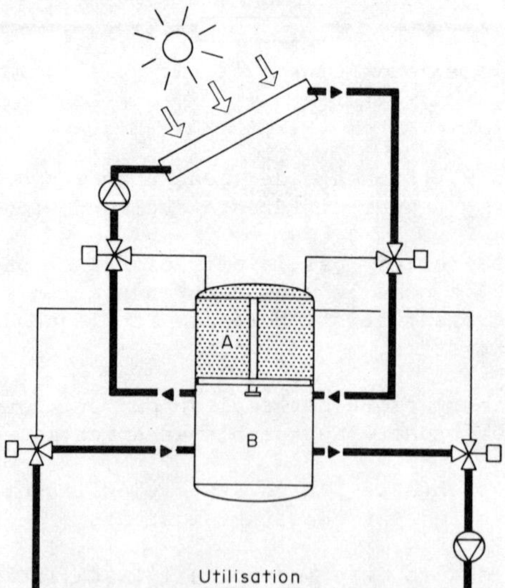

PHASE 3 : LE RESERVOIR "A" A ATTEINT SON VOLUME MAXI ET IL EST DONC HORS SERVICE.
L'ACCUMULATION ET L'UTILISATION DE L' ENERGIE SOLAIRE SONT EFFECTUEES AVEC LE RESERVOIR "B".

Fig. 7.

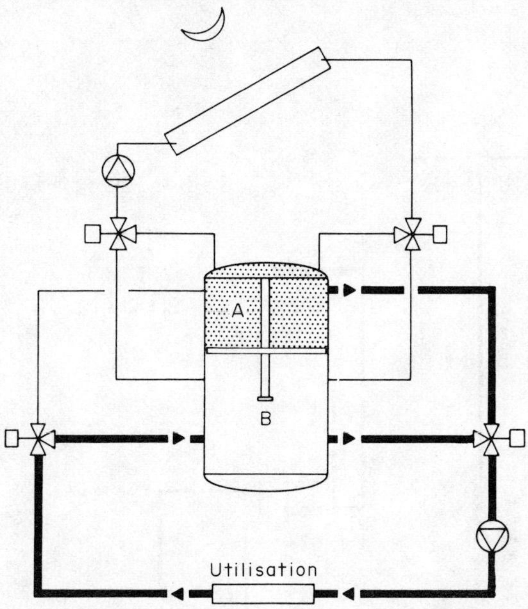

PHASE 4 : ABSENCE DE SOLEIL. LA TEMPERATURE DU FLUIDE DANS LE RESERVOIR "B" EST INSUFFISANTE. IL Y A MELANGE ENTRE LE FLUIDE PROVENANT DE "A" ET CELUI QUI PROVIENT DE "B".

LE FLUIDE RETOURNE, APRES UTILISATION EN "B".

Fig. 8.

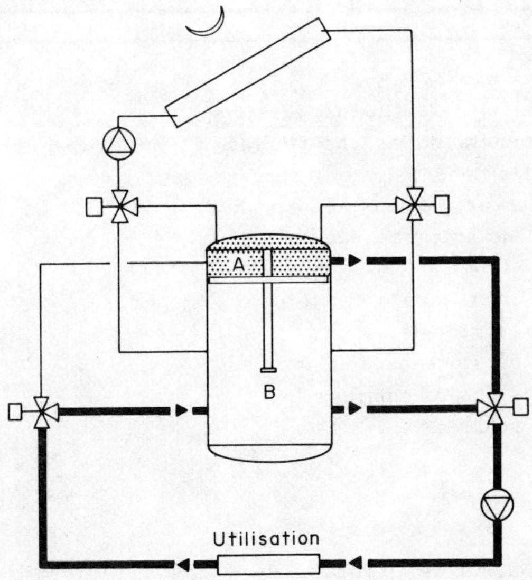

PHASE 5 : ABSENCE DE SOLEIL, FIN DU CYCLE.

LE VOLUME DU RESERVOIR "A" EST MINIMAL, SA TEMPERATURE PRATIQUEMENT CONSTANTE. LE RESERVOIR "B" AURA ATTEINT SON VOLUME MAXI ET SA TEMPERATURE MINIMA.

Fig. 9.

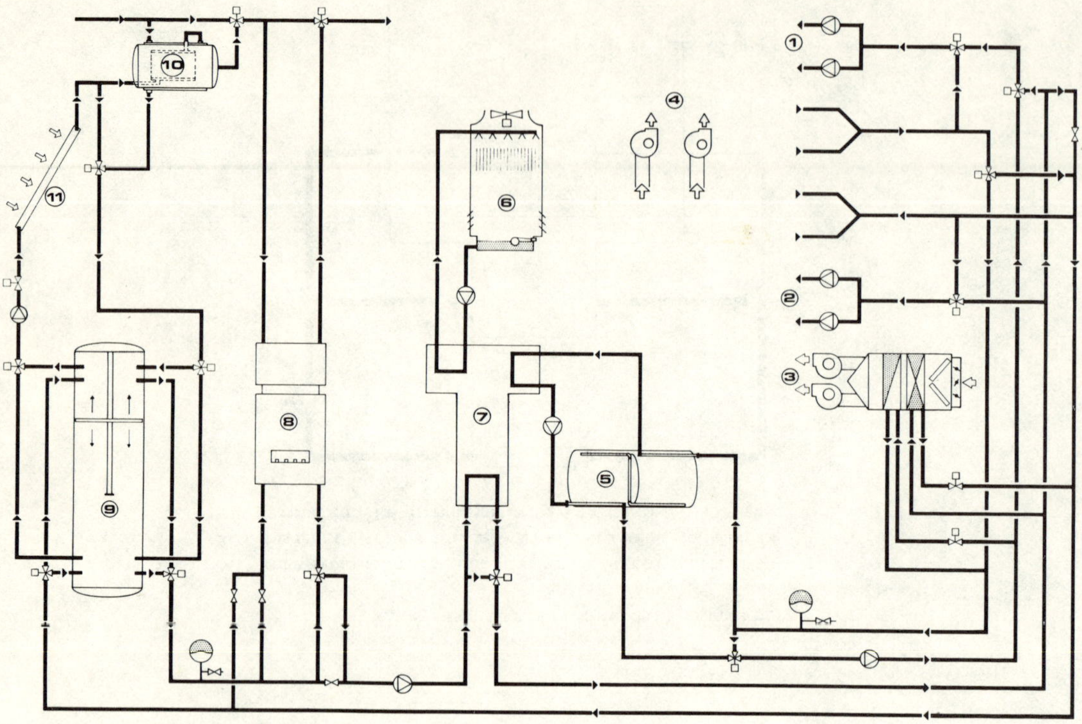

1 : Pompes circuits ventilo-convecteurs
2 : Pompes circuits dalles chauffantes
3 : Caisson traitement air de renouvellement
4 : Ventilateur extraction air vicie
5 : Reservoir de stockage eau glacée
6 : Tour de refroidissement
7 : Machine frigorifique à absorbtion
8 : Chaudière d'appoint au gaz
9 : Reservoir de stockage eau chaude
10 : Ballon eau chaude sanitaire
11 : Capteurs solaires

Fig. 10. Schéma de principe de l'installation.

CAPTEURS SOLAIRES

Le rendement d'un système solaire de chauffage et climatisation doit nécessairement être optimisé en fonction des caractéristiques des capteurs solaires, de leur surface et du cycle thermique considéré.

Au problème purement technique il faut ajouter les contraintes économiques car s'il est possible de chauffer ou refroidir une maison avec un rendement solaire de 100 % il n'est pas logique d'installer une surface de captation telle qu'une grande partie de cette dernière soit utilisée d'une façon marginale, donc sans aucune possibilité d'amortissement. Il est judicieux de réduire légèrement la surface de captation et d'utiliser un système d'appoint traditionnel.

Les études des différents systèmes de climatisation ont montré que le rendement d'un cycle thermodynamique à source froide fixée est croissant en fonction de la température disponible à la source chaude (jusqu'à 95°C) alors que le rendement du capteur solaire est décroissant en fonction de la température moyenne de fonctionnement.

Il faut donc sélectionner un système pour atteindre une performance maximum pour une combinaison "capteur/système de production de froid", en été (ou bien capteur/système de chauffage en hiver).

Dans ce cas la température optimale de fonctionnement est la température minimale qui permet l'activation de la machine de réfrigération ou le chauffage de l'ambiante. On produit donc en été de l'eau à 85/90°C pour les besoins de refroidissement pendant qu'en hiver le niveau de température est réduit à 35/40°C. Ceci permet de faire varier la température de fonctionnement des capteurs en fonction aux paramètres climatiques (rayonnement, température) et maintenir ainsi un rendement pratiquement constant du système de captation.

Les collecteurs solaires choisis sont des collecteurs Sofretès avec absorbeur en cuivre à tube et ailettes serties dont la surface sélective au noir de chrome (Black Chrome) a été mise au point par le Commissariat de l'Energie Atomique.

Les vitrages sont doubles et les collecteurs ont une longueur unitaire d'environ 9,60 m pour une largeur de 0,75 m.

Ils sont caractérisés par une très faible contenance en eau (0,5/m^2) et par un rendement élevé pour des températures de 85°C à 90°C.

La courbe de rendement de ces capteurs est présentée dans la fig. 11.

CHAUDIÈRE À GAZ

Une chaudière d'appoint pour système solaire doit répondre aux besoins spécifiques de l'utilisation:

- mise en régime instantanée;
- adaptation rapide de la puissance thermique à la demande;
- fonctionnement à puissance réduite sans dégradation du rendement;
- faible consommation électrique pour l'alimentation des accessoires (ventilateurs de tirage etc. . . .);
- limitation des pertes de toutes natures.

Les chaudières comportant une réserve d'eau consomment une quantité importante de combustible pour leur maintien en température et celà indépendamment de la puissance demandée.

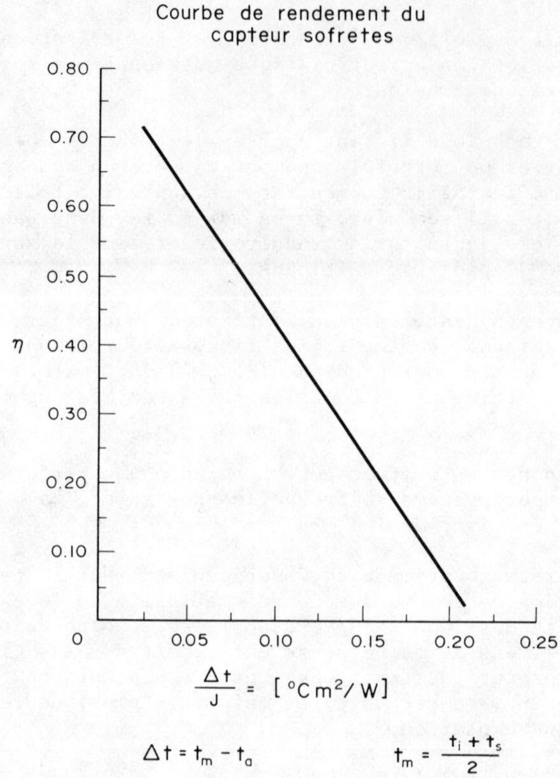

Fig. 11. Courbe de rendement du capteur SOFRETES à surface sélective.

Aussi, la chaudière et le circuit hydraulique doivent être choisis et calculés de façon à limiter au maximum la capacité thermique, mais tout en évitant de pénaliser le bon fonctionnement des appareils par de trop fréquents cycles marche-arrêt.

Ces impératifs ont pu être respectés par l'emploi d'une chaudière murale étanche deux services alimentée au gaz, avec brûleur deux allures et évacuation des produits de combustion par extracteur et micro-ventouse.

Les températures de fonctionnement doivent pouvoir varier dans une large plage. En effet, pendant l'été, la température du fluide chaud doit être comprise entre 80 et 95°C pour permettre d'activer le cycle à absorption, générateur d'eau glacée. Au contraire, en régime d'hiver, pour le chauffage, ces niveaux de température ne sont pas compatibles avec un chauffage solaire car ils sont la cause d'un gaspillage inutile d'énergie puisqu'il faut dégrader le niveau d'enthalpie de fluide avant de l'utiliser. Des températures de fluide, de l'ordre de 35°C sont suffisantes dans les systèmes dimensionnés pour l'utilisation des apports solaires.

Ce niveau de température permet d'éliminer pratiquement toutes les pertes en canalisations et améliore considérablement le contrôle des conditions de confort, évitant des phénomènes d'instabilité dans les équipements de régulation.

Pour les petites chaudières à gaz murales, les pertes d'énergie les plus importantes sont dues au tirage qui se génère entre la prise d'air extérieur de ventilation et la cheminée, en passant dans le corps de chauffe. Ces pertes sont proportionnellement plus importantes lorsque l'appoint énergétique est inférieur à 25 ou 30 % des besoins calorifiques totaux.

Pour tenir compte de ces considérations, la chaudière à gaz type mural deux services installée dans la maison CLISOTOS a été choisie pour ses caractéristiques qui sont les suivantes:

- pas de flamme de veilleuse, celle-ci étant remplacée par un système avec allumage instantané automatique;
- deux thermostats de réglage avec inversion été/hiver: en été la température de sortie est réglée à 90°C et en hiver 45°C;
- production instantanée de l'eau chaude sanitaire à 45°C;
- évacuation des produits de combustion par extracteur et système à ventouse, l'extracteur ne fonctionnant que lorsque le brûleur est allumé;
- régulation de la chaudière programmée à partir du système de gestion de l'énergie en fonction des différents paramètres mesurés et introduits dans le système de réglage.

PRODUCTION DE FROID

L'étude des différents systèmes possibles nous avait permis de constater qu'en l'état actuel de la technologie seul un système de réfrigération à absorption peut répondre favorablement à toutes les contraintes pour ce type d'application.

En effet, l'utilisation d'une machine à absorption permet d'obtenir les avantages suivants:

- possibilité d'adapter facilement des machines du commerce disponibles sur le marché;
- entretien très simple ou quasi nul;
- fonctionnement avec des capteurs à basse température avec possibilité d'améliorer le système en les changeant dès que d'autres capteurs plus performants seront disponibles;
- pas de problèmes de stockage à haute température;
- COP (à égalité de température à la source chaude) meilleur que tout autre système (jusqu'à environ 150°C).
- la machine à absorption est une machine statique donc sûre et de longue vie;
- facilité de réglage de la puissance produite;
- coût du système de production de froid très intéressant;
- machine sans nuisance acoustique;
- possibilité de fonctionner en cas d'absence solaire avec de l'eau chaude à basse température (75-90°C) produite par une chaudière murale à gaz.

La photo de la fig. 12 représente la machine frigorifique à absorption alors que les figs. 13, 14, 15 sont relatives à ces caractéristiques de fonctionnement.

BOUCLES D'ESSAIS ET SIMULATION DES DIFFÉRENTS PHÉNOMÈNES

Pour analyser le comportement des machines frigorifiques à absorption en régime variable, il a été réalisé une boucle d'essais permettant (voir figs. 16 et 17) une simulation réelle des différents paramètres qui caractérisent un système solaire. La fig. 18 représente le schéma de principe de la boucle d'essais.

Les besoins frigorifiques du bâtiment, compte tenu de son inertie etc. . . sont

Fig. 12.

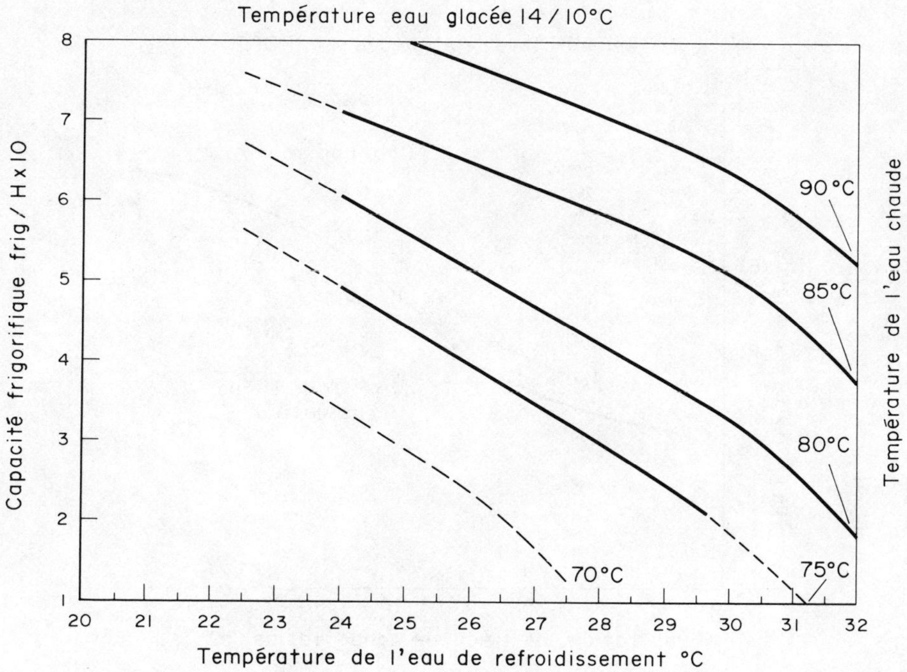

Fig. 13-14-15. Courbes caractéristiques de la machine frigorifique à absorption.

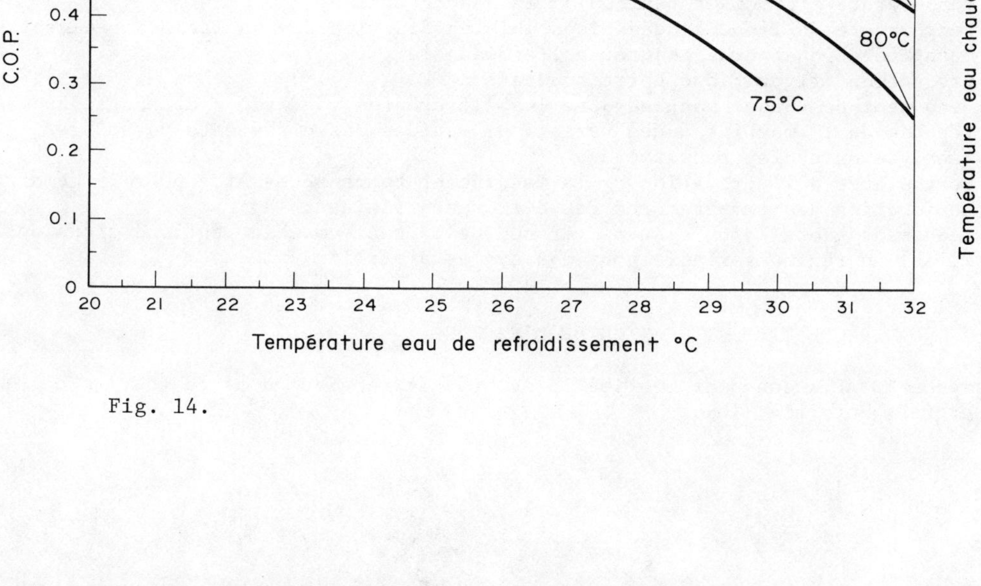

Fig. 14.

K

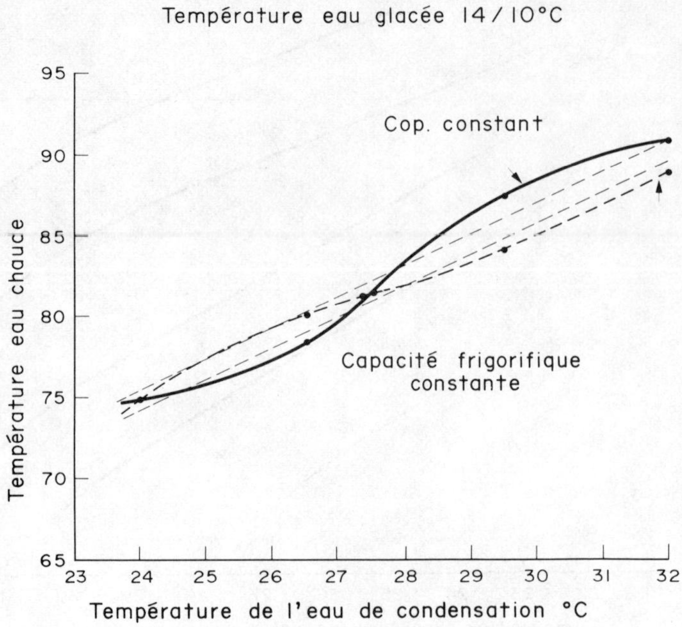

Fig. 15.

calculés d'abord théoriquement et la courbe de charge obtenue permet de définir, heure par heure, la température de retour de l'eau glacée (simulation des appareils de climatisation). La température de la source chaude est variable pour simuler la production thermique des capteurs solaires ou de l'eau stockée dans un réservoir. De même la température de la source froide peut simuler la température (variable) de l'eau de condensation sortant de la tour de refroidissement et dépendant des conditions extérieures.

Ces essais ont permis d'établir pour chaque système les paramètres suivants:

- température minimum et maximum de la source chaude;
- température minimum et maximum de la source froide;
- température de l'eau glacée disponible en fonction des variables précédentes;
- quantité de chaleur à rejeter à l'atmosphère;
- production frigorifique correspondante;
- rendement de chaque machine dans les différentes conditions de fonctionnement;
- réponse de la machine à des variations rapides des différents paramètres (analyse du cycle transitoire).
- pertes dues à l'activation de la machine et temps nécessaire pour le démarrage en fonction des températures des différents fluides;
- rendement global d'un système (et non de la seule machine en conditions optimales) en régime variable pour des cycles à établir.

RÉGULATION ET GESTION DE L'ÉNERGIE

Le groupe TOTAL a conçu et réalisé un système de régulation et de gestion de l'énergie (photo fig. 19).

Figs. 16-17-17A. Boucle d'essais et simulation.

Fig. 17.

Fig. 17A.

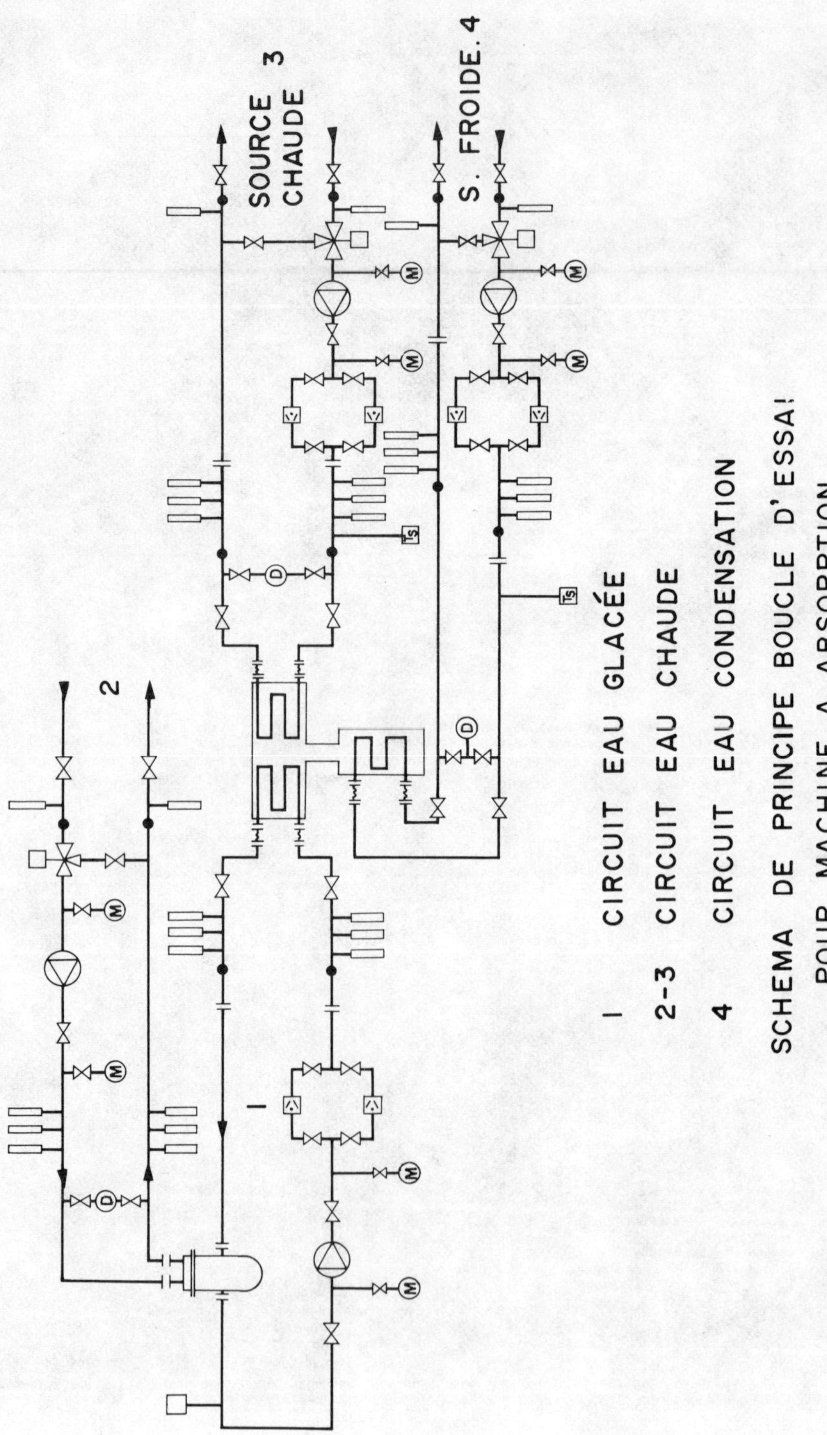

Fig. 18. Schéma boucle d'essais et simulation.

Fig. 19. Système de gestion de l'énergie TOTAL-CFP à circuits intégrés.

Ce système de type électronique à circuit imprimé ("Solid State") permet par l'utilisation d'un certain nombre de circuits imprimés avec des microprocesseurs, de faire fonctionner le système de climatisation et de production d'eau chaude sanitaire sans intervention de l'utilisateur et selon des programmes de gestion et d'optimisation de l'énergie évitant tout gaspillage.

La composition des circuits imprimés et des microprocesseurs dépend bien sûr des divers paramètres étudiés précédemment:

- conditions climatiques du lieu;
- caractéristiques physiques du bâtiment
  - inertie - transmission, etc.;
- caractéristiques de l'installation
  - rendement de la machine frigorifique,
  - rendement des capteurs solaires etc. . . .

L'évolution des conditions climatiques ainsi que des conditions de stockage et d'utlisation de l'énergie peuvent corriger automatiquement les signaux sortant du système de régulation pour tenir compte des conditions optimales de fonctionnement.

PROGRAMME DES ESSAIS

La réalisation de la maison CLISOTOS permet de:

- vérifier les études théoriques;
- tester les divers composants en site réel;
- mettre au point les systèmes de gestion et de stockage de l'énergie conçus à l'occasion du programme.

Le système de mesure utilise l'appareillage suivant:

- une station météorologique complète (mesure du rayonnement solaire direct et diffus, vitesse du vent, température, humidité) avec enregistrement sur bande magnétique;
- mesure des différents paramètres de parois, conditions ambiantes, consommation énergie solaire et d'appoint, débit des fluides, répartition des besoins énergétiques, consommation électrique, d'eau etc. . . .

L'ensemble des mesures est effectué par deux "data logger" avec enregistrement sur bande magnétique toutes les 3 minutes, et l'exploitation des données est réalisée par ordinateur. Il y a en tout 87 points de mesure. Les figs. 20, 21, 22 donnent une idée des courbes d'enregistrement après traitement par ordinateur.

ALIMENTATION EN ÉNERGIE ÉLECTRIQUE ET APPROCHE ÉCONOMIQUE

Pour une installation avec machine frigorifique à compression l'ensemble des équipements pour la production d'eau glacée nécessite environ 0,40/0,65 kW pour produire 1000 frigories/h utiles.

Le système CLISOTOS permet de réduire l'énergie électrique nécessaire à environ 0,09/0,13 kW pour 1000 frigories/h utiles. On peut alors démontrer qu'un système solaire optimisé peut limiter la puissance électrique des raccordements à environ 18-20 % de celle nécessaire avec un système de compression. Même s'il est trop tôt pour tirer un bilan énergétique de l'installation, il a été constaté que, pour le régime climatisation, la température de 23°C est maintenue avec un appoint énergétique extérieur marginal.

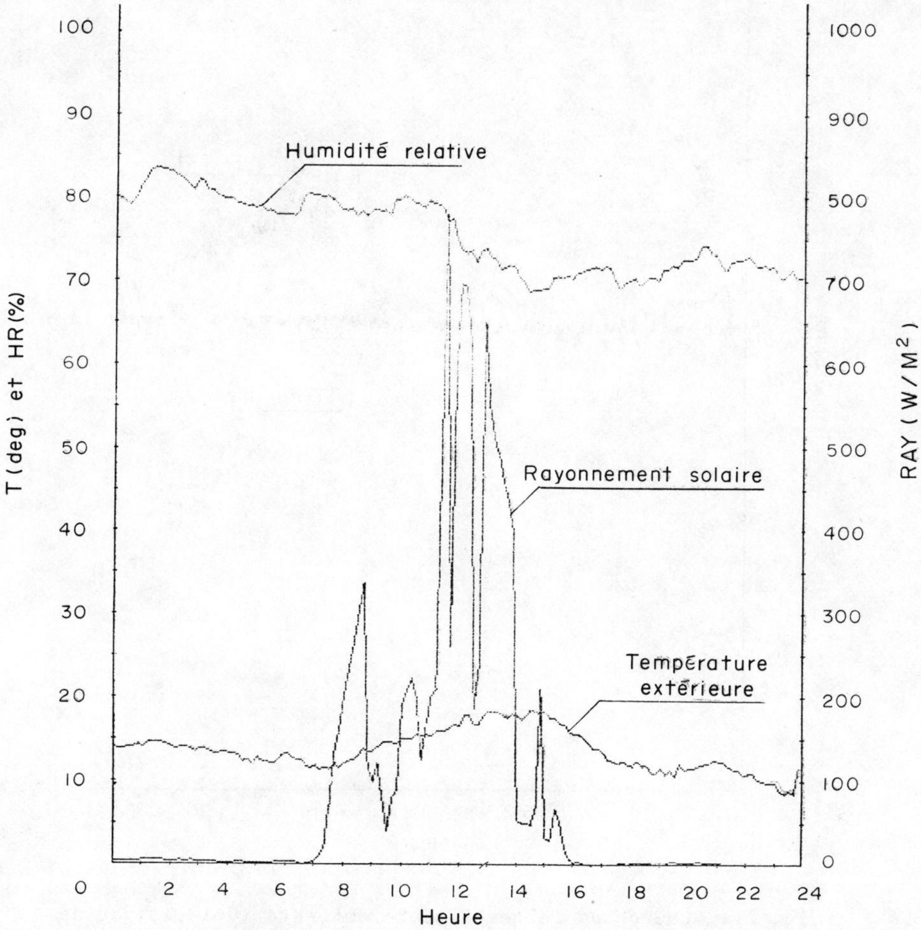

Fig. 20. Courbes type des condition climatiques journalières après traitement par ordinateur (moyenne saison).

FUTUR DÉVELOPPEMENT DE L'INSTALLATION AVEC UTILISATION D'UN CYCLE À POMPE DE CHALEUR THERMIQUE

Dans le cas d'un système solaire optimisé pour le refroidissement en été avec une machine à absorption, la surface mais surtout l'angle de montage des capteurs doivent être choisis pour satisfaire les besoins d'été. Du fait que la charge maximale est relative aux mois de juin, juillet, août et septembre, l'angle de montage optimal des capteurs est pour nos régions compris entre 20° et 30°.

Avec pareil angle de montage, l'énergie disponible en hiver est considérablement pénalisée à cause de la faible hauteur du soleil propre à cette saison.

Pour améliorer le rendement solaire du système sans pénaliser le régime d'été, nous envisageons dans un proche futur d'utiliser la machine à absorption avec un cycle à pompe de chaleur qui utilisera l'énergie solaire pour charger artificiellement l'évaporateur.

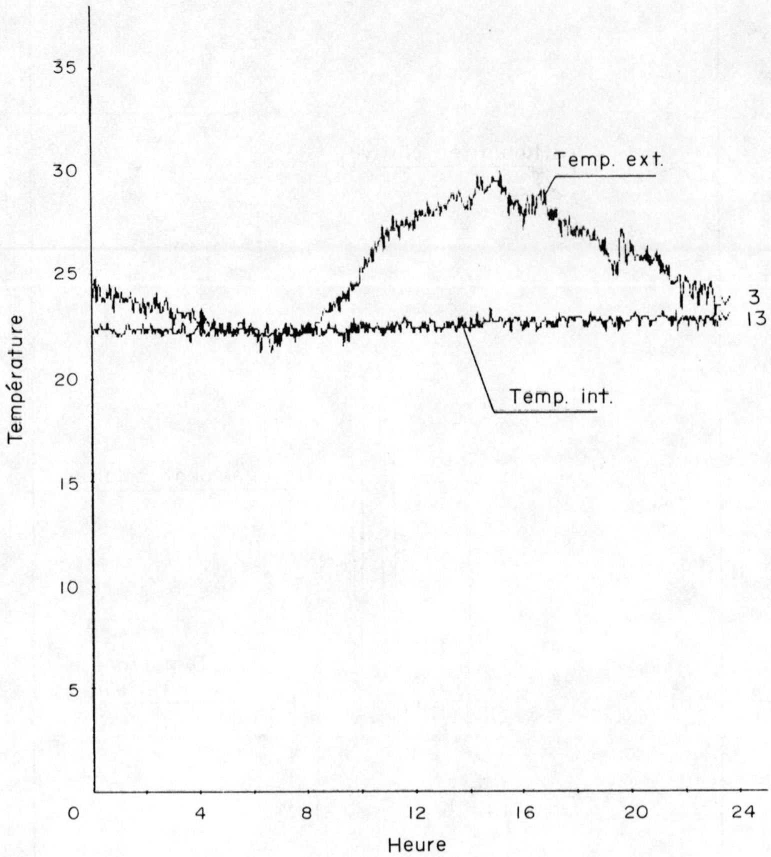

Fig. 21. Courbe de la température ambiante (voie 13) et de la température extérieure (voie 3) (été).

Cette modification a été déjà étudiée et les résultats de l'analyse sont particulièrement intéressants.

L'étude a été conduite dans le but d'utiliser sans aucune modification la machine à absorption qui produit de l'eau glacée en été comme pompe à chaleur en hiver grâce au système très souple de gestion de l'énergie.

Les figures de 23 à 31 représentent les schémas de principe et les résultats des calculs théoriques et de simulation pratique établis pour cette nouvelle expérience.

P.S.: les diagrammes et les calculs ont été établis avant l'entrée en vigueur des normes relatives aux unités de mesure, il est donc nécessaire de rappeler que 1 kcal/h = 1,163 W et qu'environ 1 atm = 1 bar.

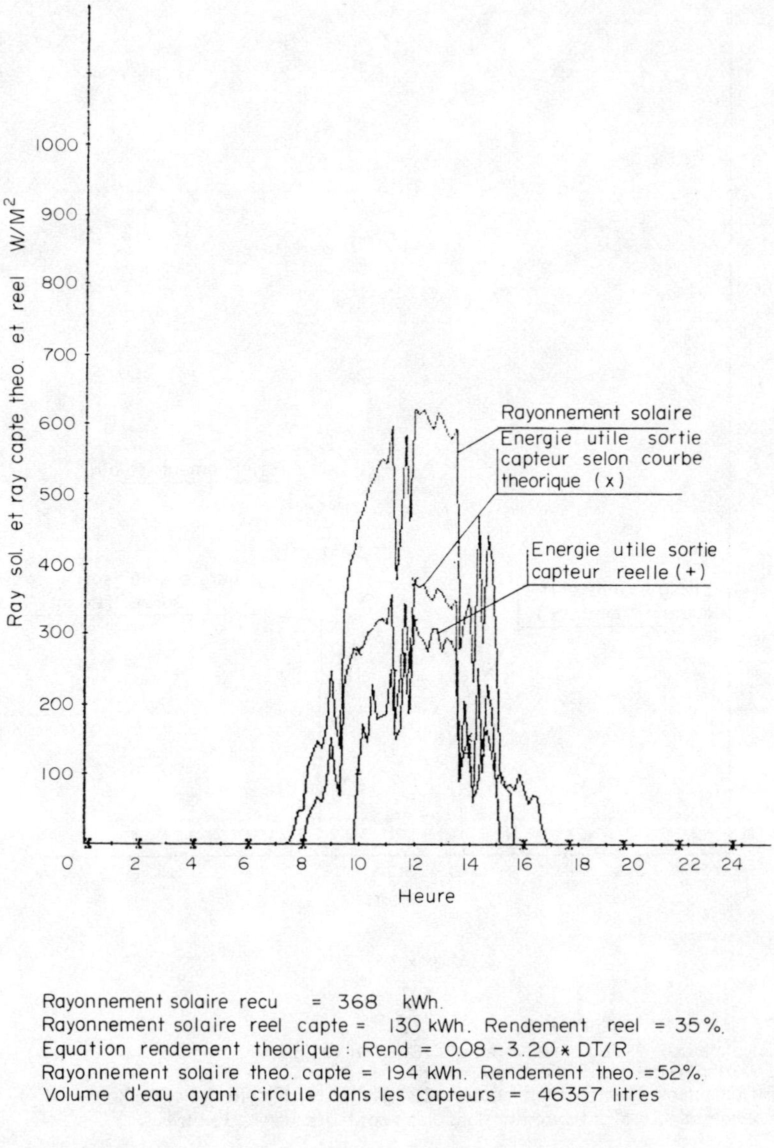

Rayonnement solaire recu = 368 kWh.
Rayonnement solaire reel capte = 130 kWh. Rendement reel = 35%.
Equation rendement theorique : Rend = 0.08 - 3.20 * DT/R
Rayonnement solaire theo. capte = 194 kWh. Rendement theo. = 52%.
Volume d'eau ayant circule dans les capteurs = 46357 litres

Figs. 22-23-24-25. Courbes de rendement théorique et réel des capteurs et rayonnement solaire.

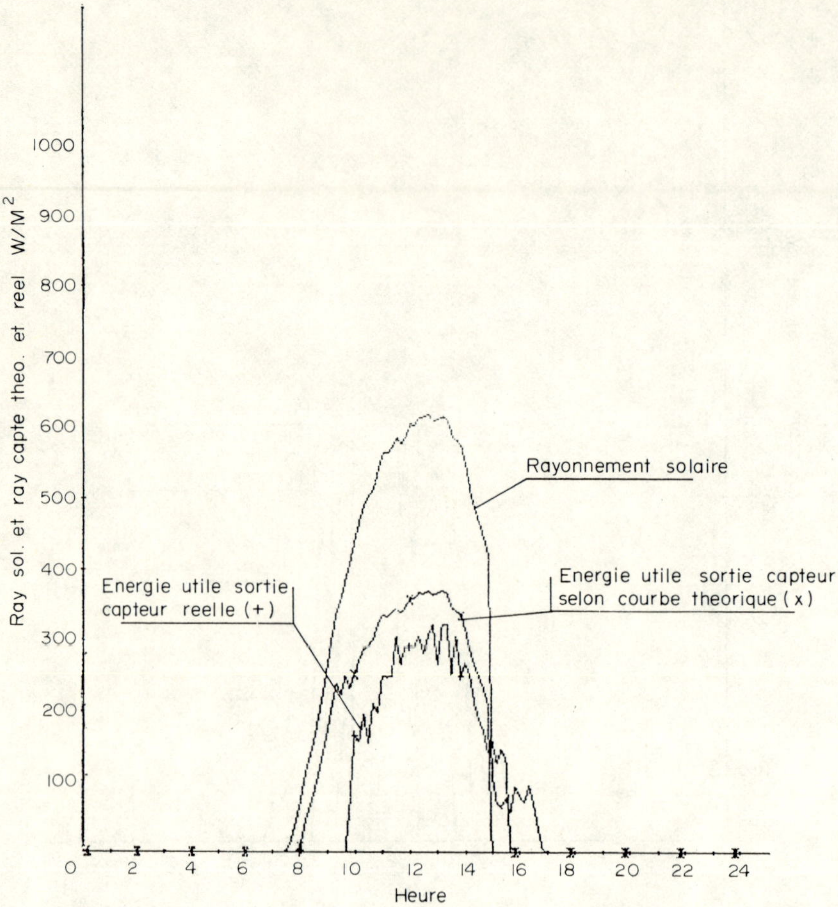

Rayonnement solaire recu = 406 kWh.
Rayonnement solaire reel capte = 160 kWh. Rendement reel = 39%
Equation rendement theorique : Rend = 0.80 - 3.20 * DT/R
Rayonnement solaire theo. capte = 227 kWh. Rendement theo. = 55%.
Volume d'eau ayant circule dans les capteurs = 47095 litres

Fig. 23.

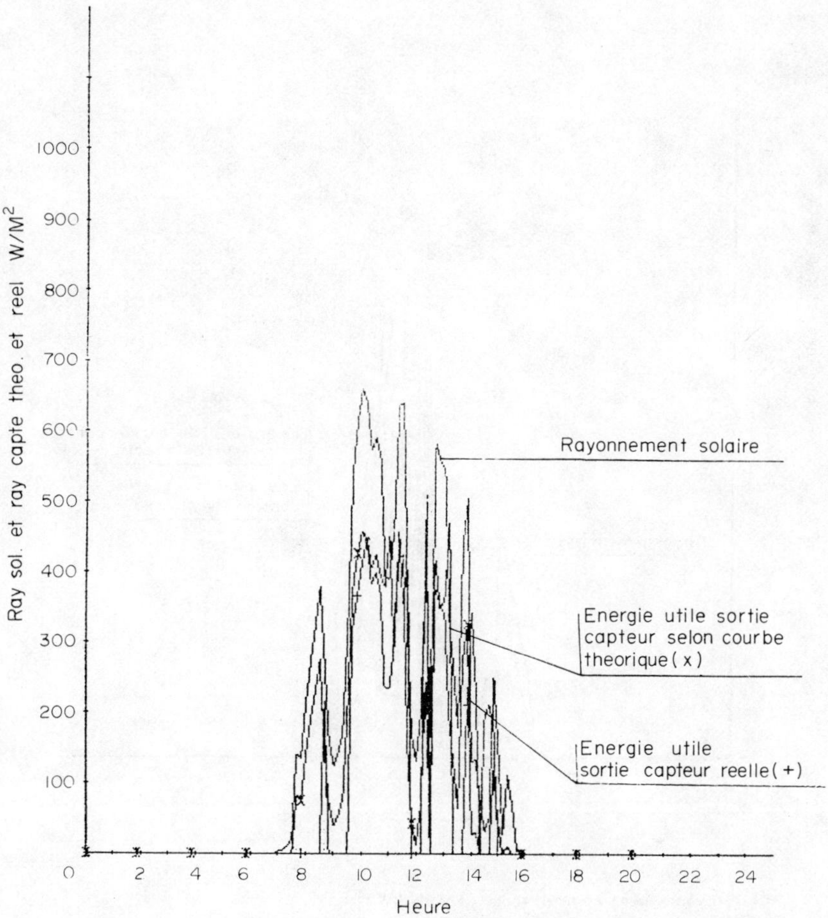

Rayonnement solaire recu = 304 kWh.
Rayonnement solaire reel capte = 175 kWh. Rendement reel = 57%.
Equation rendement theorique: Rend = 0.80 - 3.20 * DT/R
Rayonnement solaire theo. capte = 176 kWh. Rendement theo. = 58%.
Volume d'eau ayant circule dans les capteurs = 25384 litres.

Fig. 24.

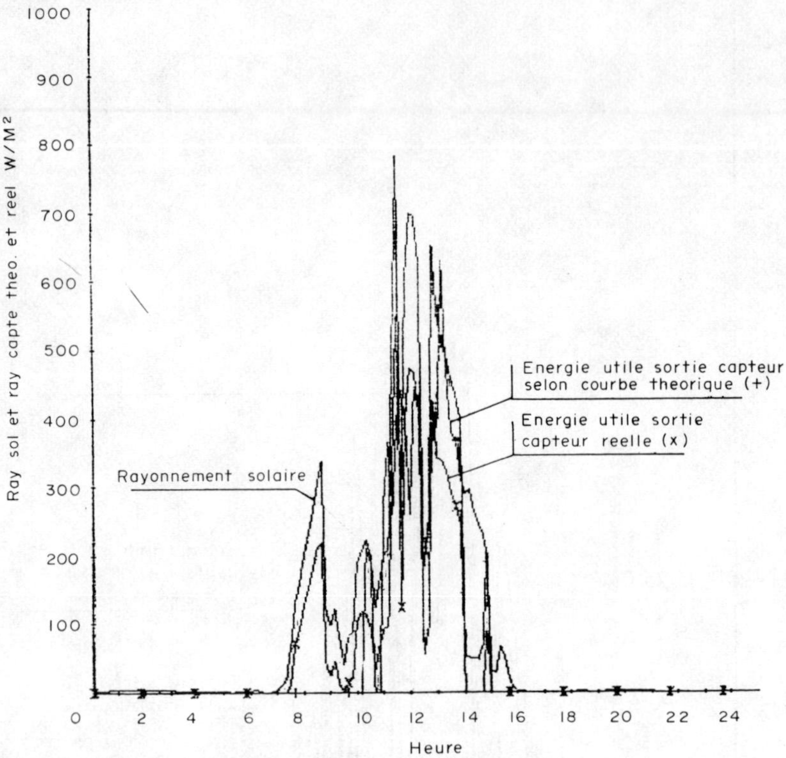

Rayonnement solaire recu = 255 kWh
Rayonnement solaire reel capte = 171 kWh, Rendement reel = 67 ‰
Equation rendement theorique : Rend = 0.80 − 3.20 * DT/R
Rayonnement solaire theo. capte = 143 kWh Rendement theo. = 56 ‰
Volume d'eau ayant circule dans les capteurs = 2658.4 litres

Fig. 25.

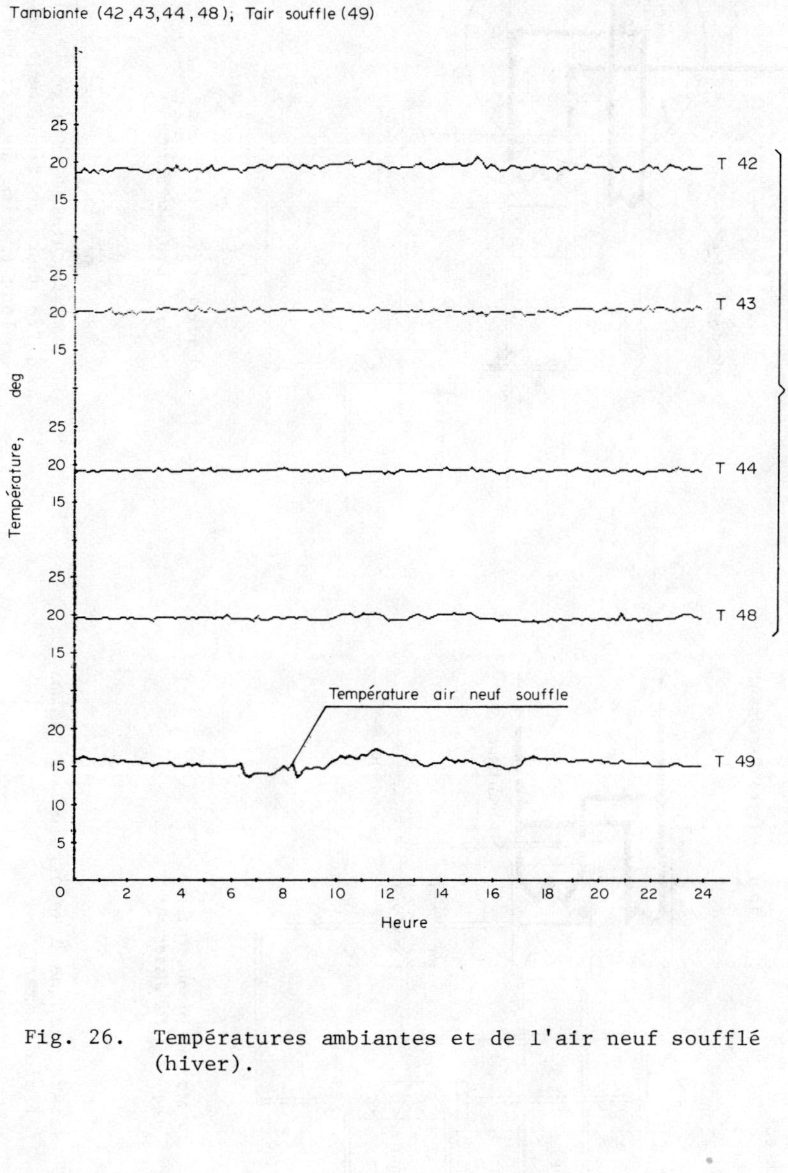

Fig. 26. Températures ambiantes et de l'air neuf soufflé (hiver).

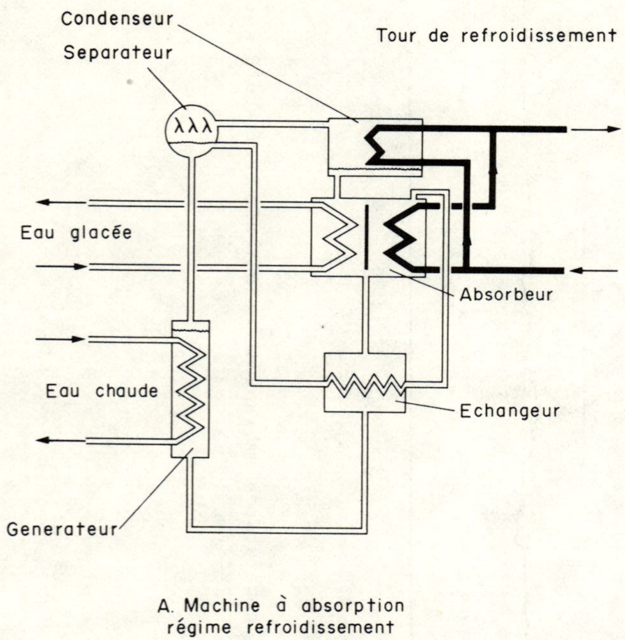

Fig. 27. Schéma machine à absorption en régime refroidissement.

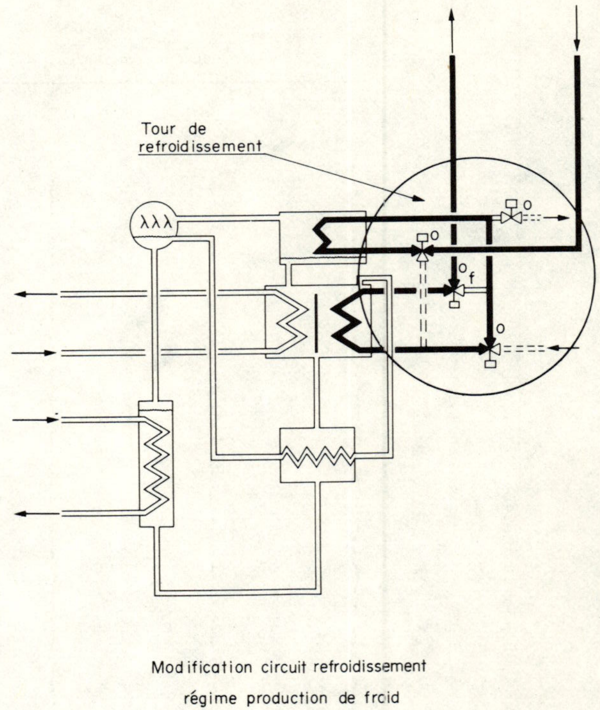

Fig. 28. La même machine mais avec le circuit modifié pour fonctionnement en pompe à chaleur (régime de refroidissement).

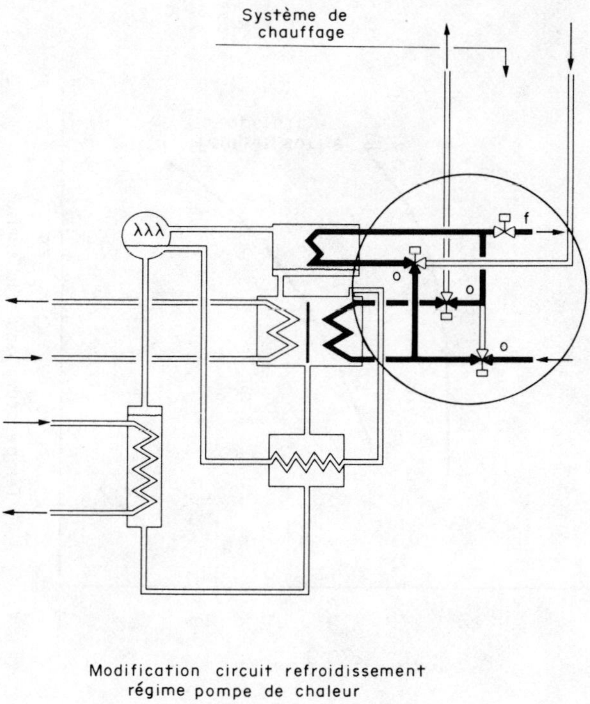

Fig. 29. Machine à absorption en régime pompe à chaleur.

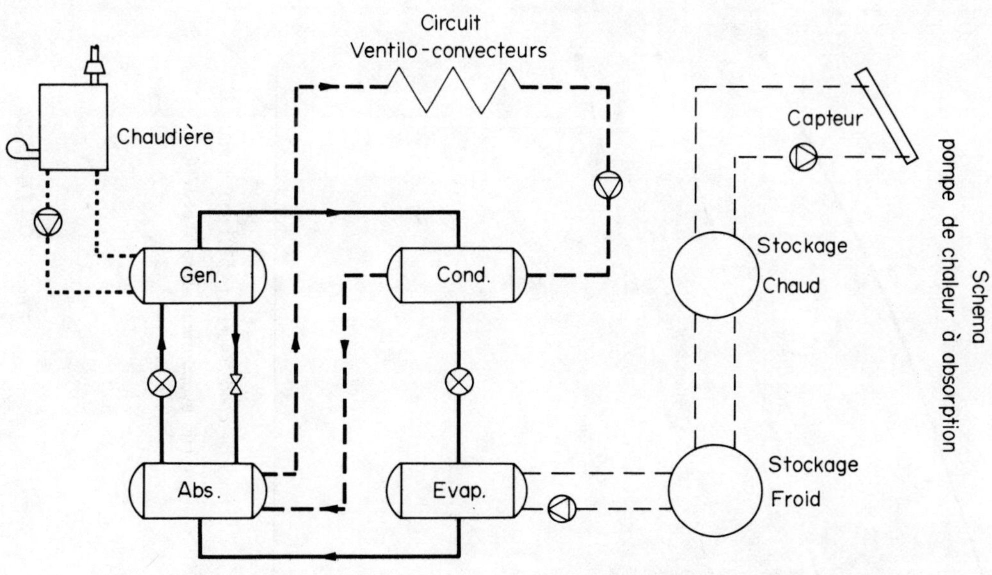

Fig. 30. Schéma de principe de schéma futur à pompe à chaleur.

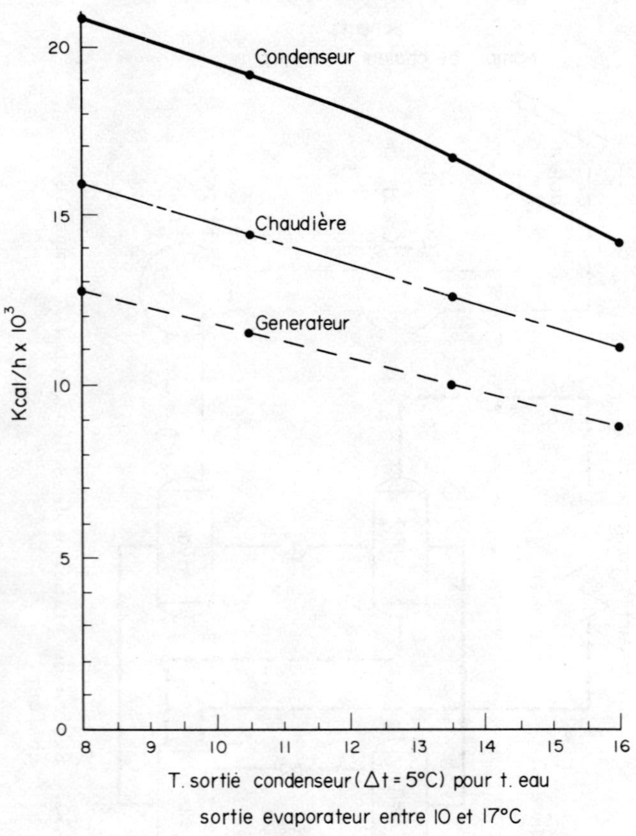

Fig. 31.

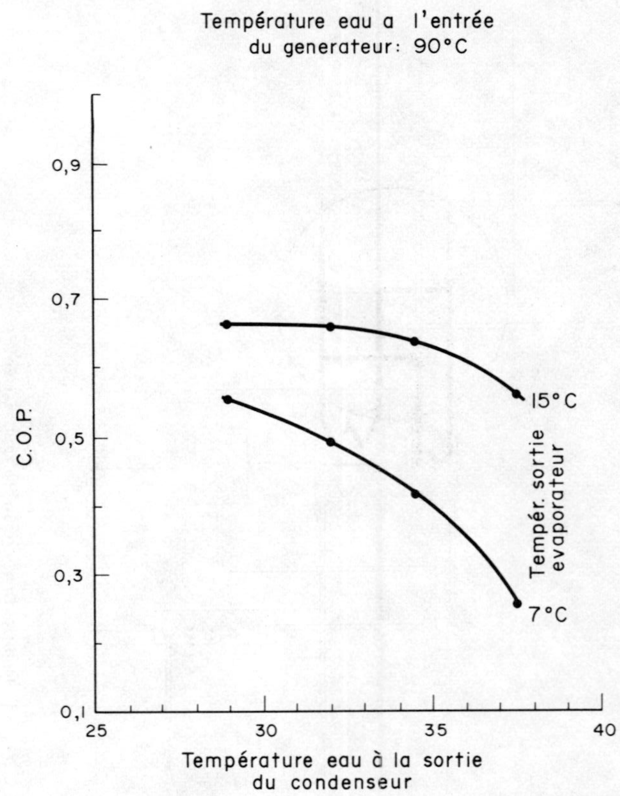

Fig. 32.

Figs. 31-32-33-34. Courbes caractéristiques de la machine à absorption en régime chauffage (pompe à chaleur).

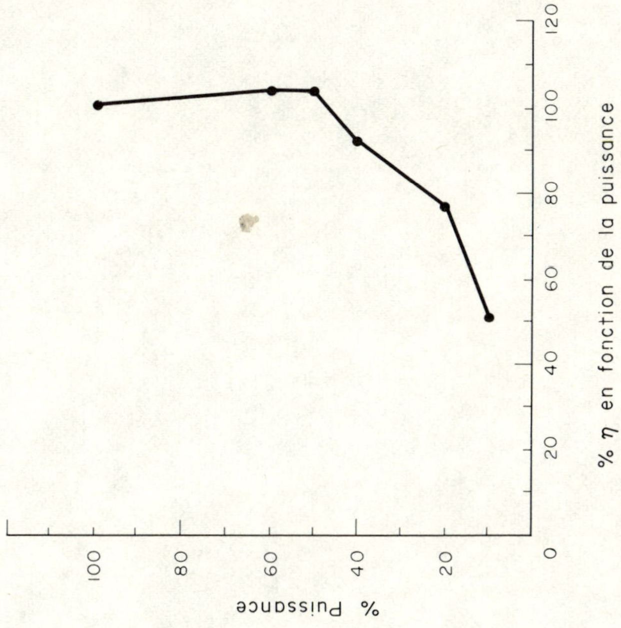
% Puissance
% η en fonction de la puissance
0
20
40
60
80
100
120

Fig. 34.

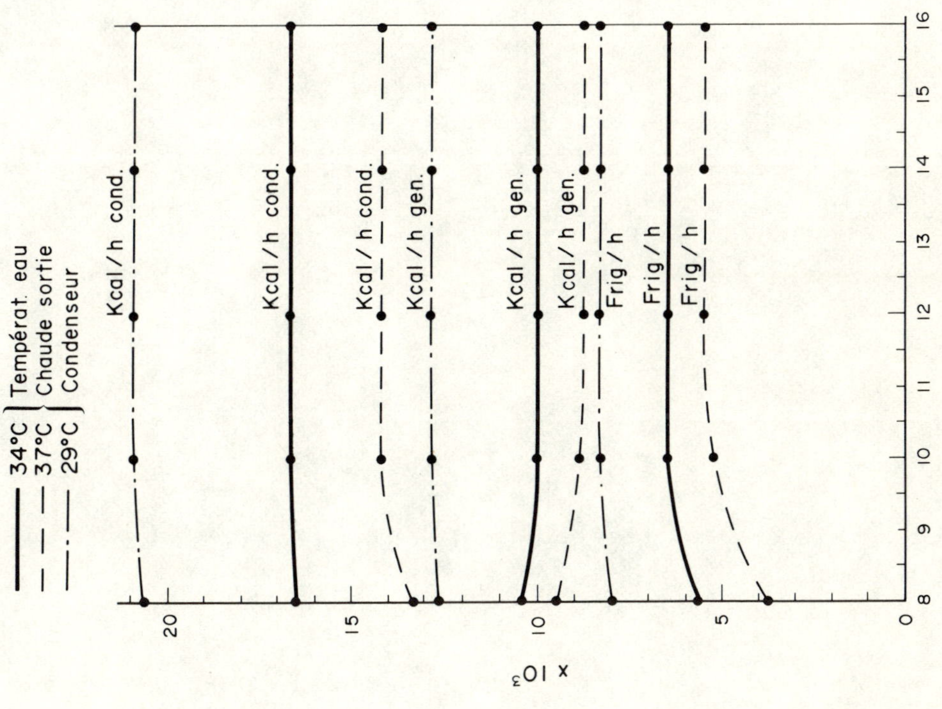
34°C
37°C
29°C
Températ. eau
Chaude sortie
Condenseur
Kcal / h cond.
Kcal / h cond.
Kcal / h cond.
Kcal / h gen.
Kcal / h gen.
Kcal / h gen.
Frig / h
Frig / h
Frig / h
x 10³
0
5
10
15
20
8
9
10
11
12
13
14
15
16

Fig. 33.

Imports and Exports

Imports and Exports

P. Cruchon (General Rapporteur)

France

INTRODUCTION

For a long time natural gas consumption developed on a world-wide scale on the basis of regional balances between production and consumption which left little room for international trade. In 1976 international trade accounted for barely 10 per cent of gas consumption (135 Gm^3 out of 1357 Gm^3), although such consumption represented a significant proportion (20 per cent) of world energy requirements. This situation was partly due to the existence of major production zones near industrial areas, but it also reflected the inherent difficulties of gas transport. The situation is changing rapidly, since around 1980 imports could cover nearly 13 per cent of total consumption. The importance of international trade in the development of the gas industry is becoming increasingly clear, and in many countries medium-term consumption will be adjusted to the level of imports which can be achieved in the context of reasonable technical and economic conditions.

Two papers of high quality and interest were presented under this topic. While reports presented under other sections occasionally contain information related to imports and exports, the fact that there were not more who dealt directly with this topic is an indication of the complexity of the subject. International trade is both the consequence of and the prerequisite for each area's balance between resources and requirements; forecasts in respect of various areas are also subject to all the uncertainties connected with production and consumption forecasts. In addition, it is particularly difficult to reconcile such forecasts with a restricted geographical approach.

Accordingly, our aim here is by no means to fill all gaps but simply to indicate a few guidelines and suggest some points which might lead to discussion. These will concern the development of technical and economic constraints on transport, the trend in international trade at the ECE regional level and the role of imports and exports in this region as a whole.

INDIVIDUAL PAPERS

The two papers are entitled:

"The Gas Industry in Algeria: Sonatrach Challenge for Investments and Financing" by C. Khelil (Sonatrach);

"Algerian Gas in the Economic Community and the Rest of Europe: An Optimization Study in Economic and Marketing Diversification" by M. Belguedj (Sonatrach).

Unfortunately, at the time this report was prepared, only a summary of the second paper was available and it was, therefore, not possible to take it fully into consideration.

A general description of the programme embarked upon by Algeria to exploit its oil and gas potential and the investments which will be necessary is outlined in the paper by C. Khelil. He notes that Algeria is potentially a major producer of gas, for according to a study by Degolyer and MacNaughton, hydrocarbon reserves at 1 January 1977 were as follows:

Oil	1100 million t
Condensate	442 million t
LPG	196 million t
Gas	3500 Gm^3

The use of these resources is provided for in the "VALHYD" programme (Valorization Hydrocarbon Development Plan). This programme aims at the expansion of hydrocarbon production based on the optimal recovery of oil reserves by the reinjection of gas, the maintenance of a high level of gas sales over a long period and the maximum recovery of LPG condensates.

The investments to be made over a 30-year period (1976-2005) concern the various stages of production: well-drilling, the construction of gathering and transmission systems, separation units, wet gas processing, reinjection, natural gas liquefaction and transport, LPG separation and loading, and refineries. The total capacity of various gas-related units will be the following:

Units for processing non-associated gas: 150 Gm^3/year.
Reinjection units: 84 Gm^3/year.
Natural gas liquefaction units: 74 Gm^3/year.
LPG separation and loading units: 9 million t/year.

Investments total $33.4 billion (1976 value), of which $25 billion are for the gas development programme. Loans will peak in 1978, 1979 and 1980 (nearly $3 billion for each of these years) and will drop to a very low level after 1985; servicing charges will amount to over $2.5 billion annually between 1982 and 1986, after which they will decline. The income anticipated by Sonatrach is more than enough to meet these obligations throughout the period in question; the supplies, which will depend mainly on fluctuations in investment costs and income, will take care of any contingencies which can reasonably be expected.

POINTS FOR DISCUSSION

The Development of Technical and Economic Constraints on the Transport of Natural Gas

The Symposium held in 1974 in Belgrade on "The Development of Gas Markets in Europe and their Supply Possibilities up to 1985" was organized to highlight such constraints by devoting its Topic D to "Developments of gas transport by pipeline, methane tankers, in the form of LNG, and in other forms". As several papers at the time dealt with technical and economic prospects in that field it does not seem necessary to deal explicitly with that question here, for to do so would, for example, result in duplication of the work done at certain specialized seminars, such as the one held at Florence in April 1976 on sea pipelines for the transport of gas.

However, possible progress in international trade depends on the constraints inherent in the methods used to transport gas, both in liquefied form and by pipeline. It might therefore be useful to review some of these constraints as a means of promoting a study of their possible development and their implications around the year 1990.

Natural gas requires much larger investments than liquid hydrocarbons, whether it is transported by pipeline (for an equivalent amount of energy, the necessary investments are twice as great as for crude oil) or by sea (particular account being taken of installations for liquefaction, storage and regasification). Moreover, the consumption of energy is greater: it is 3 per cent per 1000 km of gas pipeline and about 15 per cent at the liquefaction stage. This explains the need for strict contractual provisions and for strictly specified processing conditions, as well as the economically prohibitive nature of excessively long transport distances.

When imports are being planned, the mode of transport to be used (gaseous or liquid form) is sometimes self-evident for geographical reasons: either ocean transport is physically necessary or else the distances implied by the two methods are very different. In other cases, more detailed studies are necessary and may be related to certain theoretical considerations. For distances of over 1000 km the cost of land transport is virtually proportional to the distance. For an LNG line, on the other hand, both investments and energy consumption represent a large fixed proportion because of the liquefaction and regasification stages, whereas the marginal kilometre cost is generally lower than for the gaseous form. Transport by gas pipeline is therefore normally preferable, except perhaps over very long distances. A number of authors have tried to define this "equilibrium distance" between the two modes of transport: it is still very theoretical in nature, which explains why estimates vary over a wide range (to our knowledge, 4000 to 8000 km). In any event, this comparison also depends on the amounts transported, since the economies of scale are different in the two cases.

Economies of scale are important in transport by gas pipeline (for long distances, an increase in throughput from 3 Gm^3 per year to 20 Gm^3 per year reduces transport costs by 50 per cent). This effect becomes less marked beyond 10 or 20 Gm^3 per year, owing to the technical problems presented by large-diameter pipelines.

On the basis of past experience, it is difficult to estimate the savings which can be expected in the cost of liquefied transport. They vary according to the links in the transport chain:

At the beginning and end (between points of production or consumption and the terminals) the gas is transported by pipeline, with the economies of scale mentioned above.

Port outlay does not vary much according to the quantity transported; it differs greatly in individual cases, but is often high because of the special nature of LNG traffic.

The cost of liquefaction can probably be reduced by increasing the capacity of the lines, but in the event of an accident or outage for maintenance, the supply will be further reduced, thus penalizing both seller and buyer.

The cost of ocean transport itself depends mainly on the extent to which methane tanker capacity is used, as well as on tanker size; if used below capacity, their cost quickly becomes prohibitive. On the other hand, the number of vessels assigned to a line is of very little importance once they have reached the optimum size. Accordingly, the co-existence of parallel but independent ocean transport lines is economically feasible, although this would be absurd in the case of transport by gas pipeline. The economic urge to increase the size of vessels is undoubtedly limited by the navigational risks which giant methane tankers would create. They present very special handling and mooring problems because of the wind resistance of their freeboard area. At the present time, most operators consider a capacity of 125 000 m^3

as the optimum, and it is impossible to tell whether and when vessels of 200 000 m^3 will be used, although designs of methane tankers of 300 000 m^3 have already been studied.

The following questions might bear consideration:

Are technical developments (such as combined production of electricity at regasification terminals) likely to bring about a significant reduction in transport costs between now and 1990?

The search for economies of scale implies a general increase in the quantities involved. The most recent operations entail movements of 5 to 10 Gm^3 per year for ocean routes and more than 10 Gm^3 per year for land routes. Can an appreciable reduction in transport costs be achieved if these quantities are increased still further? Is there not a danger that this advantage will be nullified by the cost of the facilities required by the gas industry to cope with an interruption in supplies in the event of a technical hitch?

At the Belgrade Symposium it had seemed that consortiums of purchasers at the European level had almost become a necessity because of the quantities involved. That has not always been the case in liquefied gas operations because of their greater flexibility. Can it still be assumed that this approach will continue to predominate between now and 1990? If not, would it not be desirable to have more flexible procedures which can produce economies of scale at certain stages in the transport operation, and spread the technical risks among several partners?

International Gas Trade Trends at the ECE Regional Level

Agreed trade flows

Because of rigid transport requirements, international trade flows are based almost exclusively on long-term contracts concluded several years before deliveries actually take place. This time lag, which varies in individual cases, is seldom less than 5 years. A fairly accurate picture of trade patterns in 1983-1985 can therefore be obtained from a scrutiny of current contracts. A certain amount of uncertainty still persists, due in particular to the fact that certain transactions have not yet been approved, but the quantities involved are marginal. In order to simplify matters, the EEC region has been divided into three areas: North America, Western Europe and Eastern Europe. Only imports by a country from an area other than its own have been taken into account. The following figures (in Table 1) which have been obtained mainly from press reports, are sometimes imprecise and may undoubtedly contain errors.

Except for a few older contracts involving small amounts, the quantities in question are guaranteed up to dates between 1995 and 2005. The 20-year period is the most common. Flows between areas are summarized in Table 2. Figures without brackets indicate firm commitments and those in brackets are for other contracts (which in general have not yet received the necessary official approval). The table reveals the following characteristic features:

- Africa is the principal exporting area and Western Europe the principal importing area.
- Trade is mainly between contiguous areas (Africa - Western Europe; Eastern Europe - Western Europe; Middle East - Eastern Europe) or areas which are relatively close (Africa - North America; Middle East - Western Europe). Trade between areas which are much further apart (Middle East - Asia and Middle East - North America) is still very limited.
- Uncertainty about contracts which have not yet been finalized mainly involves imports from North America.

TABLE 1

International Trade in Natural Gas to the EEC Region - Current Contracts

Country of origin	Country of destination	Project	Annual volume (Gm^3)	Date started	Length of contract	Final contract
1. *Destination North America*						
Algeria	USA	Distrigaz I	0.4	1971		yes
		El Paso I	10	1977	20	yes
		El Paso II	10.3	1980	20	-
		Distrigaz IV (including Distrigaz I)	1.2	1977	20	yes
		Trunkline	4.3	1980	20	yes
		Tenneco Atlantic	11.8	1985	20	-
Indonesia	USA	Pacific Lighting	5.6	1981	20	yes
Iran	USA	Columbia Gas System	3	1982	25	-
2. *Destination Western Europe*						
Algeria	United Kingdom	BGC	1	1964	15	yes
	France	GDF (Le Havre)	0.5	1965	15 + 10	yes
		GDF (Fos)	3.5	1972	15 + 10	yes
		GDF (Montoir)	5.15	1980	20	yes
	Spain	Empresa Nacional des Gas Enegas	0.4	1974		yes
			4.5	1980	20	yes
	Belgium	Distrigaz	5	1980	20	yes
	Fed. Rep. of Germany	Ruhrgas	3.2 }	1984	20	yes
		Salzgitter Ferngas	0.8 }			
		Thyssengas-Brigitta	4		20	-
	Netherlands	Gasuni	4	1984	20	yes
	Italy	ENI	12.4	1981	25	yes
Libya	Italy	Snam Progetti ENI	3	1969	20	yes
	Spain	Empresa Nacional des Gas	1.1	1971	15	yes
USSR	Fed. Rep. of Germany	Ruhrgas (I, II and III)	9.5	1974/78	25	yes
	France	GDF	4	1976/80	25	yes
	Austria		2.5	1968/80		yes

TABLE 1 (*cont.*)

Country of origin	Country of destination	Project	Annual Volume Gm^3	Date started	Length of contract	Final contract
	Italy		7	1974/78	20	yes
	Finland		3	1974/	20	yes
Iran	Fed. Rep. of Germany	Ruhrgas	5.5	1981/83	20	yes
	France	GDF	3.6	1981/83	20	yes
	Austria	OMV	1.8	1981/83	20	yes
3. *Destination*	*Eastern Europe*					
Iran	Czechoslovakia		3.6	1981/83	20	yes
	USSR		10–15			yes
Afghanistan	USSR		2.5–3			yes
4. *Destination*	*Asia*					
Alaska	Japan	Tokyo Gas Tokyo Electric	1.3	1969	15	yes
Abu Dhabi	Japan	Tokyo Electric	3	1977	20	yes

TABLE 2

Current Contracts between Major Areas (Recapitulation) (Gm^3/year)

Exporting area	Importing area: North America	Eastern Europe	Western Europe	Far East	Total exports
North America		0	0	1.3	1.3
Eastern Europe	0		26	0	26
Middle East	0 (+3)	16-21	10.9	3	29.9-34.9 (+3)
Africa	15.9 (+22.1)	0	44.5 (+4)	0	60.4 (+26.1)
Far East	5.6	0	0		5.6
Latin America	?	0	0	0	?
Total Imports	21.5 (25.1)	16-21	81.4 (+4)	4.3	123.2-128.2 (+29.1)

Possibilities around the year 1990

The trend beyond the flows which have already been decided upon can be described only in a very general way and subject to many uncertainties. This is the approach adopted in the paper on "Possible sources of natural gas for France and the ECE area in 1990", and particularly in its Fig. 4. The authors foresee a major expansion of international trade: exports will increase from 135 Gm^3 per year (including 17 Gm^3 of LNG) in 1976 to 370-480 Gm^3 (including 130-175 Gm^3 of LNG) per year in 1990. In other words, trade by gas pipeline will double and an even greater increase will be achieved in LNG trade. They also suggest a geographic distribution of this trade, as summarized in Table 3, but confine themselves to flows between major world areas (except for imports and exports between countries in the same area).

TABLE 3

International Gas Trade between Major Areas (around 1990)*

Exporting area	Importing area: North America	Eastern Europe	Western Europe	Far East	Total exports
Eastern Europe	0		30-35	0	30-35
Middle East	0-5	20-25	10-20	5-20	35-70
Africa	40-55	0	60-75	0	100-130
Far East	5	0	0		5
Latin America[a]	35-45	0	0	0	35-45
Total imports	80-110	20-25	100-130	5-20	

[a]Mexico included in Latin America. (Figures in Gm^3 per year.)
*According to the paper "Possible sources of natural gas for France and the ECE area in 1990" in Topic A.

Table 3 is fairly similar to the one in which current contracts are recapitulated. The main trends which emerge are the following:

Exports from Latin America to North America, which are now under negotiation, will most likely materialize a few years from now.

A particularly large increase in trade with Africa, which will probably export 100 to 130 Gm^3 per year, a figure that reflects Algerian development plans (the additional export capacity envisaged in the Valhyd plan is over 70 Gm^3 per year) and development possibilities in other gas-producing countries (such as Nigeria). On the other hand, the low figure for exports from eastern Europe and the Middle East has been based more or less on only current contracts, which may be a very wise thing to do.

At one time or another, consideration has been given to major trade transactions between areas which are further away from each other (and in particular, between North America, on the one hand, and Eastern Europe and the Middle East on the other), as well as to exports of Soviet gas to the Far East. Transport costs over long distances constitute a serious obstacle to the development of such projects. It might be interesting to discuss the possibility of some of them being carried out around 1990 (taking into account, for example, possible improvements in transport techniques and reductions in transport costs) or at some later date.

This same paper helps us to compare anticipated net imports or exports and production as a means of ascertaining gas availabilities in each area.

TABLE 4

Net Imports or Exports, Production and Availabilities by Area around 1990*

Gm^3/per year	Net imports or exports[a]	Production	Domestic availabilities[b]	Consumption in 1976
North America	+ 80-100	550-800	630-910	651
Eastern Europe	10	800-900	790-890	375
Western Europe	+ 100-130	180-250	280-380	202
Middle East	- 35-70	150-200	80-165	29
Africa	- 100-130	130-160	0-60	4
Far East	0- + 15	110-130	110-145	46
Latin America	- 35-45	110-130	- 65-95	50

(a) + = net importing area; = exporting area.
(b) The range obtained by arithmetical calculation is generally greater than the actual probable range.
*Possible sources of natural gas for France and the ECE area in 1990", *loc. cit.*

There are striking differences between the three major areas into which the countries of the ECE region have been divided:

Thanks to the USSR, Eastern Europe has the largest production potential and, in view of its reserves, is the area most likely to develop its trade.

In this context, however, the level of exports to Western Europe would remain moderate and, in terms of total volume, would be practically offset by imports from the Middle East; domestic consumption, on the other hand, would increase sharply, at an annual rate of 5.5 to 6.6 per cent between 1976 and 1990.

Supplies for Western Europe would in large measure (approximately one-third) depend on imports of 100 Gm^3 or more. The Middle East would contribute a marginal amount (10 to 15 per cent) of the total and the main suppliers would be Africa (mainly North Africa and Nigeria) and Eastern Europe, i.e. the USSR.

North America would import comparable amounts, but their relative share would still be much lower (less than 15 per cent of total availabilities); they would be divided between LNG (for which Africa would probably be the main source) and imports by gas pipeline from Latin America (mainly Mexico). Since figures of possible production vary over a fairly broad range, the same is true of availabilities, the extreme assumptions being a levelling-off in consumption somewhere near current levels or moderate growth of slightly less than 2.5 per cent annually.

These differences justify a closer examination of the role of gas imports and exports in each area, although in this case it is important that the *de facto* interdependence of each area's policies should not be overlooked. Eastern Europe, for example, can satisfy a relatively large proportion of Western Europe's requirements while Africa's export capacity is divided between North America and Western Europe.

Role of Imports and Exports by Area

North America

Because of the size of the American gas market, even a relatively low level of imports can weigh heavily in the balance of international gas trade. The conclusions that can be drawn from information set out in the papers by L. D'Andrea and the AGA about possible trends in United States import policy are therefore particularly useful.

The National Energy Plan for the period up to 1985 submitted by President Carter in April 1977 was based on the assumption that, as consumption would expand very little, imports would increase from 28 Gm^3 in 1978 to 34 Gm^3 in 1985. However, as it is hardly likely that consumption will level off in this way additional sources of supply will no doubt have to be found. The following imports are therefore envisaged:

in the gaseous state from Canada and Mexico;
in the form of LNG especially from Africa and Indonesia.

The quantities covered by contracts which were still being scrutinized by ERA at the time D'Andrea's report was prepared amounted to 44 Gm^3 per year, one-half of which had already been approved:

Gm^3/per year	Algeria	Indonesia
Approved	16.8	5.5
Under study	21.9	

Further contracts may well be concluded, and the author envisages a high figure of 55 Gm^3 per year for 1985.

Although assuming increased consumption at the rate of 2.9 per cent annually between 1978 and 2000, the AGA study relating to the year 2000 expects that LNG imports will be limited to 85 Gm^3 per year, i.e. to 9 per cent of requirements — other imports amounting to approximately one-half of this figure — and that the balance of requirements will be met by the large-scale development of domestic sources.

On the basis of these forecasts for 1985 and 2000, it may be assumed that LNG imports in 1990 will amount to between 50 and 65 Gm^3 which, in the light of the general estimates of the Chapelle-Desprairies report, would not seem likely to create an imbalance in international trade. However, everything will depend on the policy that is adopted and applied. The maintenance of the wellhead price of gas at a low level, as well as the regulations applicable to its transport and distribution, have had the opposite effect on demand, which has actually been stimulated, and on the prospecting and exploitation of new deposits, which have not been promoted. Certain gas companies are therefore experiencing shortages and are consequently expanding their resources by producing SNG or importing LNG even at a particularly high price. It is not certain that this solution will be the best for the United States from a general economic point of view (it might prove less costly to develop certain domestic resources).

The adoption of an import policy (particularly in respect of LNG) will make it easier to forecast how imports will develop. Criticism of imports is concentrated mainly on their cost and the dependence they bring about, while advocates of gas imports stress the advantages they offer in comparison with oil imports which would otherwise be indispensable, namely the importance of supplies of industrial goods and the diversification of sources of energy. One point of crucial importance as regards the market's capacity to absorb LNG would seem to be its pricing, which may be either "rolled in" or "incremental". The administration seems to incline towards the second solution, and in this respect is supported by certain economists, for it seems to be more in line with actual prices and costs, as it avoids long-term sales below marginal cost.

In any case, the rules worked out will have to be evaluated from the point of view of general energy policy and all the various means used to make good the difference between gas resources and demand.

It would therefore be useful to draw the necessary conclusions from the introduction of the new energy policy with regard to gas and the rules applicable to gas imports. Another point for possible discussion is the economic validity of the two methods of gas pricing and the possible consequences that the one selected will have on LNG imports, in order to determine whether the range of 50 to 65 Gm^3 per year mentioned above is realistic.

Western Europe

The individual papers submitted to the Symposium contain hardly any information that could be used to obtain an accurate estimate of gas imports and exports by the countries of Western Europe. Conditions in these countries vary widely and, as information of this kind would be desirable, use has been made of the figures contained in the 1977 report of the International Energy Agency (IEA) on:

the total demand for primary energy;
the production of natural gas;
net imports of natural gas.

These figures, supplemented by various data pertaining to France (which is not a member of IEA), are given in Table 5. They should, in fact, be used with caution, as it is rather difficult to cross-check them against other figures because of the units used. Indeed, in some cases figures are given in cubic metres (in the case of the Chapelle-Desprairies report mentioned above), the calorific value of which is not necessarily obvious, and tonnes of oil equivalent with energy-conversion coefficients that vary from one organization to another.

Between 1976 and 1985 higher production and imports would make it possible to increase the proportion of natural gas in energy consumption from 15 to 17 per cent.

TABLE 5
Production and Imports of Natural Gas by Country (Western Europe)

Million tonnes of oil equivalent* (toe)	Total demand for primary energy			Natural gas production			Natural gas imports		
	1976	1985	1990	1976	1985	1990	1976	1985	1990
1. *Member countries of the IEA (source: IEA report for 1977)*									
Austria	23.4	33.4	39.7	1.9	1.0	0.9	2.3	5.1	6.1
Belgium	44.9	60.2	71.5	0	0	0	9.7	13	13.2
Denmark	19.3	20.6	22.9	0	2.7	2.7	-	-	0.6
Germany, Fed. Rep. of	260.3	337.8	371	14.8	14.4	12.5	21.9	46.8	48
Greece	13.2	25.3	33.5	-	0.4	0.2	-	-	-
Ireland	7	14.2	21.6	-	0.1	0.1	-	-	-
Italy	136	199.5	245	12.9	15	10	9.7	19.7	28
Luxembourg	4.6	7.5	8.6	-	-	-	0.4	0.5	0.5
Netherlands[a]	64.2	90.1	98.8	75.1	69.3	51.2	-41.2	-37.7	-16
Norway[b]	20.9	27.5	31	0.3	27	17	-	-27	-17
Spain	65.7	98.4	122.8	-	-	-	1.3	7	10.8
Sweden	49.6	60.5	65	-	-	-	-	-	-
Switzerland	21.4	26.7	29.4	-	-	-	0.6	1.7	1.9
United Kingdom[c]	206.9	244	268	33.2	40	42.5	0.9	10	10
Western Europe (member countries of IEA)	937.4	1245.7	1428.8	138.2	169.9	137.1	5.6	39.1	85.4
2. *France*									
France	174.5	230	270	6.6	6	4.5	13.1	31	39.5

(a) 1990 figures: estimates of the IEA secretariat.
(b) 1976 and 1990 figures: estimates of the IEA secretariat.
(c) Average ranges transmitted by the United Kingdom.

*Conversion coefficients:

IEA: 1 toe = 10 000 therms net calorific value (41.86×10^{15} J) - 1 tce = 0.7 toe, - 1 TWh = 0.086 million toe.

France: 1 toe = 10 000 therms gross calorific value - 1 tce = 0.66 toe, - 1 TWh = 0.222 million toe.

That proportion would then decline slightly to 16 per cent in 1990. Western Europe's net imports still will continue to increase considerably (at an annual rate of 13 per cent) but two-thirds of them simply offset declining production, particularly in the three main producing states, namely, the Netherlands, Norway and the United Kingdom. In fact, the amount available for export from those three countries will increase from 40.3 million toe in 1976 to 56.7 million toe in 1985 and then drop to 23.7 million toe in 1990,* assuming that production, after reaching a peak around 1985 (135 million toe), will fall back to its 1976 level (approximately 110 million toe)** in 1990. However, this assumption, which has been used by official bodies, possibly errs on the side of caution, which is the view expressed by P. Odell in his paper.

Around 1985-1990 seven countries, namely, Austria, Belgium, the Federal Republic of Germany, Italy, Spain, Switzerland and France, will depend on imports. The rate at which their imports increase will vary considerably and between 1976 and 1985 (a period when domestic production in these countries will be relatively stable) will make it possible for gas to maintain its share in the demand for primary energy where such demand exceeds 15 per cent, and to increase it elsewhere. After 1985 this share will at best be maintained and more likely it will be necessary to make up for declining production. The annual imports of these seven countries will thus rise from 2.64 EJ gross calorific value in 1976 to 5.59 EJ gross calorific value in 1985 and to 6.448 EJ gross calorific value in 1990; imports will therefore increase, mainly between 1976 and 1985, as shown in Table 6.

TABLE 6

	Austria	Belgium	Fed. Rep. of Germany	Italy	Spain	Switzerland	France (estimate)
Share of natural gas in total demand for primary energy:							
1976	18%	22%	14%	17%	2%	3%	10.8%
1985	18%	22%	18%	17%	7%	6%	16%
1990	18%	18%	16%	16%	9%	6%	16%
Average annual rate of increase of imports:							
1976-1985 period	9.2%	0.3%	8.8%	8.2%	20.6%	12.3%	13%
1985-1990 period	3.7%	0.4%	0.6%	7.3%	9%	2.3%	4.2%

*In EJ gross calorific value: 1.86 in 1976, 2.37 in 1985 and 0.99 in 1990.

**In EJ gross calorific value: 4.60 in 1976 and 1990 and 5.65 in 1985 (on a basis of $\frac{\text{gross calorific value}}{\text{net calorific value}} = 1.1$ for gas).

Compared with other estimates, the production forecasts supplied by the member countries of IEA may appear cautious: 1990 figures revert to the 1976 level — a trend which corresponds to the low hypothesis in the Chapelle-Desprairies paper. It may well be that the decline in production between 1985 and 1990 will be less pronounced. In this case, countries in which production is of secondary importance in their availabilities (mainly Italy, the Federal Republic of Germany and France) might increase their consumption of gas or restrict their search for new supplies. For their part, the major producing countries (the United Kingdom, the Netherlands and Norway) would be in a position to prevent a decline or levelling-off in their own consumption or else to prolong their export flows. The respective advantages of these various solutions might be the subject of a discussion.

Eastern Europe

The importance of the gas potential of the USSR is casting the other countries of Eastern Europe in a twofold role:

as importers, since the natural gas from their domestic fields is no longer sufficient to meet their needs;
as transit countries for gas exports to the West.

Czechoslovakia provides an illustration of this situation. Since 1967 the USSR has been its main source of natural gas, and subsequently the transport of natural gas to the countries of Western Europe and to the German Democratic Republic led to the construction of high-capacity gas pipelines; the existence of this infrastructure is encouraging the expansion of imports (contract with Iran).

Similarly, the paper prepared by Hungary predicts that, even assuming a stable level of domestic production thanks to new discoveries, a considerable increase in imports is desirable. Imports, which amounted to only 600 000 Gm^3 in 1975, would be 7 Gm^3 in 1990 (or approximately one-half of domestic requirements), of which 4 Gm^3 will be supplied under contracts already concluded. The paper also envisages the development of underground storage in a depleted gasfield (Hajduzobozlo) in connection with the transit pipeline system through Czechoslovakia. In view of the distance between the site and the pipeline (estimated at 100 km) and the general importance of having stocks close to points of consumption, the economic advantage of this project does not seem obvious. It might be a subject for discussion.

The export capacity of the USSR will obviously play an important part in the development of international trade in this product in Europe. In 1976 exports amounted to approximately 26 Gm^3, of which 12 Gm^3 were exported to Western Europe and 13 Gm^3 to Eastern Europe. The map in Fig. 4 in the Chapelle-Desprairies paper assumes total exports of 70 to 80 Gm^3, of which 30 to 35 Gm^3 will go to Western Europe, i.e. exports will increase about threefold. In view of the size of reserves in this area and the amounts covered by contracts already concluded (26 Gm^3 to Western Europe), it would be interesting to discuss the possibilities of expediting the development of these supplies.

Conclusion

What has been said above offers a picture of sorts of the international gas trade in the ECE region. By way of conclusion, it should be pointed out that this picture — which is rather indistinct — can be used only as a basis for discussion; it is flawed by so much uncertainty that it cannot be described as a forecast. As was noted in the introduction, it is necessary to proceed cautiously in this field, which is subject to many constraints, only a few of which have been mentioned. Gas contracts also fall into a more general trade framework which, whether bilateral or multilateral, can greatly influence the policy pursued by a country. Various specific problems — such as methods of financing or counterpart industrial deliveries — may arise and increase the complexity of negotiations. The availabilities of an exporting country are determined by comparing its total resources with its domestic requirements. And as its requirements may vary with the policies it adopts, future prospects can change quite rapidly.

Each country in the ECE region has to ensure that its gas balance, reflecting production, consumption and international trade is in equilibrium. Current domestic production indicates a variable and often very low degree of elasticity. There are considerable possibilities of expanding international trade, subject to a considerable amount of uncertainty; once concluded, trade agreements remain rigid. In the short and medium term, consumption is still the principal factor of adjustment, and an adequate amount of flexibility must be maintained in forecasting consumption trends.

The Gas Industry in Algeria: Sonatrach Challenge for Investments and Financing

C. Khelil

Algeria

INTRODUCTION

Having the fourth largest reserves of gas in the world, Algeria has decided to develop them and make them available to the world markets. Sonatrach embarked on the study of gas reserves 5 years ago with the help of several consultants; while the study was going on, the development of certain areas was undertaken.

At the time Sonatrach was negotiating the possibility of selling gas to El-Paso in 1970, arguments were raised that Algerian gas reserves at that time may not be sufficient to cover this contract. From 1970 to date alone Sonatrach has developed additional reserves which will more than cover the El-Paso contract. Since then Sonatrach has met the challenge to find money to develop these reserves, and meanwhile gained tremendous experience. Today Sonatrach appreciates the opportunity to present to you the Valorization Hydrocarbon Development Plan herein after referred to as "VALHYD" which consists of a review of the progress achieved in the development of its large oil and gas resources with an insight into what remains to be done in the next few years.

This presentation will cover:

- objectives of the VALHYD programme;
- programmed actions;
- gas and liquid hydrocarbon reserves and the planned production potential;
- investment requirements;
- financial projection over the period 1976-2005;
- total gas and oil cash-flow analysis;
- sensitivity analysis to variation in revenues, costs of capital and escalation.

OBJECTIVES OF THE VALHYD PROGRAMME

The VALHYD programme aims at achieving the following:

- recovery and utilization of gas associated with oil production;
- processing of associated gas and non-associated gas from gas condensate fields to recover the highest quantity possible of condensate and LPG (liquefied petroleum gases) within acceptable economic constraints;
- reinjection of gas especially associated gas whenever this will lead to a better oil-reserves recovery;

maximum recovery of condensate by total or partial cycling in gas condensate fields;
maintaining a level of sales gas volume as high and as stable as possible during the longest period of time while taking into account gas needs for cycling operation, reinjection in oil fields and gas lift;
systematic production of LPG at the level of gas processing units in the fields using gas from condensate stabilization units, and intermediate pressure units; in addition to this production in the fields, LPG extraction will be achieved at the level of gas-liquefaction units.

PROGRAMMED ACTIONS

In order to fulfil these objectives Sonatrach has engaged the following actions and programmes:

Reservoir engineering studies, surface facility studies, infrastructure and manpower-needs studies.
Drilling of various types of wells for extension, development, injection in gas condensate fields, dry gasfields and oilfields. Water wells for the needs of injection in oilfields, human consumption and industrial use.
Laying gathering lines and reinjection lines between wells and fields.
Construction of separation units for processing oil and associated gas.
Construction of compressor stations for gathering and delivery of associated gas to gas-processing units.
Construction of wet gas-processing centres.
Construction of recompression booster stations whenever necessary.
Construction of gas-reinjection stations besides gas-processing units on gas condensate fields and oilfields.
Construction of:
gas pipeline for the delivery of gas to the coast;
condensate pipelines for delivery to storage and loading facilities at the coast;
LPG pipelines for delivery to LPG separation units at the coast.
Construction at the coast of:
installations for storage and loading of condensate;
refrigerated storage installations of LPG for separation into propane and butane and loading;
gas liquefaction units and their associated units for LPG extraction.

GAS AND LIQUID HYDROCARBON RESERVES AND THE PLANNED PRODUCTION POTENTIAL

According to a study made by Degolyer and MacNaughton* Algeria's hydrocarbon reserves have been evaluated as follows:

Oil	= 1.1	billion tons
Condensate	= 442	million tons
LPG	= 196	million tons
Gas	= 3500	billion m^3

These represent the remaining reserves as of 1 January 1977 and represent the best estimates based on the latest technical data available as of that date.

*_Report on the Oil, Gas, Condensate and LPG Reserves of Algeria_, January 1977, Degolyer and MacNaughton, Dallas, Texas.

These reserves exclude:

- potential reserves in several basins of Algeria where practically no exploration has been carried out yet;
- reserves classified in the possible categories in basins containing producing fields;
- growth of reserves of fields recently discovered and which should extend due to further development drilling.

In order to develop these reserves and deliver their output to world markets, the following programme will be realized during the period from now to 2005:

- drilling of around 2760 development wells;
- laying of appropriate gathering and reinjection lines to connect wells to processing installations;
- construction of eleven processing units for non-associated gas, five of them for Hassi-R'mel only. The total capacity of these processing units is around 150 billion m^3 per year;
- eight gas reinjection units for a total capacity of 84 billion m^3 per year, out of which 60 billion m^3 per year are planned for Hassi-R'mel alone;
- laying about 4800 miles of transmission pipelines;
- constructions of seven LNG liquefaction units with a total capacity of 74 billion m^3 per year.
- construction of LPG separation units, storage and loading with a capacity of 9 million tons per year.
- construction of four new refineries for a total capacity of 29 million tons per year.

Figures 1 and 2 show the production of hydrocarbon liquids and gas respectively to be derived from the implementation of such a programme.

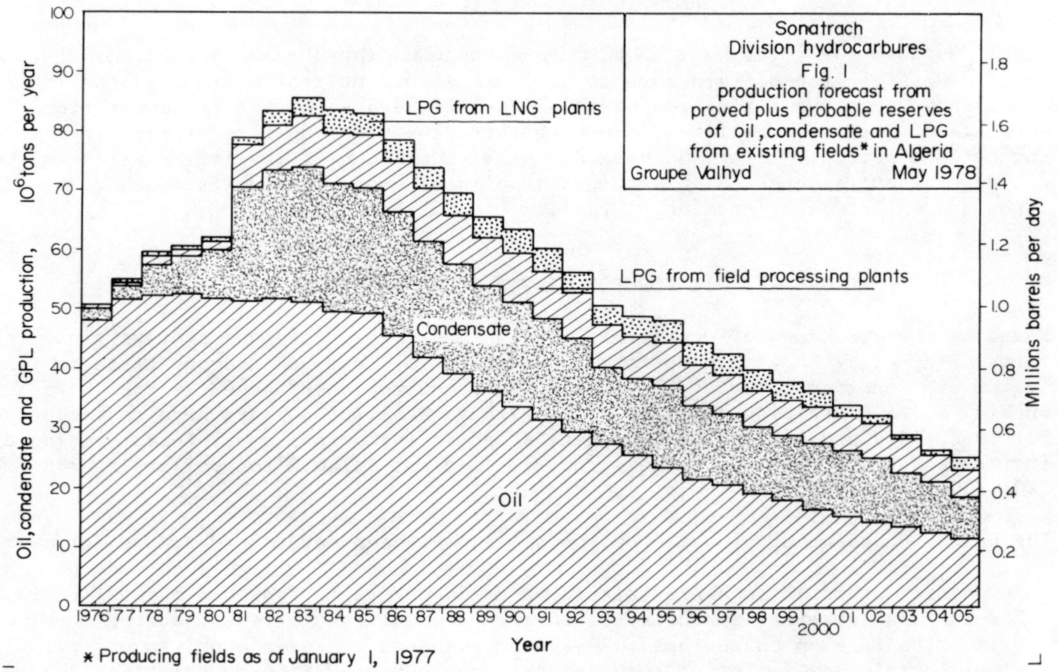

Fig. 1.

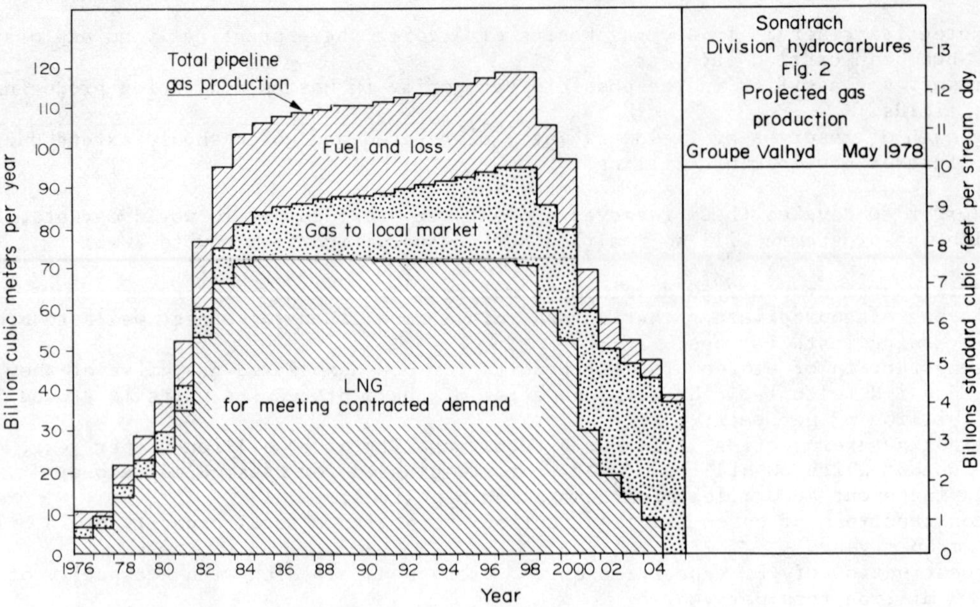

Fig. 2.

INVESTMENT REQUIREMENTS

Figure 3 shows the investment to be made throughout the 30-year period. In 1976 US dollars 33.4 billion dollars have to be invested during that period. These investments show a peaking during the period 1976-1985 because of the implementation schedule of various facilities requiring that 28 billion dollars will be invested during that period. Figure 3 shows also the proportion of investments tied to the gas development scheme (LNG units, gas pipelines, processing units, gathering lines and wells) as a function of time.

TOTAL GAS AND OIL FOREIGN DRAWINGS AND PAYMENTS

Based on the breakdown of costs into foreign and local costs and average financing terms obtained by Sonatrach on foreign component of cost and average terms of financing on the local component. The foreign debt and repayment schedule are shown on Fig. 4. Also in Fig. 4 it can be seen that 3 billion dollars have to be borrowed for 1978, 1979 and 1980 while repayment schedule (principal and interest included) increases from 250 million dollars in 1979 to about 2.7 billion dollars in 1983, 1984, 1985 and 1986.

The bases and assumptions on which the above have been determined are as follows:

New equity contributions are calculated at 25 per cent of annual capital costs.
Foreign loan drawings includes all foreign capital even interest during constructions on foreign borrowings. Repayment is based on seven equal annual principal instalments following a 4-year drawdown period during construction. Interest is capitalized during construction, financing costs are calculated at an interest of 8 per cent.

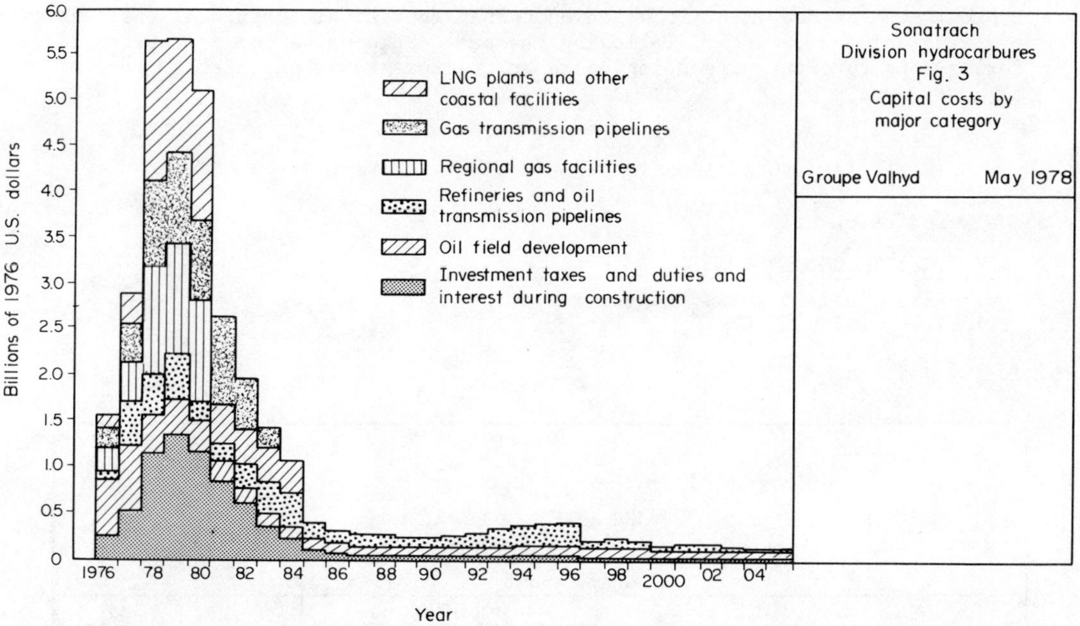

6.0
5.5
5.0
4.5
4.0
3.5
3.0
2.5
2.0
1.5
1.0
0.5
0
Billions of 1976 U.S. dollars
LNG plants and other coastal facilities
Gas transmission pipelines
Regional gas facilities
Refineries and oil transmission pipelines
Oil field development
Investment taxes and duties and interest during construction
Sonatrach
Division hydrocarbures
Fig. 3
Capital costs by major category
Groupe Valhyd
May 1978
1976
78
80
82
84
86
88
90
92
94
96
98
2000
02
04
Year

Fig. 3.

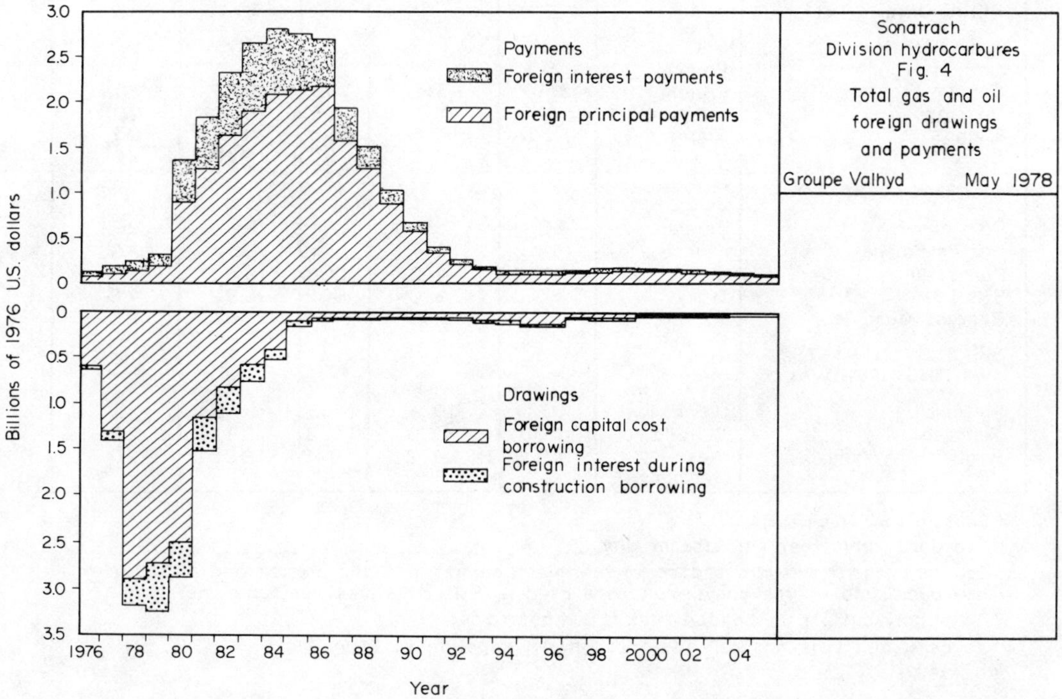

3.0
2.5
2.0
1.5
1.0
0.5
0
Billions of 1976 U.S. dollars
Payments
Foreign interest payments
Foreign principal payments
Sonatrach
Division hydrocarbures
Fig. 4
Total gas and oil foreign drawings and payments
Groupe Valhyd
May 1978
0.5
1.0
1.5
2.0
2.5
3.0
3.5
Drawings
Foreign capital cost borrowing
Foreign interest during construction borrowing
1976
78
80
82
84
86
88
90
92
94
96
98
2000
02
04
Year

Fig. 4.

Local loan drawings are calculated as the amount necessary to complete financing capital costs. Repayment is based on twenty four equal semiannual instalments of principal and interest, following the same drawdown period as foreign loans during construction. Financing costs are calculated at an interest rate of 5.75 per cent.

FINANCIAL PROJECTIONS 1976-2005

Based on prediction of:

1. Oil, condensate and LPG sales.
2. Gas sales (Table 1 shows the contracts).
3. Prices for 1976.

SONATRACH
Division hydrocarbures

TABLE I

LIQUEFIED NATURAL GAS SALES AGREEMENTS

Groupe Valhyd May 1978

CUSTOMER	DESTINATION COUNTRY	VOLUMES (a) Thermies Billion/Yr	Cubic Metres Billion/Yr	scfd Million	START OF DELIVERY	TERM OF CONTRACT (Years)
OPERATIONAL CONTRACTS						
British Methane	U.K.	10.5	1.1	116	1965	15
Gaz de France	France	5.3	0.6	64	1965	25
Gaz de France	France	35.0	3.7	392	1973	25
Distrigas	U.S.A.	12.0	1.3	138	1976	20
Enagas	Spain	45.0	4.8	508	1976	23
El Paso	U.S.A.	103.4	10.9	1153	1978	25
CONTRACTS APPROVED						
Distrigas	Belgium	50.0	5.3	561	1980	20
Gaz de France	France	51.5	5.4	571	1980	20
Panhandle	U.S.A.	45.0	4.8	508	1980	20
ENI (c)	Italy	85.0	9.0	952	1982	20
Ruhrgas Gasunie	West Germany	75.6	8.0	846	1982	20
SIGNED CONTRACTS AWAITING APPROVAL						
El Paso	U.S.A.	103.4	10.9	1153	1982	20
Tenneco	Canada (b)	1000	10.6	1122	1981	20
Brigitta-Thyssen	West Germany	40.0	4.2	448	1985	20

Source : SONATRACH

Standard cubic feet per stream day

(a) Contract quantities are stated in thermies volumes in cubic metres and scfd shown are approximate, the conversion rate used is 945 thermies per cubic metres

(b) To supply outlets in Canada and U.S. East Coast

(c) To be supplied as natural gas through 48" pipeline to Italy

Figure 5 shows the total gas and oil cash-flow analysis as well as the level of revenues to be expected covering out-of-pocket costs of operations, income taxes levied by the Algerian State and repayment of the debt service. It is to be noted that the buffer between revenues and debt is very large providing a margin for any reasonable variations in prices, costs or other parameters.

If we were to include future exploration activity for the years 1976-2005 the return on equity to Sonatrach would decrease from 35 to 32 per cent and the debt coverage would fall from 8.2 to 7.6 as shown in Table 2.

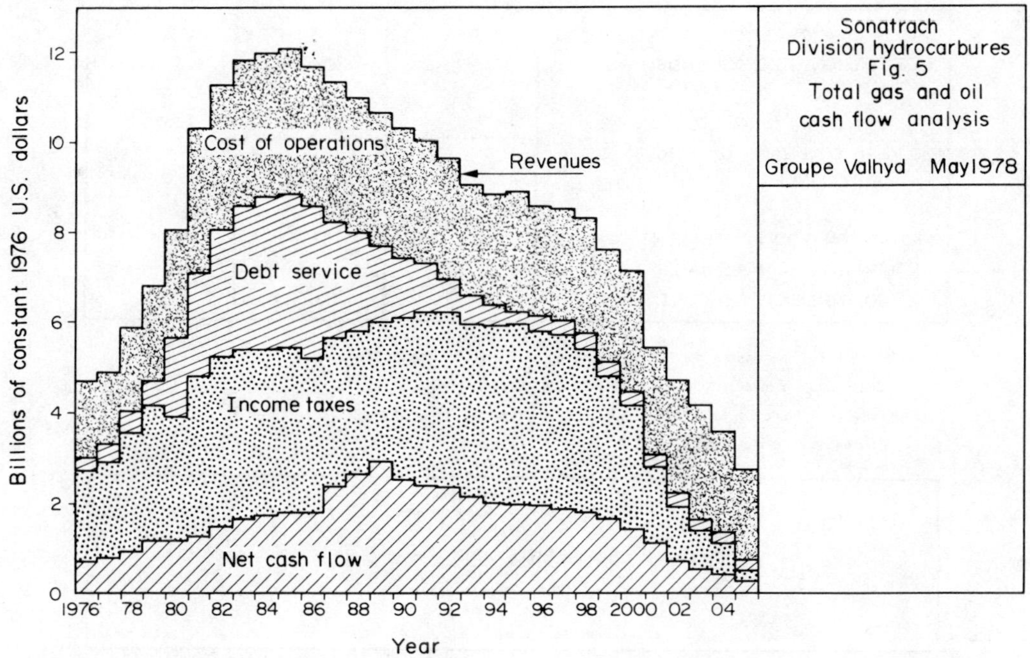

Fig. 5.

Sonatrach Pro Forma Income Statement and Pro Forma Net Foreign Exchange Inflows to Algeria

Table 3 shows the Sonatrach income statement for the oil and gas operations comprising the VALHYD project over the period 1978-1982.

To be noted is the decrease in net income in 1980 due to completion and starting of various installations and the increase in depreciation.

Table 4 shows the net foreign exchange in flows to Algeria during the period 1978-1982 and the coverage of the foreign debt service.

Sensitivity Analysis to Variations in Revenues, Capital Costs, Operating Costs and Interest Rates

In order to proceed with this sensitivity analysis, it has been decided to take as a parameter for comparison the discounted cash flow rate of return on equity to Sonatrach after tax of 35.42 per cent for the base case.

SONATRACH
DIVISION HYDROCARBURES
TABLE - 2
FINANCIAL PROJECTIONS INCLUDING
FUTURE EXPLORATION ACTIVITY
TOTAL FOR THE YEARS 1976-2005
(BILLIONS OF 1976 U.S. Dollars)
GROUPE VALHYD MAY 1978

	CASE 1 A BASE CASE	CASE 1-B*	CASE 1-C**
GROSS FOREIGN EXCHANGE EARNINGS	221.3	250.1	232.6
NET FOREIGN EXCHANGE INFLOWS	179.1	204.6	187.8
FOREIGN LOAN DRAWINGS	17.4	20.0	20.0
30 YEAR COVERAGE OF FOREIGN DEBT SERVICE	8.2X	8.2X	7.6X
RETURN ON EQUITY, PERCENT	35	32	32
RETURN TO THE ALGERIAN GOVERNMENT, PERCENT	181	176	175

* Case 1-B Expenditures for future exploration and revenues from future projected discoveries are taken into account for both gas and oil

** Case 1 C: Same as Case 1-B except that no revenues from the sale of gas from future discoveries are included

SONATRACH
DIVISION HYDROCARBURES
TABLE - 3
PRO FORMA INCOME STATEMENT
(BILLIONS OF 1976 US. DOLLARS)
GROUPE VALHYD MAY 1978

	1978	1979	1980	1981	1982
REVENUES					
FOREIGN	5.34	6.19	7.41	9.62	10.33
LOCAL	0.53	0.59	0.65	0.74	0.96
TOTAL REVENUES	5.87	6.78	8.06	10.36	11.29
COSTS AND EXPENSES					
OPERATING COSTS	0.73	0.82	0.96	1.47	1.65
ROYALTIES	0.95	1.09	1.23	1.44	1.43
INTEREST EXPENSE	0.18	0.19	0.63	0.78	0.92
DEPRECIATION	0.36	0.42	1.36	1.61	1.95
TOTAL COSTS AND EXPENSES	2.22	2.52	4.18	5.30	5.95
INCOME BEFORE TAXES	3.65	4.26	3.88	5.06	5.34
INCOME TAXES	2.61	3.02	2.74	3.53	3.71
NET INCOME	1.04	1.24	1.14	1.53	1.63

SONATRACH
DIVISION HYDROCARBURES
TABLE - 4
PRO FORMA NET FOREIGN
EXCHANGE INFLOWS-ALGERIA
(BILLIONS OF 1976 US DOLLARS)
GROUPE VALHYD MAY 1978

	1978	1979	1980	1981	1982
SOURCES OF FOREIGN EXCHANGE					
FOREIGN REVENUES	5.34	6.19	7.41	9.62	10.33
FOREIGN LOAN DRAWINGS	3.19	3.25	2.89	1.54	1.12
DIRECT PAYMENTS BY FOREIGN ASSOCIATES	1.33	1.31	1.24	1.21	1.20
TOTAL SOURCES	9.86	10.75	11.54	12.37	12.65
USES OF FOREIGN EXCHANGE					
FOREIGN OPERATING COSTS	0.18	0.20	0.25	0.36	0.43
FOREIGN INTEREST EXPENSE	0.12	0.13	0.48	0.60	0.70
FOREIGN CAPITAL COSTS	3.19	3.25	2.89	1.54	1.12
FOREIGN DEBT REPAYMENT	0.12	0.19	0.88	1.24	1.62
TOTAL USES	3.61	3.77	4.50	3.74	3.87
NET FOREIGN EXCHANGE INFLOW	6.25	6.98	7.04	8.63	8.78
COVERAGE OF FOREIGN DEBT SERVICE	26.21	22.77	6.18	5.71	4.78

Figure 6 shows the results when only one parameter, either revenues or capital costs, or operating costs or interest rate was varied.

The discounted rate of return on equity is sensitive to revenues and costs as shown in Fig. 6 but rather insensitive to other parameters.

Figure 7 shows the sensitivity to rate of return on escalation.

Were the investment cost to increase by 5 per cent compounded annually, the operating cost to remain unchanged and the revenue to escalate by 5 per cent compounded annually the rate of return on Sonatrach equity would increase from 35 per cent (Base Case) to 47 per cent.

CONCLUSIONS

With the fourth largest reserves of gas in the world, with more than sixty gasfields, Sonatrach is carrying out the VALHYD project to develop its oil and gas resources and make them available to the world market.

As determined by the consultant, De Golyer and MacNaughton, the estimates of the hydrocarbon reserves and production potential of Algeria are the best that Sonatrach could have as of 1 January 1977 on the latest technical data available then.

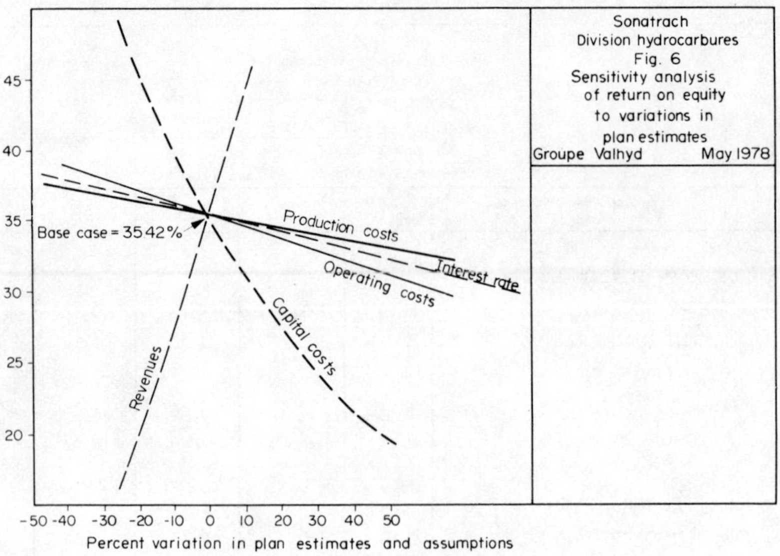

Fig. 6.

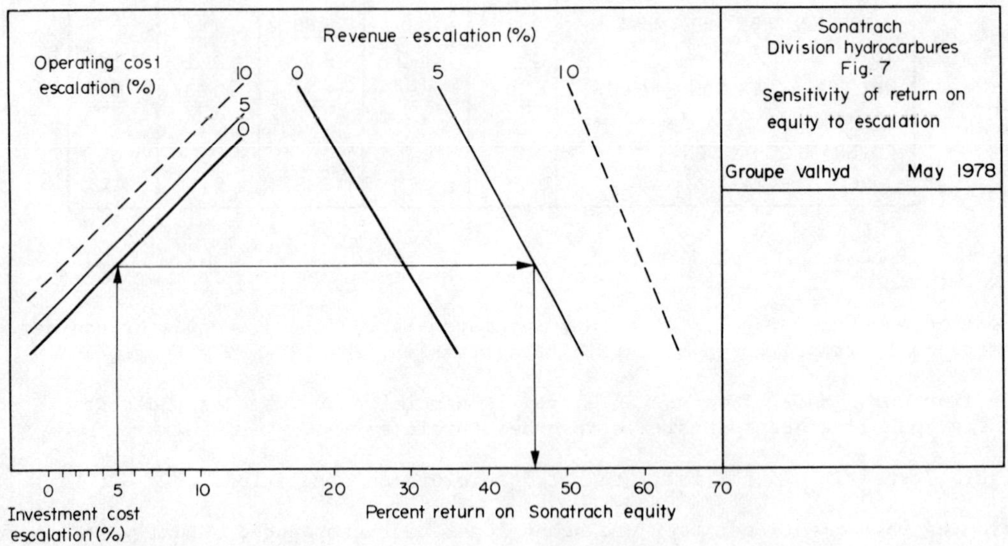

Fig. 7.

The recoverable reserves we have used in the VALHYD project do not include either:

the potential reserves in either known producing basins or other basins;
or the future reserves to be derived from the growth during the development of existing producing fields.

The gas reserves are large by any standard and would require big investments in order to develop them and make them available to the world market.

Sonatrach, being an active borrower in the international financial markets in the last few years, was able to raise financial resources for quite a few projects including that of GNL 1, SKIKDA, GNL 2, various gas pipelines to the coast and gas-processing and reinjection plants of Hassi-r'mel, Alrar and Rhourde Nouss.

However, Sonatrach needs to raise funds of the order of 3 billion $ US for each of the years 1978, 1979 and 1980.

Once developed, the reserves mentioned above should result in large revenues to Sonatrach and Algeria and should provide:

a quick pay out on investments;
an excellent coverage of the debt service;
and a good buffer against any reasonable swings in costs or other parameters underlying the plan;

No difficulty is to be expected for raising the necessary funds in the international markets.

Sonatrach is convinced that this is a sound and reasonable plan and is open to co-operation in carrying it out with all its partners in the international financial community.

Algerian Gas in the European Economic Community and the Rest of Europe: An Optimization Study in Economic and Marketing Diversification

M. Belguedj

Algeria

ABSTRACT

The bringing on stream in September 1977 of natural-gas production from two of the North Sea gasfields, Frigg and Ekofisk, has apparently begun the transformation of the gas industry of the European Economic Community. In fact this transformation started in the early 1960s with the first shipments of liquefied natural gas from Algeria being delivered to France and the United Kingdom.

The purpose of this paper is to investigate, through the development of this industry on the Continent, the penetration of Algerian gas.

This study must necessarily evolve through a preliminary analysis of existing reserves in the Community and their rapid depletion and to this end, it will show how during the first part of the twentieth century and until the late 1950s, natural gas accounted for only a relatively small share of the total energy consumption in Europe.

This situation evolved rapidly after the discovery in the early 1950s and 1960s, of the Lacq field in southern France and the Groningen field in the Netherlands. These were followed later by those of the southern part of the British sector of the North Sea and finally the offshore Norwegian fields of Cod, Ekofisk and Frigg.

Another underlying dynamic this paper will investigate is how, from a general point of view, the overall European gas picture does not clearly show: either the United States' characteristic of large reserves depleted rapidly due to extensive consumption patterns with few new reserves added; or that of Japan of near absolute lack of domestic reserves and undeveloped consumption potential.

Europe had a more balanced development model based on substantial domestic resources developed around manufactured gas plants extending into a wide gas-distribution network.

Then, it is assumed that if present strong market pressures for growth in gas consumption remain, they will have to be supplemented in growing proportion by natural gas and, increasingly so, by imported gas.

The fact that the Netherlands, hitherto the only exporter, is beginning to import small amounts of gas and will increase these volumes in the next decade means that the Community's internal supply, reaching an asymptotic level, will necessarily have to be complemented by other sources of imports, especially other than Algerian gas.

Thus the purpose here is to analyze through a detailed study of the supply-demand imbalance of these countries, the overall demand for Algerian and other gas in the Community. The basic concept is centred around the limited absorptive capacity on this side of the Mediterranean through the study of present and projected internal consumption patterns in Algeria. This leaves the country with a large balance for export calling for an optimization model based on marketing and economic concepts already proven in the field of petroleum.

This is further reinforced by the relative proximity of Algerian gasfields which lead to an innovative trade, as the technological gap were being filled, both by way of LNG technology and deep undersea pipelines. The concluding remarks will demonstrate how Algerian gas perking only at first into the Community's energy sector will necessarily become, by the next decade, its largest non-European supply source.

Other Reports

*Overall Economic Perspective for the ECE Region up to 1990 A Summary**

United Nations Economic Commission for Europe Secretariat

The overall economic perspective is based on the past and likely future development of the several factors and conditions of growth, *inter alia:*

investments,
basic products and energy,
environmental improvement aims, etc.

The overall growth rates both in market and centrally planned economies are expected to slow down up to 1990. The following factors could be sighted as being at the root of this slow down.

Labour supply is expected to slow down in the centrally planned economies up to 1990 and beyond. For the market economies, the opposite is true in the 1980s, with a likely slow down by the end of the century.

Capital requirements or, in other terms, investment requirements are likely to grow in most countries as compared with output in general. This is mainly due to the growing need for investments in sectors with low returns (energy, basic products, agriculture, infrastructure, improvement of environment, housing, etc.).

Energy and basic products are becoming scarcer and/or relatively expensive. This tends to result in foreign-trade imbalances for countries heavily dependent on imports of these materials. It imposes additional low-return investment requirements on countries endowed with these resources but extracting them under adverse conditions (USSR, Canada and the North Sea).

During the 1980s in the market economy countries there will be an excess of labour supply, while in the centrally planned economies demographic factors would tend to tighten it. In the 1990s, demographic factors would tend to reduce the growth of labour supply in the region as a whole.

---

*The complete version of the "Overall economic perspective for the ECE Region up to 1990" was endorsed by the Commission at its XXXIII session and recommended for circulation to the principal subsidiary bodies interested in long-term issues. The study is available as a United Nations document under the symbol ECE/EC.AD/17. The summary prepared by the ECE secretariat pursuant to paragraph 51 of ECE/GAS/38, is a selection of relevant parts of the study made in consideration of the work programme of the Committee on Gas and the Symposium.

In the centrally planned economies, the energy intensiveness of consumption is expected to decrease or, at least, to adjust to the slow down of supply. The shift in employment from agriculture to industry may tend to decrease and, in some countries, even to level off. The growth rate of fixed assets will be affected by the need to increase the share of investments in energy and investments for improvements in environmental conditions.

Because of the link between policies for domestic output and investment patterns on the one hand and for enlarging international co-operation on the other, the adjustment of economy to changing conditions is likely to be one of the main concerns of the centrally planned economy countries in the period up to 1990.

In the market economies, some saturation is beginning to appear with respect to "home equipment" and prospects are poor for growth in the housing service sector. In urban or very densely populated areas, energy saving may set limits on expenditure for heating and transport.

The overall annual growth rate of productivity is likely to be much lower than in the past for North-western Europe (2.8 per cent up to 1990) as compared to past growth rates of 3.8 in Northern Europe and 4.0 in Western Europe on account of a slowing down of industrial growth and investment and the changing structure of the economy, in particular the shift from sectors with high-productivity growth to low-productivity growth.

The projected level for 1990 of the gross fixed capital formation as a percentage of GDP might stabilize around 21-22 per cent in North-western Europe. In the non-farm sector, declining capital-output ratios were observed in the past 20 years but evidence points to likely upward trends in the future, especially considering such sectors as energy and utilities.

In the public sector, capital requirements are high but with a relatively low depreciation rate for the existing stock. Gross investment requirements for the public sector should therefore decelerate, at least in North-western Europe.

Annual demand for houses might stabilize at 20-30 per cent less than in the past in North-western Europe.

Regarding overall growth projections for Western Europe as a whole, a range of 3.5 to 4.1 per cent as a potential GDP growth rate up to 1990 may be expected, i.e. a figure close to the range estimated for North America.

In the supply of energy, conditions have fundamentally changed. Costs have generally risen and the structure of international price formation has been modified. These changes are likely to affect the future economic growth pattern in the countries of the ECE region. The following prospective in the period up to 1990 may be given special consideration.

Readjustments in energy-*supply* patterns, decrease in imported fuels; increase in R and D activities and investments for the development of renewable resources; rational supply patterns (reduction of loss and waste when transforming primary energy).
Increases in efficiency of the use of energy.
Increases in investments of energy as compared with other sectors.
Means of reducing demand for energy, while maintaining a satisfactory growth rate for the economy as a whole.

New approaches to the problem of supply and use of energy may also bring about important changes in the pattern of foreign trade of ECE countries. These changes might be summarized as follows:

- the complementarity between countries well and poorly endowed with energy resources is expected to call for the establishment of long-term and stable international trade arrangements together with other forms of international co-operation;
- future international trade patterns are expected to be less energy-intensive in general.

In view of the foregoing the adequacy of institutions and mechanisms for a widened and deepened co-operation between countries is being examined.

Future economic co-operation implies a long-term vision of economic development which goes beyond trade in the traditional sense and comprises problems relating to investment policies, etc.

With regard to region-wide co-operation, especially between countries having different economic and social systems and at different stages of economic development, this co-operation in the years ahead seems likely to give, *inter alia*, special emphasis to:

- joint projects for the development of energy resources;
- common efforts for the improvement and extension of the regional intrastructure;
- international financing of investments.

Within the framework of the CMEA, the following main forms of co-operation are expected to continue in future years:

- joint investments;
- co-operation and co-ordination in building up medium- and long-term national plans.

The preparation of long-term programmes of co-operation is under way in the fields of energy.

The market economy countries may be expected to continue to make special efforts towards:

- a harmonization of national economic policies, particularly with respect to energy, finance and external economic relations;
- co-operation in large-scale investment and production projects.

Countries developing from the economic point of view seem likely to attach special importance in the decades ahead to the means of financing investments and imports of capital equipment, especially arrangements for repayment in products.

At the end of the study, the Senior Economic Advisers to ECE Governments suggest subjects as important areas for further consideration or work, some of which are:

1. current and projected energy-investment requirements for ECE member countries and projected changes in the energy-supply capacity;
2. estimation of long-run marginal energy-supply costs and energy-price policy for changes in energy-supply and demand patterns;
3. forms of long-term investment and trade arrangements between energy-resource-surplus and energy-resource-deficit countries of the ECE region, with particular emphasis on co-operation.

Natural Gas Reserves and Resources in Europe (General Review)

W. Schott*

Federal Republic of Germany

SUMMARY

Natural gas is a mixture of gaseous hydrocarbons, and sometimes of other gases (nitrogen, carbon dioxide, hydrogen sulfide), as found in the subsurface. In many regions, natural gas is associated with the formation of petroleum, in other places it is related to the coalification of organic matter in fossil sediments. This review of natural gas reserves in Europe, especially the development of these reserves during the last few years, stresses the intensive efforts of the natural gas industry to explore for accumulations of natural gas in Europe. Various methods are used to determine the reserves of natural gas in a gas field (e.g. material balance method, volumetric method). The amount of natural gas in a gas field is dependent on the geological factors existing in the reservoir rock (e.g. porosity, permeability, connate water, etc.). The geological factors present in the gas reservoir, the state of exploitation technology, and economic factors influence the volume of production from a gas field. The ultimate amount of gas recovered fluctuates between 50 per cent and 90 per cent of the gas existing originally in the reservoir. Different terms are used in European countries for the classification of natural gas reserves. The main centres of gas reserves in European sedimentary regions can be seen in the geographical distribution of the natural gas fields. Natural gas resources and further natural gas prospects are expected to be developed in the onshore sedimentary regions, especially with the help of ultra-deep wells. There are several possibilities in the European shelf regions and also beyond the shelf edge. New discoveries of natural gas accumulations in these areas will influence the gas situation around 1990.

INTRODUCTION

A general review of natural gas reserves and resources in Europe is given here as an introduction to this symposium on the situation with respect to natural gas in the ECE region around the year 1990. Following a general introduction, the problems connected with the estimation of the reserves in a gas field will be discussed and, finally, the European regions where natural gas resources can be expected from the geological point of view.

Natural gas is a mixture of gaseous hydrocarbons and sometimes other gases (e.g. nitrogen, carbon dioxide, and hydrogen sulfide) as found in the subsurface. In many

*In accordance with para. 11 of the report of the Preparatory Meeting for the Symposium held in April 1977 (GAS/SEM.4/AC.1/2).

Prof. Dr. W. Schott, c/o Bundesanstalt für Geowissenschaften und Rohstoffe, Alfred-Bentz-Haus, Stilleweg 2, D-3000 Hannover 51.

parts of the European sediment basins, the natural gas is associated with the formation of petroleum, e.g. in the Orenburg field (Fig. 1). In other places the gas is related to the coalification of organic matter in fossil sediments. Thus, the gas accumulation at Groningen was generated by the further coalification of the coal in the Upper Carboniferous below the Rotliegend (Fig. 1, Ref. 39).

Natural gas can be found in nearly all European regions that contain a thick sequence of sediments that are not strongly folded (Fig. 1*). The distribution of the gas fields is dependent on the general geological and tectonic situation in these sedimentary basins. This can be recognized on the International Map of Natural Gas Fields in Europe. This map has been compiled by the Group of Experts on Natural Gas Resources and published 1972 by the ECE Committee on Gas and the Bundesanstalt für Geowissenschaften und Rohstoffe at a scale of 1:2 500 000. An explanatory note to this map has been elaborated in the three official languages of the United Nations (English, French, and Russian).

NATURAL GAS RESERVES

Very large gas fields with reserves of more than 100×10^9 m^3 have been discovered in Europe during the last thirty years, e.g. the fields Orenburgskoe and Shebelinskoe in Eastern Europe and Groningen and Leman in Western Europe (Fig. 1). The main gas-condensate reservoir at Orenburgskoe is a limestone sequence of the Middle Carboniferous up to 530 m thick. The gas was discovered in a large structural uplift along a 100 km profile at depths of 1300-1750 m. The gas reservoir is sealed by a thick salt deposit, thus the gas could not escape (Ref. 37, 38).

The Groningen natural gas reservoir is situated in a gently folded and faulted dome structure in the northern part of the Netherlands. The pay zone is the Slochteren sand of the Rotliegend. The thickness of this sand ranges from 100 m in the south part to 200 m in the north part of the field. The gas-bearing Rotliegend is covered by Zechstein salt (Ref. 1, 51).

The intensive efforts of the European gas industry in the exploration for natural gas accumulations have been successful in causing a rapid increase in the natural gas reserves in Europe (incl. Asiatic USSR) from about 6808.5×10^9 m^3 in 1967 to about 26083.4×10^9 m^3 on 1 January 1978 (Fig. 2, Table 1). These amounts are only estimations, because several countries have not submitted their annual figures for natural gas reserves to the ECE, Committee on Gas.

It is very important to determine the amount of reserves in a newly discovered natural gas accumulation immediately after its discovery. When figures are available from the first production tests, it is possible to decide from the point of view of the economics whether a gas discovery should be developed.

The amount of exploitable reserves of natural gas in a gas field is dependent on the geological structure which contains the gas accumulation and on the geological and physical parameters of the reservoir rock, e.g. porosity, permeability, water saturation of the pore space, pressure, temperature, etc. The lithofacies of the reservoir rocks influences the porosity and permeability of the pay zone. Porosity and permeability do not generally fluctuate so intensively in sandy strata as in limestone. The average porosity in the sand of the Groningen gas field is for instance 18.3 %, the permeability ranges from 0.1 to 1.0 Darcy. Calcareous reservoir rocks, especially those of reef formations, are often very inhomogeneous in comparison with sandy rocks. Therefore, porosity and permeability may be different from well to

*Non prospective areas in the Figs. 1, 3 and 4 are preponderantly areas with basement and highly metamorphic rocks outcropping.

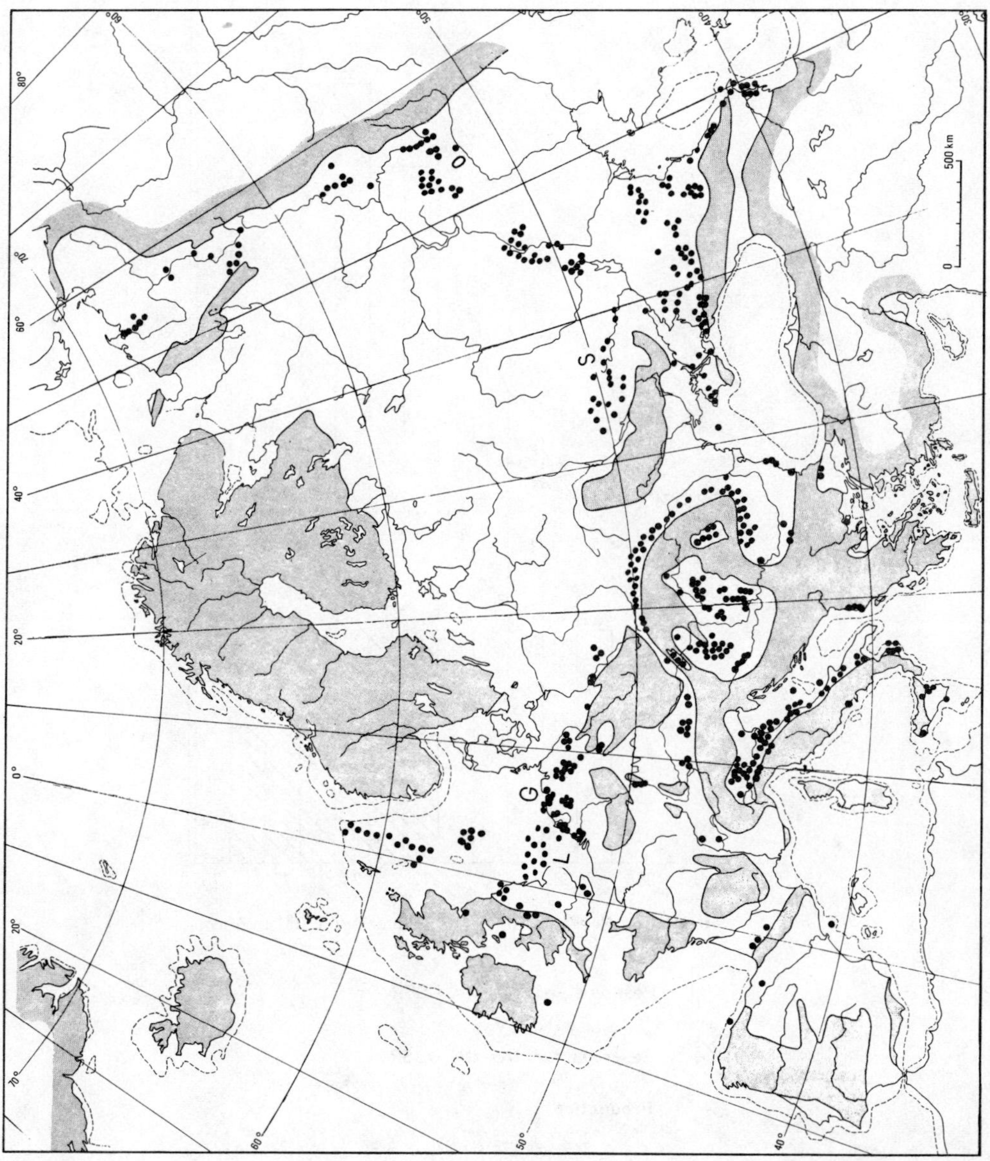

Fig. 1

Natural gas in Europe

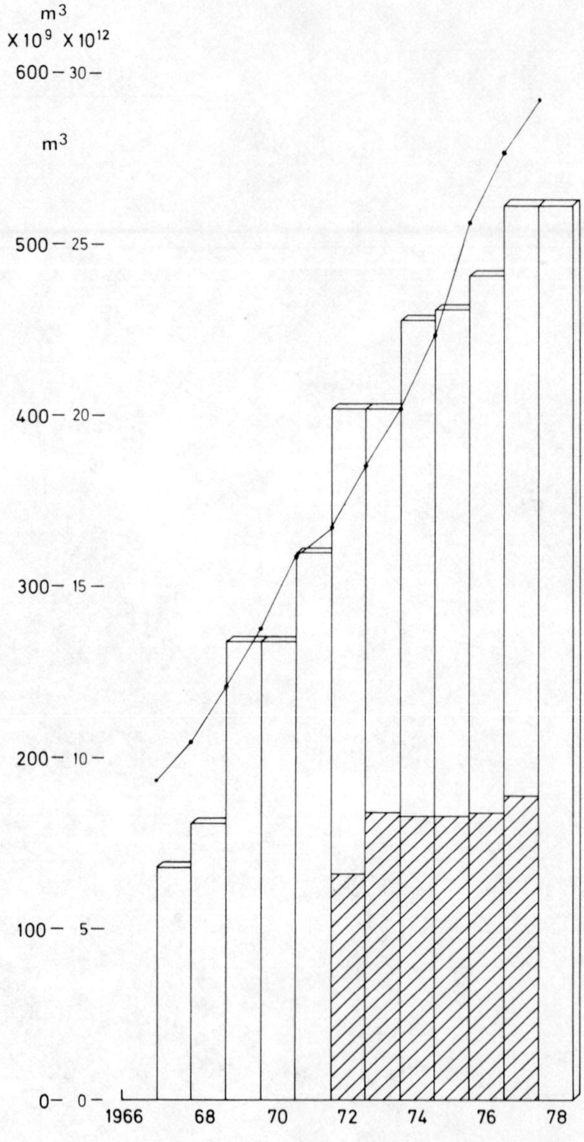

Fig. 2 Natural Gas Reserves (10^{12}) and Production (10^{9}) in Europe

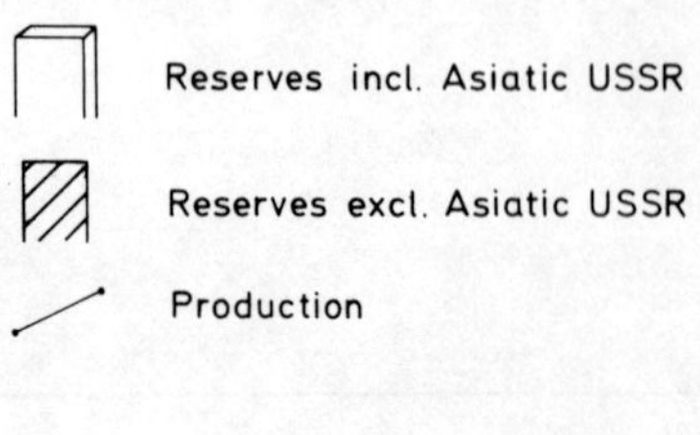

TABLE 1

Reserves and Production of Natural Gas in Europe
(in 10^9 m^3)

Year	Reserves incl. Asiatic USSR	Reserves excl. Asiatic USSR	Production incl. Asiatic USSR
1966			187.4
1967	6808.5		209.0
1968	8056.4		242.4
1969	13 448.5		275.2
1970	13 436.5		317.8
1971	16 016.5		333.0
1972	20 221.1	6607.5	369.9
1973	20 246.1	7389.2	404.2
1974	22 785.4	7273.8	447.4
1975	23 069.8	7311.7	512.5
1976	24 088.9	7379.4	552.3
1977	26 060.6	7925.0	582.9
1978	26 083.4		

well in calcareous pay zones. On the other hand, the permeability of the fissures in limestones may influence the flow of gas to a well. Not only these factors present in the gas reservoir but also the state of exploitation technology and economic factors decide the amount of gas production from a field and thus the amount of exploitable reserves.

There are two methods used for determining the amount of reserves in a gas field: the volumetric method and the material balance method. After the discovery of a gas accumulation, only the volumetric method can be used to determine the amount of gas in place in the reservoir rock of a structure so long as there is only one well in the deposit. The parameters of the pay zones obtained by well logging and/or investigation of cores and from seismic results are necessary for calculating the amount of gas in place in the deposit. After extension wells have been drilled, the amount of gas in place can be calculated more accurately with this method because more is known about the inhomogenity of the reservoir rock and sometimes the gas/water contact, if present.

The amount of gas in place may be calculated more exactly with the help of the material balance method after the production of gas has begun in the field. The decrease in pressure in the reservoir is dependent on the ratio between the original amount of gas in place and the amount of gas produced. This method can be used even if water intrudes into parts of the reservoir that had been occupied by gas before production was started.

If the amount of gas in place is known for a reservoir, the amount of exploitable reserves can be determined. A recovery factor is used to calculate the exploitable reserves in the reservoir. The recovery factor for a gas deposit is dependent on the parameters of the pay zone mentioned above, on the present technological and he economic situation in the gas industry and last but not least on the quality of the gas. Sweet gas and gas condensate are more economic than sour gas with its higher hydrogen sulfide content. Sour gas must be chemically stripped before transport. Therefore, it has a relatively lower value; the same is true for gas with a high nitrogen content. For these reasons the ultimate amount of gas recoverable from a gas field fluctuates between 50% and 90% of the original amount of gas in place in the reservoir.

The members of the Group of Experts on Natural Gas Resources of the Committee on Gas have not only collaborated in the preparation of the International Map of Natural Gas Fields in Europe but have discussed for several years all problems connected with the determination of gas reserves. One aim of these discussions has been to correlate the different terms used for the classification of the natural gas reserves in the different European countries (see Table 2). Only thus is a comparison of the published reserves possible from country to country.

Tab. 2 TERMS USED IN DIFFERENT COUNTRIES FOR CLASSIFICATION OF NG RESERVES

Category of Reserves	France	Federal Republic of Germany	Netherlands	Poland	United Kingdom	United States	USSR	Yugoslavia
Proven	Proven	Proven	Proven	A+B	Proven	Proven	A+B	A
Probable	Probable	Probable		C_1	> 50 % chance	Probable		B
Unproven			Unproven				C_1	
Possible	Possible	Possible		C_2	< 50 % chance	Possible		C_1
Prospective				D_1+D_2		Potential	C_2	C_2
Undiscovered	Potential	Undiscovered	Potential		Undiscovered			
Speculative	Speculative					Speculative	D_1+D_2	D_1+D_2

NATURAL GAS RESOURCES

Natural gas resources and further natural gas prospects are to be expected in the sedimentary regions of the European continent. There are likewise several possibilities in the European shelf areas and also in deep water regions beyond the continental shelf.

Resources on the Continent

Exploration has not yet been finished in many onshore sediment basins even if large gas fields have already been discovered there. This is the case in several regions of Europe, especially in the European part of the USSR (Ref. 33). The exploration will no doubt be generally more difficult after the main tectonic structures have been investigated.

Some examples:

In the southeast European Caspian basin, hydrocarbon prospects are connected with reef-building zones of the subsaliferous sediment sequence. In elevated marginal regions of the basin, carbonate rocks of the subsaliferous sequence may be gas-bearing, similar to the Orenburg gas condensate field (Ref. 15). The results of a detailed tectonic investigation in the southern part of the Ural foredeep permit the assumption that further hydrocarbon prospects can be discovered in this over 2500 km long foredeep although several fields are already present there (Ref. 24).

The Pechora basin in the northern part of the USSR is a new hydrocarbon province that will be very important for the future. This basin is filled with Paleozoic, Mesozoic and Cenozoic sediments that are up to 15 000 m thick in some places. Since the discovery of the Vuktylskoe field in 1966, with more than 100×10^9 m^3 gas condensate in a Permian-Carboniferous reservoir, geophysical investigations in the basin have found several anticlinal structures that contain hydrocarbon accumulations (Ref. 57). An onshore well, deviated under the Pechora Bay, on the arctic coast tested natural gas (Ref. 25, 1975, p. 1831). This new Vasilkovo gas-condensate field is important for the future exploration on the Barents Sea shelf (Ref. 25, 1978).

Detail prospecting for hydrocarbons, directed to a large extent at stratigraphic traps of various origins and types, will also be successful in the large sedimentary basins of the European USSR (Ref. 61). Such intense exploration of stratigraphic or combined structural-stratigraphic traps can be successful even in small sediment basins in which hydrocarbons have already been produced for several years, as for example in the Vienna Basin. Four new gas accumulations have been explored during the last few years in the Tertiary (Neogene) fill of this basin (Ref. 28, 1977). A similar success is to expect from the Pliocene strata of the Po basin of Italy (Ref. 4). The wildcat well Cancello 1 was the first gas producer on the coast of the Tyrrhenian Sea about 25 km northwest of Naples in an area where hydrocarbon accumulations had previously not been found.

The deepest parts of the gas-bearing sedimentary basins, as well as the strata beneath them, have generally been regarded as prospective for gas accumulations for several years, but especially since the discoveries in the Vienna and Po basins. The first gas reservoirs to be opened up below the Tertiary sediments of the Vienna and Po basins consist of dolomite strata of primarily Triassic age at depths of about 5000 and 5400 m, respectively (Ref. 17, 23, 29).

A large number of deep and ultradeep wells have been drilled during the last few years in all parts of Europe, some in the margins of the sedimentary basins, some in the deeper parts of the basins (Fig. 3). The pre-Permian rocks of the Cambrium to the Carboniferous are regarded as prospective for hydrocarbons in the northwestern and northern European basins (Ref. 5, 7, 8, 12, 26, 27, 46, 47, 56, 62, 63).

Deep and ultradeep wells have also been drilled or are planned in the northern frontal zone of the Pyrenees and the Alps, others within the Alpine-Carpathian overthrust zone to test Cenozoic and Mesozoic rocks below the overthrust. For instance, an oil and gas accumulation was found in Liassic sandstone in Austria (well Klement 1) (Ref. 28, 1975). Several ultradeep wells have been drilled or are planned also in other European countries, some of them outside of the gas-bearing sedimentary basins (Ref. 25, 1976, pp. 1710, 1712, 1719; 1977, pp. 1649, 1650, 1652, 1653; Ref. 49, pp. 43, 51, 147; Ref. 50, p. 354, etc.).

Many deep wells have been drilled to investigate the deep reservoir rocks in the hydrocarbon-bearing sedimentary basins in the European part of the USSR. New pay zones have been explored at depths of more than 4000 m (Ref. 14, 16, 35, 41). The Saatly well in the Kura basin southwest of Baku has a target depth of 15 000 m. Oil and gas shows have been found at a depth of 7000 m by the deep test well Shevchenkovo 1 in the Dolina field in the Ukraine (Ref. 25, 1975, p. 1831).

Other ultradeep well projects being carried out are Vorderriß 1, which is supposed to investigate primarily the sediments below the overthrust of the northern "Kalkalpin" south of Munich (final depth 6468 m) (Ref. 58), and Zistersdorf ÜT 1, which is

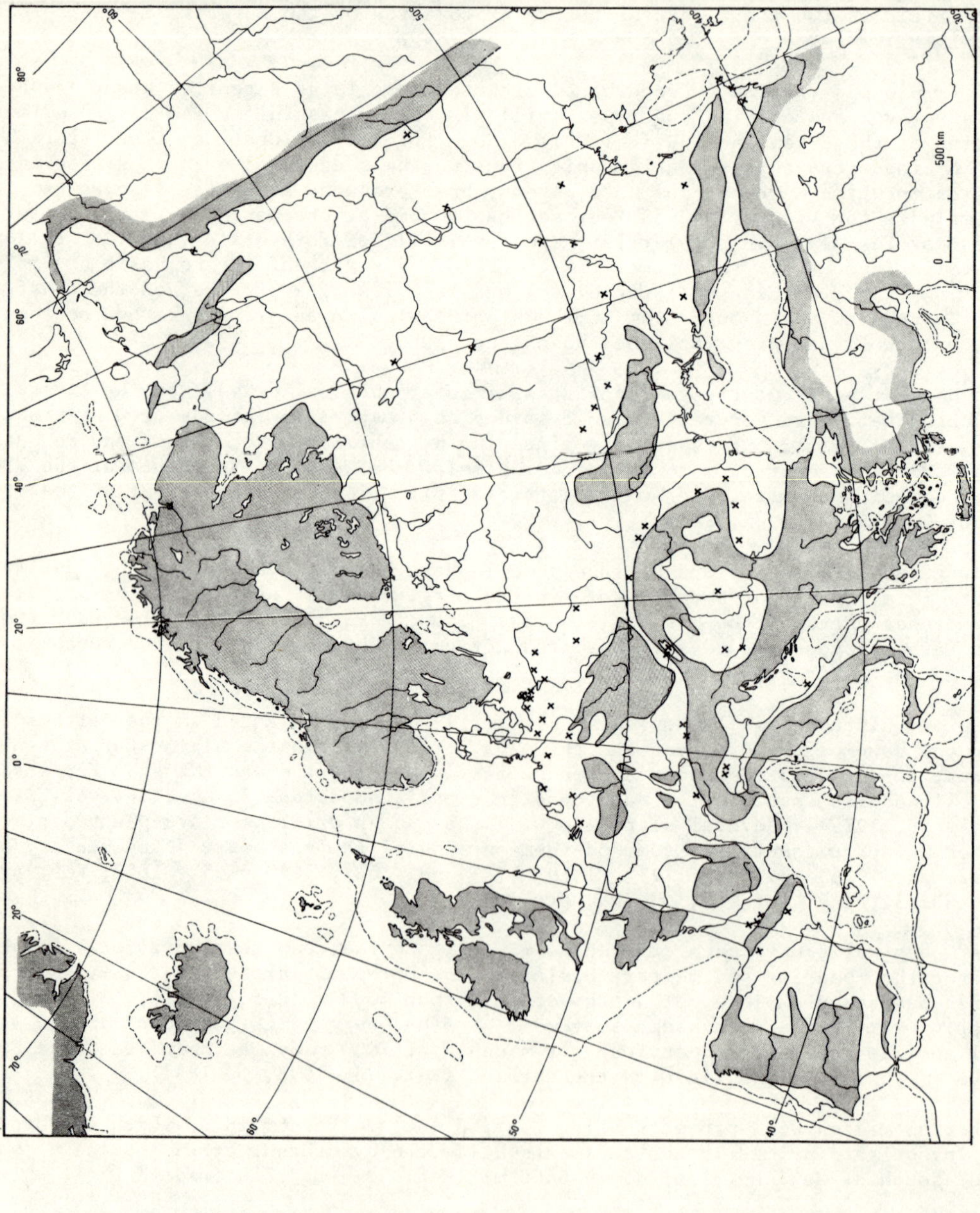

Fig. 3

Deep or ultradeep wells in Europe

to explore the deepest part of the Tertiary in the Vienna basin and the Mesozoic rocks below (target depth 7500 m).

Exploration results and the geological and tectonic situation in the basins clearly indicate that natural gas resources are still present in the onshore European sedimentary regions.

Resources in the Shelf Regions

Several European gas-bearing basins extend into the bordering shelf regions, for instance, the northwest German basin extends into the North Sea, the Po basin into the Adriatic Sea. After the discovery of enormous gas reserves in the North Sea, interest in offshore prospects extended immediately to all of the European shelf regions where sedimentary basins with possible hydrocarbon reservoirs are expected (Fig. 4). This intensive search has been successful in the last few years in discovering new areas with natural gas occurrences. Mesozoic and Cenozoic rocks are the reservoirs (Ref. 20; 25, 1975, 1976, 1977, 1978; 49, pp. 148-150).

New hydrocarbon accumulations were discovered in the Mediterranean off Tarragona Province in 1975 and 1976. Further occurrences of natural gas can be expected on the southern shelf of Sicily, now that wildcat Mila 1 has found natural gas and condensate in the Upper Triassic Taormina dolomite southwest of Ragusa. Most of the shelf in the Adriatic Sea is now prospective after gas observed in Yugoslavian offshore wells (Ref. 25, 1978); new discoveries are also to be expected on the shelf in the Italian Adriatic. Exploration in the offshore Tertiary basins of Greece has been intensified as a result of the first commercial development of a natural gas and condensate occurrence in the northern Aegean Sea.

The Black Sea is also very prospective for hydrocarbons. The offshore Golitsina gas field is about 160 km southeast of Odessa in the western continuation of the Odessa – Sivaskoje graben, which is gas-bearing on the Crimean Peninsula. Activity has also begun in the Black Sea off Romania. The first well, about 130 km offshore, has found oil in this area (Ref. 25, 1978). The Tjulenovo oil field has a gas cap and extends seaward from the Bulgarian coast. Cenozoic and Mesozoic reservoirs are prospective primarily for gas in these shelf areas of the Black Sea. The Strelkovskoe gas discovery initiated intensive exploration in the Sea of Asov. The gas-bearing Odessa-Sivaskoje and Indol-Kuban depressions extend under the Sea of Asov (Ref. 20, 1977; 25, 1977; 40, 49, p. 148). Several further hydrocarbon accumulations have been discovered recently below the European and Asiatic shelf regions of the Caspian Sea, e.g. the Shakhovo More gas field and the Bakhar gas condensate field south of the Apsheron Peninsula. The latter contains eight reservoirs, the deepest at 5200 m. Numerous prospects indicated by seismic surveys in the Caspian Sea have remained untested for lack of offshore drilling rigs. These are excellent prospects for the future (Ref. 25, 1977).

In the Atlantic shelf regions, a new hydrocarbon province in the Spanish part of the Bay of Biscay was found in 1976 in an area where future activity will be very interesting (Ref. 59). Exploration activity in the Irish Sea will be intensified as a result of the discovery of natural gas in the Triassic sandstone of the Morecambe field (Ref. 6; 25, 1975, 1977; 60). The search for hydrocarbons on the Atlantic continental margin west of the Shetland Islands has been of special interest during the last few years. Wildcats have been drilled there in the troughs flanking the ridge at the shelf edge. Three petroleum occurrences were discovered in 1977, one with natural gas in a Lower Cretaceous/Upper Jurassic reservoir (Ref. 13; 25, 1978).

The North Sea shelf has been the most significant European offshore exploration area for natural gas for several years. The natural gas in the Permian pay zones

Fig. 4

Offshore exploration in Europe

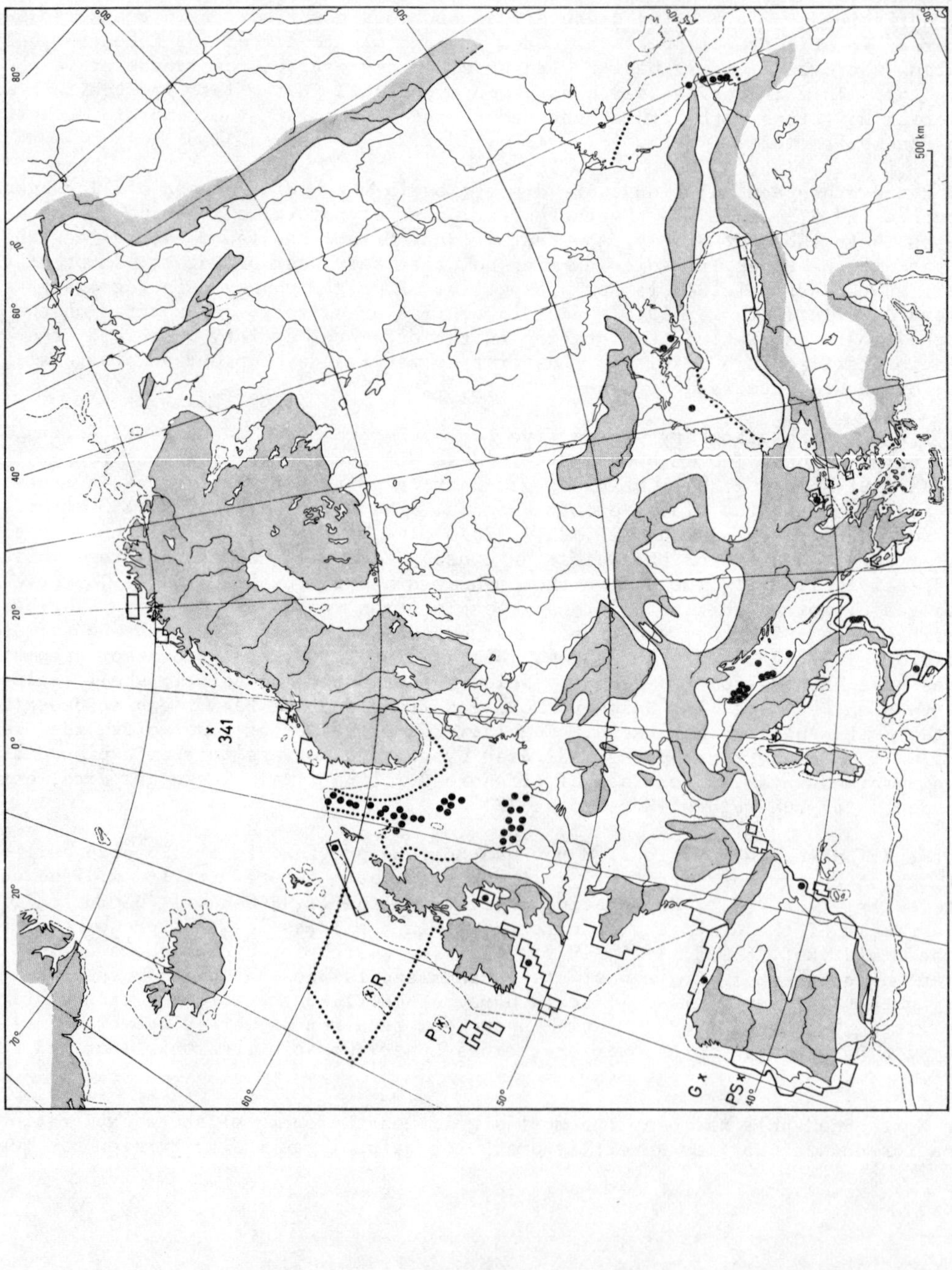

of the southern North Sea was generated by the further coalification of the coal in the Upper Carboniferous. The gas in the Mesozoic and Cenozoic rocks below the middle and northern North Sea shelf is associated gas. Further natural gas resources are to be expected in the North Sea, but the "giant" gas accumulations have probably already been discovered. A gas discovery was made in 1974 in the Netherlands sector in a Carboniferous sandstone that had not been previously found to be productive in the North Sea. Exploration activity will continue for several years, but the intensity of exploration depends on several economic factors. The proven and probable natural gas reserves in the North Sea amounted to about 2250×10^9 m^3 at the beginning of 1978 (Ref. 3, 19, 25, 1975; 32).

Natural gas resources can be expected along the continental shelf off central and northern Norway. It is assumed that Cenozoic and Mesozoic sediments are present with reservoir rocks similar to those of the northern North Sea. Gas shows in a shallow borehole on the island of Andøya (Vesterålen) are good indications. Seismic measurements during the last few years show that the structural features are in general similar to those in the northern North Sea, but there are also clear differences. The question as to whether the success in the North Sea will continue along the Norwegian Shelf can be answered only by exploration wells. A first wildcat is planned in the northern permit area of the Norwegian State Oil Company northwest of Finnmark (Ref. 18, 1975, 1976, 1977; 19, 21, 22, 49, pp. 101, 149; 52).

The Barents Sea is the most prospective area for hydrocarbon resources in the European part of the Arctic Sea. The Barents Sea region is probably an old epicontinental platform which developed after the Caledonian orogeny. Thick sediments were deposited during periods of subsidence. This can be assumed mainly from the onshore geology of northern Scandinavia and Svalbard. Several geological structures (basins and ridges) have been recognized in the western Barents Sea by refraction and reflection seismics.

There are rock outcrops of different ages on the sea floor. Tertiary sediments up to 4000 m thick have been observed. The thickness of the Mesozoic rocks varies between 3000 and 8000 m. A thicker sequence of Paleozoic rocks is to be expected. Two evaporite sediments are probably of Upper Paleozoic age (Ref. 21, 45). Gas shows were reported from the exploration well on the Tromsøbreen anticline in the southernmost part of Spitsbergen (Ref. 25, 1978).

As far as is known, the geological structures under the eastern part of the Barents Sea are similar to those in the western part. In the southeast, the Barents Sea basin passes into the gas and oil producing Pechora Depression. In addition to the well in the Vasilkovo gas-condensate field (deviated from the coast under Pechora Bay) that tested natural gas, oil and bitumen indications have been observed in Paleozoic rocks on Novaya Zemlya (Ref. 10, 11, 18, 1977; 25, 1975; 30, 49, p. 149). The hydrocarbon-bearing Pechora basin passes in the Barents Sea shelf area in the same manner as the onshore, northwest European basin passes in the North Sea.

Oil and gas accumulations have also been discovered in the coastal area of the Baltic Sea. Reservoirs are Paleozoic rocks in the Baltic and Gotland region, Jurassic sandstone in Schleswig-Holstein. It is questionable whether natural gas occurs in commercial quantities in the Baltic Sea area (Ref. 25, 1975, 1976, 1977; 34, 42, 49, p. 149; 55).

Resources Beyond the Shelf

Geoscientific investigations of the continental margins have shown that sediment basins can be present in the deep water regions of the oceans beyond the shelf edge. These basins may also be prospective for natural gas. Therefore, several permits have been granted for areas beyond the continental shelf and some wildcats have already been drilled there (Fig. 4, Ref. 9, 31, 48).

Only three examples from the European deep-water regions:

A basin with sediments 5000-6000 m thick parallels the coast of the Iberian Peninsula beyond the shelf edge between the city of Porto and Cape Finisterre (Fig. 5). The basin is bordered in the west by the Porto and Vigo seamounts and the Galicia Banks. These structures are isolated, sunken continental blocks (microcontinents). The fill of the basin consists of Tertiary, Cretaceous, and Jurassic sediments, an interesting sequence that probably contains natural gas resources (Ref. 36). Northwest of the shelf edge off Ireland and Scotland is another major microcontinent, the Rockall Plateau with Rockall Island. This microcontinent is bordered by the Hatton-Rockall basin in the northwest and the Rockall Trough in the southeast. The thickness of the sediments in these troughs are about 2000 m and more than 2000 m, respectively. Discordances have been observed in the Tertiary sediments. The Porcupine Bight basin between the southwest Irish shelf and the Porcupine Bank microcontinent contains a prospective thick sedimentary section above the Precambrian to Carboniferous basement indicated by seismic surveys and boreholes. Uneconomic hydrocarbon indications were observed by the first two wildcats in this area. Additional drilling is planned (Ref. 2, 25, 1975, 1978; 43).

The offshore areas of central and northern Norway, extending from the shelf seawards into the deep water regions, especially in the Vøring Plateau area, are very prospective for natural gas. The basins along this part of the Norwegian continental margin contain sediments from 2000 to more than 4000 m thick. Very interesting hydrocarbon indications were discovered in 1974 in Pleistocene and Tertiary sediments in the DSDP well 341 drilled at a water depth of 1439 m southeast of the Vøring Pleateau escarpment (Ref. 44, 45, 52, 53, 54).

CONCLUSIONS

This review of natural gas resources in the onshore and offshore sedimentary regions of Europe has shown that further natural gas accumulations of economic value can be explored. It is understandable that in areas where exploration and exploitation activities have been taking place for some time, the search for natural gas occurrences will be more difficult than before and often more expensive; deeper strata will frequently have to be investigated. The exploration and exploitation in arctic and offshore areas, especially in deep water regions, will be very expensive and require new technologies. But these economic and technical efforts are necessary for the exploration for natural gas accumulations of economic value there. The gas fields that will be discovered in Europe in the near future will influence the natural gas situation in the ECE region around 1990.

ACKNOWLEDGMENTS

I am indebted to several colleagues for assistance and useful suggestions. Members of the Group of Experts on Natural Gas Resources have sent material needed for preparing this paper. I must especially mention Dr. K. Brüning, Wintershall, Kassel, the Mining Engineers H. J. Meyer and H. Schöneich, Niedersächsisches Landesamt für Bodenforschung, Hannover.

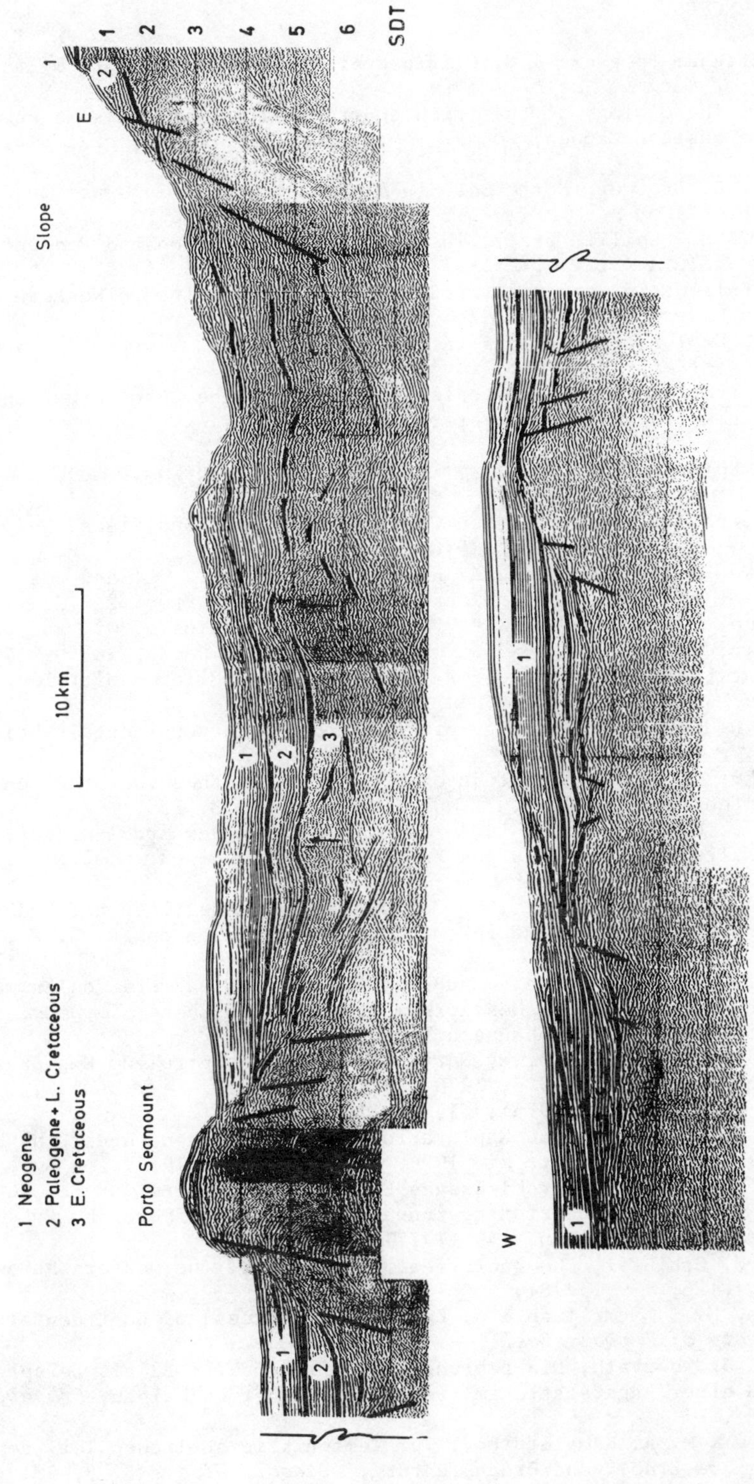

Fig.5 Seismic profile of the Portugal margin
(Montadert et al.)

REFERENCES

(1) H. A. van Adrichem Boogaert & G. C. Brouwer, in *Explanatory Note of the International Map of Natural Gas Fields in Europe*, 85 (1976).

(2) R. J. Bailey, The Geology of the Irish continental margin and some comparisons with offshore eastern Canada, *Canad. Society of Petrol. Geologists*, 4, 313 (1975).

(3) H. Boigk & G. C. Brouwer, North Sea, in *Explanatory note International Map of Natural Gas Fields in Europe* (1976).

(4) D. Bongiorni & M. Pupilli, Italy, in Petroleum Exploration and Production in Europe 1977, *Bull. Amer. Ass. Petrol. Geol.*, 62, 10, 1913 (1978).

(5) A. Brouwer, Palaeozoic Structures and Post-Variscan Basins in Western Europe, *2nd Meeting of European Geological Societies*, 29 (1978).

(6) V. S. Colter, Exploration for Gas in the Irish Sea, *2nd Meeting of European Geological Societies*, 34 (1978).

(7) J. Dadlez, Petrography of the Devonian sediments in the Goscino-Czluchow Zone of the Pomerania Trough (North-Western Poland), *Kwartalnik Geologiczny*, 19, 533 (1975).

(8) S. Depowski, Poland, in *Explanatory note of the International Map of Natural Gas Fields in Europe*, 102 (1976).

(9) W. G. Dow, Petroleum Source Beds on Continental Slopes and Rises, *Bull. Amer. Ass. Petrol. Geol.*, 62, 9, 1584 (1978).

(10) N. A. Eremenko, A. A. Geodokian, A. H. Alikhanov *et al.*, Oil and gas prospects of the shelf of the USSR, *8th World Petroleum Congress* Proc. 2, 121 (1971).

(11) N. A. Eremenko, G. P. Ovanesov & V. V. Semenovich, Status of Oil and Gas Prospecting in USSR in 1971, *Bull. Amer. Assoc. Petr. Geol.*, 56, 9, 1711 (1972).

(12) D. Franke, Entwicklung und Bau der Paläozoiden im nördlichen Mitteleuropa, Teil 1 *Z. Geol. Wiss*, 6, 1, 5 (1978).

(13) J. P. Gerard & G. Boillot, Geology of the North Irish Continental Shelf, *Marine Geology*, 23, M 1 (1977).

(14) W. W. Gluschko, G. Ch. Dikenschtejn *et al.*, Das Aufsuchen von Erdöl und Erdgas in Großen Tiefen (über 6000 m), *Z. angew. Geol.*, 19, 3, 135 (1973).

(15) A. A. Golow, L. G. Kirjuchin & B. A. Solowjow, Zonen der Erdöl-Erdgas-akkumulation des subsalinaren Komplexes des Kaspibeckens, *Z. angew. Geol.*, 23, 2, 49 (1977).

(16) L. J. Grdzelov, I. M. Siryk & V. V. Jagodin, Die Perspektiven auf Erdöl-Erdgas in tiefliegenden Horizonten und ihre Erkundung im Rahmen des R. G. W., *Z. Geol. Wiss.*, 4, 3, 411 (1976).

(17) G. Groppi, A. Muzzin & G. Vaghi, Erdölfeld Malossa, Die erste Kohlenwasserstoffgewinnung aus dem tiefen Mesozoikum unter der Po-Ebene, *Compend. 1976/77*, 1, 50, Industrieverl. v. Hernhaussen (1976).

(18) F. Hagemann & F. Al-Kasim, Norway North of 62°N., in Petroleum Exploration and Production in Europe 1974, 1975, 1976, *Bull. Amer. Ass. Petr. Geol.*, 59, 10, 1827 (1975); 60, 10, 1718 (1976); 61, 10, 1647 (1977).

(19) H. U. Hark & H. Schöneich, The Exploration for Hydrocarbons under the European continental shelf from 1973 to mid 1976, *Anep 1976*, 78 (1976).

(20) H. U. Hark & H. Schöneich, Erdöl-Erdgas-Explorationsgebiete vor den Küsten Europas, in Mineralische Rohstoffgewinnung aus dem Schelfbereich, *6. Aufbauseminar Meerestechnik*, Berlin, 146 (1977).

(21) K. Hinz & H.-U. Schlüter, The geological structure of the western Barents Sea, *Marine Geology*, 26, 199 (1978).

(22) G. L. Johnson, N. J. McMillian & J. Egloff, East Greenland Continental Margin, *Canadian Society of Petrol. Geol.*, 4, 205 (1975).

(23) J. Kapounek & Sz. Horvath, Die Bohrung Schönkirchen Tief 32 als Beispiel für den Aufschluß einer Lagerstätte im tiefen Anteil der Kalkalpen, *Erdoel, Erdgas-Z*, 84, 396 (1968).

(24) J. V. Kazancev & M. A. Kamaletdinov, Zur Tektonik im südlichen Teil der Uralvorsenke und ihre Erdöl- und Erdgasführung, *Z. geol. Wiss.*, 5, 5, 581 (1977).

(25) R. E. King, Petroleum Exploration and Production in Europe in 1974, 1975, 1976, 1977, *Bull. Amer. Ass. Petr. Geol.*, 59, 10, 1814 (1975); 60, 10, 1704 (1976); 61, 10, 1636 (1977); 62, 10, 1906 (1978).
(26) W. Krebs, Geologische Aspekte der Tiefenexploration im Paläozoikum Norddeutschlands und der südlichen Nordsee, *Erdoel, Erdgas-Z.*, 91, 9, 277 (1975).
(27) W. Krebs, Aspekte einer potentiellen Kohlenwasserstoff-Führung in den devonischen Riffen Nordwestdeutschlands, *Erdoel, Erdgas-Z.*, 94, 1, 15 (1978).
(28) A. J. Kröll & K. F. J. Kollmann, Austria, in Petroleum Exploration and Production in Europe, *Bull. Amer. Ass. Petr. Geol.*, 59, 10, 1816 (1975); 61, 10, 1638 (1977).
(29) A. Kröll & G. Wessely, Neue Ergebnisse beim Tiefenaufschluß im Winer Becken, *Erdoel, Erdgas-Z.*, 89, 11, 400 (1973).
(30) L. I. Lebedev, N. A. Eremenko, L. O. Grdzelov & O. D. Korsakov, Sedimentary basins of the submarine margin of North Europe and their oil and gas prospects, *Canad. Soc. Petrol. Geol.*, 4, 257 (1975).
(31) P. Lehner, Atlantic Margins of Europe, *2nd Meeting of European Geological Societies*, 44 (1978).
(32) H. Lögters, Das Entwicklungspotential der Nordsee und dessen wirtschaftliche Aspekte (Exploration Potential of the North Sea and its Economic Aspects), *Interocean 1976*, 1, 7 (1976).
(33) M. S. Lwow, Das Erdgas in der Sowjetunion, *Rohstoffwirtsch. Internat.*, 3 (1973).
(34) S. P. Maksimow & T. A. Botnewa, Die Bildungsbesonderheiten der Erdöllager in kambrischen Schichten der Baltischen Syneklise, *Z. angew. Geol.*, 19, 4, 164 (1973).
(35) S. P. Maksimov, E. W. Tschaikowskaja, R. G. Samwelow & M. I. Lodshewskaja, Grundlegende geologische Ergebnisse von Tiefbohrungen (über 4000 m), *Z. angew. Geol.*, 19, 7, 359 (1973).
(36) L. Montadert, E. Winnock, J. R. Deltiel & G. Grau, Continental Margins of Galicia-Portugal and Bay of Biscay, in *The Geology of Continental Margins*, 323, Berlin-New York (1974).
(37) V. E. Orjol & B. A. Solowjow, Die Gasführung der Analoga des Rotliegenden in der Dnepr-Donez-Senke, *Z. angew. Geol.*, 23, 1, 1 (1977).
(38) V. J. Orjol & B. A. Solovjov, Grundzüge der Erdöl-Erdgas-Führung in der USSR, *Z. geol. Wiss.*, 5, 11/12, 1301 (1977).
(39) R. J. H. Patijn, Die Entstehung von Erdgas infolge der Nachinkohlung im Nordosten der Niederlande, *Erdöl u. Kohle*, 17, 2 (1964).
(40) B. A. Rahmer, Soviet Union, offshore prospects and problems, *Petroleum Economist*, 44, 191 (1977).
(41) G. E. Rjabuchin, Betrachtungen zum Aufschluß neuer Erdöl- und Erdgaslagerstätten in großen Tiefen, *Z. angew. Geol.*, 22, 3, 101 (1976).
(42) G. E. Rjabuchin, Die Erdöl- und Erdgasführung in altpaläozoischen und praekambrischen Schichtenfolgen, *Z. angew. Geol.*, 22, 2, 56 (1976).
(43) D. G. Roberts & V. N. D. Caston, Petroleum Potential of the Deep Atlantic Ocean, *Ninth World Petroleum Congress Proc.*, 2, 281 (1975).
(44) H. Rønnevik, E. I. Bergsager *et al.*, The Geology of the Norwegian Continental Shelf, in *Petroleum and the continental shelf of North-West Europe*, 1 (Geology), 117, Appl. Science Publ. Ltd. (1975).
(45) H. Rønnevik & E. Bergsager, On sedimentary Basins off Norway, North of 62°N, *2nd Meeting of Europ. Geol. Soc.*, 60 (1978).
(46) K. Schmidt & D. Franke, Zur lithologisch-faziellen Entwicklung des Praeperms im Nordteil der DDR, *Z. angew. Geol.*, 23, 11, 541 (1977).
(47) K. Schmidt, G. Katzung & D. Franke, Zur Entwicklung des praepermischen Untergrundes und des Magmatismus im südwestlichen Vorfeld der Osteuropäischen Tafel, *Z. angew. Geol.*, 23, 9, 426 (1977).
(48) W. Schott, Kohlenwasserstoffvorkommen jenseits der Schelfkante der Ozeane? *Erdöl u. Kohle, Erdg., Petrochem.*, 30, 251 (1977).
(49) W. Schott & D. Stoppel, editors (1976) *Explanatory Note International Map of Natural Gas Fields in Europe*, Bundesanst. f. Geowiss. u. Rohst., Hannover.

(50) L. Schröder, Die Erdöl- und Erdgasexploration in der Bundesrepublik Deutschland 1976, *Erdöl u. Kohle, Erdg., Petrochem.*, 30, 8, 349 (1977).
(51) A. J. Stäuble & G. Milius, The Geology of the Groningen gas field, Netherlands, *Amer. Assoc. Petrol. Geol.*, 14, 359 (1970).
(52) E. Sundvor & E. Nysaether, Geological outline of the Norwegian continental margin between 60° and 68° North, *Canad. Soc. Petr. Geol.*, 4, 267 (1975).
(53) M. Talwani & O. Eldholm, Margins of the Norwegian Greenland Sea, in *The Geology of Continental Margins*, 361, Springer Verl., Berlin - Heidelberg, New York (1974).
(54) M. Talwani, G. Udintsev *et al.*, *Initial Reports* of the Deep Sea Drilling Project, XXXVIII, 1256 (1976).
(55) D. A. Tugolesov, J. J. Kusnezov *et al.*, Probleme der Erdöl- und Erdgassuche im Ostssebecken, *Z. angew. Geol.*, 17, 3, 73 (1971).
(56) N. J. Uspenskaja, Untersuchungen zur Tektonik und Erdöl-Erdgasführung der nordwesteuropäischen Senke, *Z. geol. Wiss.*, 4, 3, 425 (1976).
(57) B. J. Wassermann, N. T. Sabrodozki, L. D. Tschurilov, O. O. Scheremeta & W. E. Leschtschenko, Der europäische Norden der UdSSR - eine neue und bedeutende Erdöl-Erdgasprovinz, *Z. angew. Geol.*, 23, 12, 633 (1977).
(58) J. Wildelau, G. Bachmann, G. Dohr, K. Koch, M. Müller & K. Weggen, Die Erdgas-Tiefenaufschlußbohrung Vorderriß 1, ein erster Schritt zur Erschließung der Kalkalpen im Freistaat Bayern, *Compendium 1978/79*, 1, 14, Industrieverl. v. Hernhaussen (1978).
(59) C. A. Williams, The opening of the Bay of Biscay, *Canad. Soc. Petrol. Geol.*, 4, 363 (1975).
(60) J. E. Wright, The Geology of the Irish Sea, *Canad. Soc. Petrol. Geol.*, 4, 295 (1975).
(61) I. P. Zhabrev, I. P. Zubov, N. A. Krylov & V. V. Semenovich, Comparative Evaluation of the Oil and Gas Potential of the Epi-Paleozoic Basins of the USSR, *9th World Petroleum Congress Proc.*, 2, 83 (1975).
(62) P. A. Ziegler, Northwestern Europe: Tectonics and Basin Development, *2nd Meeting of European Geol. Soc.*, 69 (1978).
(63) H. J. Zwart, The Tectonic Framework of Central and Western Europe, *2nd Meeting of European Geol. Soc.*, 70 (1978).

Investments in the Gas Industry and the Possibilities to Finance them within the ECE Area

F. Gläser

Federal Republic of Germany

THE DEVELOPMENT OF THE GAS MARKET

So as to be able to evaluate the probable demand for investments of the gas industry within the ECE region, it appears necessary to give an approximate picture of the development of the gas industry within the ECE countries. Considering the difficulties of giving a prognosis of the development of the international gas industry up to the year 1990, taking into account the additional factors of insecurity influenced by the monetary sector (e.g. the parities of the rates of exchange) it is of greater importance to show the probable tendencies of the development of the gas industry and the resulting problems of investments and the possible solutions.

The Probable Development of Gas Demand to 1990

In recent times some surveys and studies have dealt with estimates of the future requirements of energy and thus with the demands for gas, above all stating a long-term lower growth rate of the economy than prognoses in the past.[1]

As to present opinion we have to expect consequently a value of just 18 000 million t SKE (hard coal equivalents) for the world energy consumption in the year 2000 (until 1973/74 we still had 22 000 to 24 000 million t SKE). The share of natural gas steadily increased in the past.[2] Even if there is weakening growth of the total energy consumption, a continuation of this tendency must and need not be expected for the future. The consumption of natural gas, however, will grow in real values to the year 2000. Its share, however, will stagnate and slightly sink.

---

[1]e.g. Workshop on Alternative Energy Strategies (WAES). Growth-rate for the world (but for communist countries):

	High	Low
up to 1985	6	3.5 p.c./a
1985 to 2000	5	3 p.c./a

(For this field national studies often reach too low a scope.)

[2]United Nations Study 1976: 25 years (1949-1974) the share of natural gas of the total energy consumption of the world increased by 9.8 per cent to 21 per cent.

The latest estimates for a world demand of natural gas for 1990 are based on estimates of about 2400 million t SKE.[3] The study of the Workshop on Alternative Energy Strategies (WAES) sets a demand of natural gas for Western Europe in 1990 of 342 000 million m^3; in more pessimistic evaluations it comes to a volume of demand of 305 000 million m^3 at the utmost.[4] As the progressive development of nuclear energy, on which the latter figures are based, does not quite seem to be realized in the countries planning for this development, a volume of demand of about 350 000 million m^3 appears absolutely realistic. This statement is supported to a certain extent by the BP study, *Energy 2000*, which shows a demand for natural gas for Western Europe of a volume just under 400 000 m^3 for 1990.[5]

For the COMECON states figures of demand vary from 456 000 million m^3, as the lowest figure, up to 632 000 million m^3, as uppermost figure.[6]

Differing Situation of Supply and Demand

In this context it is of importance to point to the differing situation of supply and demand in individual countries. Whereas for the European region there may be expected a regressively offered quantity out of their own resources,[7] the produced quantity may double in the field of the COMECON states from the period 1985 to 2000. Two tendencies in particular result as a consequence of this:

The first tendency will be the necessary endeavour of Western Europe to import more (WAES study) to a volume of from 117 000 to almost 151 000 million m^3 in the year 1990;[8]

This will be followed by considerable endeavours in the field of exploration and production, on the part of the COMECON states above all but also by states bordering the North Sea, in order to meet expected demand in natural gas exports in competition with other offering countries, especially OPEC countries.

Expansion of Production Activity

Within the ECE region further exploration of natural gas reserves will naturally not be confined only to the USSR. Every country with natural gas reserve volumes worth mentioning will try to extend its own production (see Table 1) or, if possible, even to increase it for several reasons — such as advantages of the national balance of payments and national security so that home production satisfies the total demand. A correspondingly broad spectrum of demand for investments for exploration and production activities is evident. To this must be added drillings into deep-lying levels and the development of new drilling procedures (Frac drilling). The production activities in the off-shore area extend more and more to deep-sea areas. For these reasons, the expenses of "normal drilling" increase by amounts of about 15 million DM — as, for instance, on the German North Sea shelf — and even by double that amount.

The expenses for the development of the North Sea natural gas have been steadily increasing. Until 1982, at which point a certain level of the North Sea region so

---

[3]Economics Institutes of the Federal Republic of Germany: *OECD World Energy Outlook, Workshop on Alternative Energy Strategies* (WAES), Exxon.

[4]Calculated as to WAES study, idem., p. 158.

[5]BP study, *Energy 2000; Tendencies and Perspectives*, Hamburg, 1977, p. 10.

[6]Evaluated as to the statements of the Economics Institutes.

[7]As to Organization for Economic Co-operation and Development (OECD) estimates as of about 1985.

[8]Interpolated from the statements given on p. 158 of the WAES study, *idem.*

TABLE 1

Status 31 December 1976

	Proved world natural gas reserves (10^9 m^3)		Production (10^9 m^3)		Quota reserves/ production	
Denmark	20		–		–	
Fed. Rep. of Germany	306		18.8		16.3	
France	140		6.6		21.2	
United Kingdom	850		37.6		22.6	
Italy	190		14.8		12.8	
Netherlands	1780		96.0		18.5	
Norway	520		–		–	
Western Europe		4090		176.0		23.2
Algeria	3560		7.3		487.7	
Libya	730		6.0		121.7	
Nigeria	1245		0.2		622.5	
Africa		5910		14.5		407.6
Iran	9345		22.0		424.8	
OPEC countries		14 550		42.2		344.8
Canada	1590		89.4		17.8	
United States	6320		560.3		11.3	
North America	7820			649.7		12.0
Middle and South America		2550		45.0		56.7
Far East/ Australia		3400		29.1		116.8
World subtotal		38 320		956.5		40.1
USSR	26 000		320.8		81.1	
Communist World		27 310		415.3		65.8
World total		65 630		1371.8		47.8

Source: Esso, Öldorado, 1976.

far exploited and explored might have been reached, there is a stated volume of investments of about US $9000 million (dollar basis as of 1973)[9]

At present we can hardly estimate the demand for investments that will ensue from production activities north of latitude 62° in the North Sea as well as from further exploration of the USSR reserves. The USSR has about 30 per cent of the world reserves of natural gas which are not at all exploited today. This is

[9]Organization for Economic Co-operation and Development, *Energy Prospects to 1985*, Vol. II, p. 141.

particularly valid for the reserve areas of the province Tjumen (West Siberia) and Jakutsk (East Siberia). The former is of interest for the West European market representing a share of about 55 per cent of the secured natural gas reserves of the USSR. For this purpose, however, enormous endeavours in the technical field will be necessary that, for their part, require corresponding investments.

Increasing Import Quota

A further, decisive factor determining the demand of investments is the increase of import quotas to be expected for the West European countries. This will ensue from the growing demand for natural gas with a simultaneous decrease of the share of self-supplying. West Europe will by the year 1990 purchase about 33 000 million m^3 each year from the USSR.[10] To be added are approximately 50 000 million m^3 in the form of LNG imports[11] particularly from Africa and the Near East. This is about a quarter of the expected total LNG trade of the world. For this about US $7160 million (value 1975) had to be raised which covered conditioning and liquefaction plants, thirty-five LNG tank ships and six terminal and regasification plants.

Perfection of the Infrastructure of Transport and Distribution

The requirements for investments up to 1990 will be essentially decided by the extension of the existing infrastructure of transport and distribution. We have to add to this the internationalizing of pipeline transports, which are at the same time an essential factor for the increase of the security of supply. In this context we have to mention particularly the north-south and the east-west lines which connect Italy with the North Sea and France with the USSR and Iran respectively. To this we have to add the already mentioned LNG supply chains. Moreover, the gas industry will have to meet particular technical measures to maintain the supply standards; for instance, the adaptation of the various use-structures of demand as well as of supply and the adaptation to the various qualities of natural gas, just to mention some of them.

For obvious economic reasons the main pipeline (e.g. of the North Sea or of the USSR) will be run with relatively high charge. Equally, the LNG terminals work round the clock. This is offset by the considerably lower total demand charge on the part of consumers, influenced above all by the gas heating consumption patterns of households which is dependent on the temperature. Balancing this requires considerable future storing capacities (aquifers, cavern storing, or exploited gas fields) and, if possible, close to consumption centres.

We should not underestimate the fact that, at least in some countries, there have to be taken into account increases in centres of low-pressure distribution and area supply. This, however, might be countered by 1990, in the countries with a developed gas economy, by a relatively regressive demand for investments in the field of distribution.

New Techniques of Gas-production and Application

It is difficult to determine quantitatively the investments of the gas industry in the development of new techniques, from the development of thermal pumps, bivalent installations, and solar energy for the low-temperature range, to the production of

[10] WAES study, p. 159, *idem*.
[11] British Petroleum, *LNG in the Next Ten Years*, 1976.

synthetic natural gas (SNG). Just this latter process may have become operational in some countries by 1990 to a certain degree through the application of the well-known pressure vaporizing process and under exploitation of high process heat, as for instance, in high-temperature reactors, so that from this time onwards continual investment amounts might be expected for these developments.

DEVELOPMENT TENDENCIES OF THE CAPITAL MARKET TILL 1990

General Price Development

In the foreground is the annual price increase rates for investment goods. These increases have the effect of inflating, in terms of money, the required volume of investments. Moreover, they influence the structure of the money and the capital market in so far as it will become more and more difficult for the investor to place his capital for long terms. When they do so, it is mostly at variable rates of interest which makes long-term calculation of capital cost increasingly difficult for them. Thus borrowers will be dependent for the most part on medium-term loans.

Considering the necessary long-term capital fixation in the energy and gas economy respectively, such a development is not satisfying on a long-term basis. Adequate financing – i.e. adapting time of validity and height of interest on the capital to the individual investment project, thus making it calculable on a long-term basis – necessitates striving to maintain stability of the monetary value an absolute precondition. However, difficulties encountered in this respect are world-wide and, in fact, too well known.

Setting of Objectives for the Capital Market Policy

For the future development of long-term financing, the respective aims of the states will play a role concerning the stability, price and capital market policy (see Table 2). Depending upon the application of the financial-political means at the disposal of the governments, the capital market policy can influence (particularly in cases of a liberal ordering of the capital market) the ability of the companies concerned, as well as the state, to take up liabilities. Further, depending upon the economic aim, there arises the question of the distribution of the streams of capital, particularly when this distribution made by the state itself or is, at least, influenced by the state. Then it depends on what priority has been assigned to the energy and gas industry within the given order scale. As the ranking in the scale of importance concerning the distribution or even the allotment of capital may be different, for whatever reasons a long-term prognosis of the covering of the capital requirements of the gas industry will thereby be complicated. Further factors (that can be estimated in this context only with difficulty), apart from functionability and efficiency of the national capital markets, are the future mobility of the international capital streams. Equally decisive will be the changes in the structure of the world capital market. In this context, I wish to recall the accumulation of capital on the part of the OPEC countries since the November 1973 oil price increase – so that it may be seen that today it is hardly possible to form an exact picture of the scope, structure, terming, and costs of capital by the year 1990.

I shall therefore mainly base the discussion of the possibilities of financing on what the possible demand for capital will look like on the basis of the peculiarities of the gas industry. Whether or how the industry will find the required capital cannot be forecast with any degree of certainty at present.

Covering the demands for capital thus remains one of the main risks of the gas industry.

TABLE 2

International Discount-rates

Country	Rate per 03.01.1978		Last rate	
	% p.a.	current per	% p.a.	current per
I. *European countries*				
Belgium-Luxembourg	9	14.12.77	7	02.12.77
Fed. Rep. of Germany	3	16.12.77	$3\frac{1}{2}$	15.07.77
Denmark	9	03.03.77	10	07.12.76
France	$9\frac{1}{2}$	31.08.77	$10\frac{1}{2}$	23.09.76
United Kingdom	7	25.11.77	5	14.10.77
Ireland	$6\frac{3}{4}$	08.12.77	$5\frac{1}{10}$	11.10.77
Italy	$11\frac{1}{2}$	29.08.77	13	13.06.77
Netherlands	$4\frac{1}{2}$	11.11.77	$3\frac{1}{2}$	06.05.77
Island	10	08.77	$7\frac{1}{4}$	08.74
Norway	6	06.09.76	5	06.10.75
Austria	$5\frac{1}{2}$	10.06.77	4	10.06.76
Portugal	13	25.08.77	8	28.02.77
Sweden	8	04.10.76	6	04.06.76
Switzerland	$1\frac{1}{2}$	15.07.77	2	08.06.76
Finland	$8\frac{1}{4}$	01.10.77	$9\frac{1}{4}$	01.07.73
Greece	11	03.01.77	10	27.09.75
Spain	8	26.07.77	7	10.08.74
Turkey	9	01.10.74	$8\frac{3}{4}$	26.11.73
II. *Countries out of Europe*				
Japan	$4\frac{1}{4}$	05.09.77	5	19.04.77
Canada	$7\frac{1}{2}$	09.05.77	8	01.02.77
Iran	10	16.06.77	9	11.76
United States	6	26.10.77	$5\frac{3}{4}$	31.08.77

Source: Monatsberichte der Deutschen Bundesbank, 30. Jahrgang Nr. 1, Jan. 1978.

Even in case of a temporary detention we shall have to devote our full attention to this problem. If it is possible for the gas industry in some countries to be put in a position considering profits that depends on a minimum of state assistance, then this problem has lost quite a lot of severity. These considerations become obvious in some countries that have joined our investment research, more or less clearly.

THE PROBABLE DEVELOPMENT OF INVESTMENTS TILL 1990

In connection with investment development I would like to refer to the work of the *Ad hoc* Meeting "Investments and their Financing" convoked by the ECE's Committee on

Gas in its 12-14 November 1973 meeting. The idea that the financial sector may become the limiting factor for the growth of the gas industry was a fundamental issue. A questionaire was developed to obtain annual investment and financing data for the gas industry. This was the third one. Data of twenty-five countries has been collected, however, not always for every year. For the following countries no data whatsoever was received and no basis of information is available: Albania, Bulgaria, Cyprus, Democratic Republic of Germany, Iceland, Romania and Spain.

As concerns the exactitude of the research, it would be welcomed if these countries, too, could answer the questionaire. There are still difficulties in the exact evaluation of data which affects the content and results of the analyses. However, such difficulties could be reduced greatly for countries that regularly participated in the evaluation.

Total Investments in the Gas Industry

The development of the total investment is marked, above all, by the respective situation of the development of the natural-gas economy in the individual countries. On the basis of data in the questionaire mentioned above, not considering the USA, investment figures in the gas industry — distributed for the first time in 1974 — yielded for 1975 a total sum investment of US $8700.2 million (dollar value as of 30 June 1976) (see Table 8). For 1976 the data of two important countries (the Netherlands and Canada) was not received; however, it is relatively certain that, compared to 1975, a slight decline of about 3 to 4 per cent of the total volume of investments has set in.

Nevertheless, an increase of investments for the future is expected, although to a lesser degree than was expected in the past. This expectation is supported by the fact that in 1976 the investments projected originally were not reached.

For one thing, this was an effect of a dampening of demand as endeavours were made to save energy, a valid tendency for all countries with a developed gas industry. However, the slowing of demand for investments is the result of contrasting tendencies. On the grounds of the investment survey information gathered so far, investments for subsequent years may be increasing in some countries, to name just France, the Federal Republic of West Germany and Poland, but here, too, we have to expect that the growth rates will flatten off until 1990. This tendency can already today be realized for some countries; e.g. Austria, Belgium, Italy, Great Britain. A decisive factor for this diverging development of investments in the West European countries is the fact that, in comparison with the past, the investments for domestic production may decrease while investments for imports may increase.

The development to be expected for the USSR cannot be defined exactly at the present time. The 1976 volume of investments was slightly down as compared to 1975 (by 5 per cent); however, it was still far beyond the preceding year. There are no reliable indicators for a statement of trends to 1990.

Altogether the volume of investments in the countries reporting has doubled between 1970 and 1976 (from about US $3700 million to about $8400 million in 1976). As already mentioned, this rapid development will not continue. Regarding a trend, it must be noted that the investments for production and exploration were so far at the disposal of only three countries. To deal with such gaps of information, equally of importance for a statement on the infrastructure, we may have to agree upon a new version of the questionaire.

Investments Structures

Considering the future investments structures it must be said that the majority of countries will hardly offer worth while or important shares of the production sector in the total investments. This is particularly valid for the countries that essentially provide traditional gas production. On the basis of available data, it may perhaps be the countries Denmark, the Federal Republic of Germany, Great Britain, the USSR, Hungary and perhaps Italy that will raise 10 per cent of their total investments after 1980. This will be guided by the extension of hydrocarbon production whereby, according to present knowledge, in the Western European countries there is expected to be a flattening off after the period 1990 to 1995.[12]

Furthermore, the share of distribution investments will decrease at least in the developed gas economy of some West European countries; at the same time additional expenses for exploration will play a minor role. The main share of investments will then be eaten up by imported volumes for transport and other infrastructural sectors. In the USSR and in some other COMECON countries, however, the volume of investments for distribution will have a share equal to production.

FINANCING POSSIBILITIES

Traditional Ways of Financing

Financing out of own resources

To begin with I should like to speak about the traditional forms of financing out of domestic resources as well as other resources. It will be mere guesswork in speaking about the future to weigh these two forms of financing. This is particularly true because with increasing demand for investment, the rate of domestic capital investment will shrink if it continues to be difficult to pay a corresponding risk benefit for the capital bearing liability.

This is above all applicable if expenses rise faster than profits as the earning of interest for the capital invested, as well as a proper and sufficient reserve, will not leave sufficient means. Depreciation financing is to be considered an important means of financing and an evaluation, adapted to economic development, is of decisive importance. We know that, *inter alia*, by varying degrees of intensity of inflationary development as well as tax laws, depreciation financing is more negatively influenced or less negatively influenced depending upon the country.

Usually in times of a difficult capital market companies will increasingly use financing by reserves and depreciation so as to finance an increase of the capacity on a long-term basis.

It can hardly be foretold what the respective situation will be in 1990. For the present, however, we may state that the situation of the international capital market does not give a clear incentive for an intensified effort to finance out of domestic resources. However, from the analyses so far one can form the impression that in addition to financing out of domestic resources by building up reserves, depreciation financing has increased in several countries during recent years. Contrasting to former methods there has been more and more adapting of prices to economic rather than political considerations, naturally without infringing on competitiveness of the gas which should never suffer under such actions.

Principally it can be said that, by joint endeavours to reach and maintain stability of money and by cost-covering, market-oriented pricing, financing out of domestic resources can be maintained in the end with all its importance for the gas industry. Stability of money supply and cost/market conscious prices are to be considered

---

[12]BP *Energy 2000*, p. 10, *idem*.

simultaneous conditions for a sufficient supply of domestic capital which, for its part, determines the scope of possibilities to increase indebtedness and is thus the measuring-stick for the volume of financing based on other resources.

Financing out of other resources

The difficulties of financing out of other resources and the applicable basis of calculation considering rates of interest, terms, etc., have been already discussed. The gas industry can but be interested in long-term capital from other sources. To cover this long-term capital requirement with medium- or even short-term capital or to cover it by a revolving process is a finance attitude that we should judge negatively.

At present the credit terms on the international money markets may be viewed altogether as favourable. In 1977 the medium-term as well as the long-term international capital market placed a volume of approximately 150 000 million (= US $63 500 million).[13] During the years 1976 and 1977 in particular medium-term Euro-credits showed a lively demand, practically from all quarters: from petroleum exporters; from developing countries, from OECD countries — both Europe and America — but in particular from the COMECON states. Considering the abundant offers, the terms are more than favourable because the competition among investors has led to a brittling away of interest rates as well as to an extension of the terms of validity. The future development of the volume of Euro- and other international loans is difficult to judge in view of the insecurity resulting from the dollar fluctuations.

The dependence on the finance and the capital markets of the USA will further remain a decisive factor for the European capital markets. The movement of the parities of the rates of exchange has increased in most countries with hard currencies, the attraction being lower rates of interest. However, it is generally expected that, as a result of endeavours, particularly of the USA, to release the currency from the pressure of the exchange markets, there will be an increase of interest rates.

These few highlights for 1978 naturally do not suffice for a long-term prognosis. For this it will essentially depend on the influence of OPEC countries on the international money flows particularly under the aspect of the recycling possibilities in the field of energy considering the enormous requirements of some countries (Iran, Algiers) on other markets.

Of the total Euro-credit volume of 1977 according to preliminary statements, about US $5600 million was spent on the energy sector (in 1976 it was just US$ 6400 million).[14] That is just 20 per cent of the total volume of Euro-credits which, for its part, makes just 50 per cent of the total capital volume. In the majority of cases individual loans ranged from US $10 to 50 million; i.e. more than half of all demands were within this scope. The terms of validity, as a rule, were between 5 and 8 years.

The demand for investments has greatly increased in the field of exploration and LNG imports during recent years. Long starting times delay the essential flow of liquidity for financing the capital costs and covering the risks. Liquidity is possible only after many years; this has led to the development of the new, rather unconventional forms of financing which has been practiced.

---

[13]OECD, *Financial Market Trends*, No. 2, Dec. 1977, whereby US $1.00 = DM 2.3217 (Ø 1977).
[14]OECD, *Financial Market Trends*, pp. 92 ff.

Unconventional Financing

Particularly in the field of energy two forms of such new financing have been increasingly used. One is leasing-financing. Another is the "off-balance-sheet financing", i.e. financing outside the scope of the balance-sheet and balance methods to relieve them.

Leasing in the most customary form comprises the renting of installations, e.g. transport lines, underground storages, whereby at a specified time property rights pass over to the lessee (the so-called hire-purchase).

By the capital being put up by a third party, however, financing is made step by step through payment of the lease instalments. Leasing therefore represents a form of financing restitution.

A further possibility of financing off the balance-sheet is particularly of importance for the gas industry. Such financing may occur if, for instance, a natural gasfield is offered to the investor as capital security after suitable geological surveys and if the investor, by a certain pricing of the gas sold from this field, gets a guarantee for an adequate yield and repayment of his investment.

A further example could be seen in that a gas pipeline is so secured that all contract-secured and agreed conducting and transport charges guarantee the payment of the interest and repayment at any time.

This way of financing, moreover, may also be entrusted to particularly established financing corporations that execute the respective project financing; for instance, finance corporations for certain gas pipelines, LNG supply chains, LNG-terminals, etc.

A further possibility is that of barter which is already being exercised (natural gas for pipes). Here, however, the difficulty is to conform to the interest of all parties involved, those concerned with the selling price of the natural gas and those concerned with the pipe price or the credit interest rates, those concerned with the purchase price and suitable ways of delivery to third parties (e.g. by pipes), the needs of the investor regarding interest rates and security, as well as those of the further suppliers. In this connection there is a certain disadvantage in the possibility that influences may play a role, e.g. those of the steel and pipe market, respectively, that have nothing to do at all with the original energy specific peculiarities of the business.

A further solution is the so-called production payment. This, as we know, is used in case of the North Sea exploration and exploitation, whereby the investors, as a rule a group of bankers, have a working gasfield and interest and repayment are paid from the yields. This, of course, calls for a change of the security assessment on the part of the investors, as far as bankers are concerned.

In the past there have been various suggestions regarding the establishment of an energy banking institute. Such a specialized bank, in my opinion, could only start the transactions done already by the existing general trading banks unless this bank were bestowed with special privileges by the respective government or set in an international framework.

However, those who grant or facilitate such special terms for the availability of the capital would wish to have a say in the selection of the projects to be financed. The extent to which this should and could be accepted is a question of taste. I am of the opinion that thus, too easily unjustified factors and considerations may play a role in the proper requirements and demands of the gas industry.

STABILITY OF MONEY AND COMPETITIVE PRICE POLICY AS DECISIVE FACTORS OF FUTURE POSSIBILITIES OF FINANCING

All these methods of financing listed suggest that the search for suitable financing adequate to the long-term development of the gas industry is not at all exhausted. In considering the unexploited reservoir of possibilities, whether that of an entrepreneurial or state decision, one may set out basic conditions for financing the demand for investments expected for 1990. These conditions are: the international endeavour for a liberal order of the capital markets, at least for a certain flexibility; common striving for stability of monetary values; and an internationally adjusted, competition-oriented price policy.

If the gas industry can build upon international success in financing it will then be in a position to hold or to extend its own share in the investments and able to furnish its enterprises or organization with its own capital supplied by the owners. With a self-sufficiency economy the gas industry, together with investments from outside quarters, may be able to finance and to carry through its planned long-term undertakings and investments, which, as you know, require a very long time before realization. Consumers may then benefit optimally from the advantages that gas offers as a modern and environmentally safe energy.

While definite statements about the respectively suitable form of financing cannot be given for a time reaching so far into the future, I hope I have shown the extent to which it is fundamentally important to have sufficient, economical financing for the optimal development of the gas industry in the ECE countries. Therefore, we shall all be well advised not to lose sight of this objective and not to delay long in finding solutions, should difficulties in the realization of our big, gas-economic tasks arise, be they big investment leaps or particular difficulties in procuring the capital required.

TABLE 3

Total Investments Broken Down by Sub-areas in the Years 1970, 1975, 1976 and 1980
(in per cent)

Country	Production and conditioning				Transmission				Distribution			
	1970	1975	1976	1980	1970	1975	1976	1980	1970	1975	1976	1980
Austria	20	–	–	–	80	–	–	–	–	–	–	–
Belgium	–	3	7	?	16	32	22	?	84	65	72	?
Canada	–	27	–	–	–	36	–	–	–	37	–	–
Czechoslovakia	2	6	2	–	57	62	61	63	24	37	37	37
Denmark	71	22	–	85	–	–	–	–	29	78	–	15
Finland	–	–	4	–	–	–	47	–	–	100	49	–
France	5	3	3	4	28	52	24	54	67	45	72	42
Federal Republic of Germany	3	13	14	14	22	32	29	86	75	55	57	–
Greece	–	23	–	–	–	–	–	–	–	77	–	–
Hungary	4	1	46	1	56	74	31	47	40	25	23	52
Ireland	28	63	–	–	21	12	–	–	51	25	–	100
Italy	10	3	–	9	–	–	–	–	90	97	–	91
Malta	–	–	–	–	–	–	–	–	–	–	–	–
Netherlands	–	–	–	–	–	59	–	–	–	41	–	–
Poland	12	12	3	6	64	56	61	71	24	32	36	23
Portugal	–	–	–	–	–	–	–	–	100	100	100	100
Sweden	–	–	–	–	–	–	–	–	–	–	–	–
Switzerland	23	3	3	2	13	45	8	50	64	52	89	48
USSR	–	–	47	–	–	–	12	–	–	–	41	–
United Kingdom	4	2	3	–	43	68	57	50	53	30	40	50
Yugoslavia	37	3	–	17	63	97	–	83	–	–	–	–
Luxembourg	–	–	–	–	–	–	–	–	–	100	–	–
United States			25			–	75				–	

TABLE 4 CONT'D

TABLE 4

Development of Investments in the Gas Industry, 1970-1980
Terms expressed in 10^6 \$US (rates on 30 June 1976)

	Country	1970	1971	1972	1973	1974	1975	1976	1977	1978	1979	1980
Group A	France 1 FF = \$0.203	163.1	193.0	219.1	225.9	287.2	459.4	485.2	446.6	483.1	674.0	708.0
Group A	Fed. Rep. of Germany 1 DM = \$0.427	337.3	540.2	625.6	774.2	780.1	807.5	851.0	1044.9	1141.8	1244.3	1350.9
Group A	USSR 1 Rub. = \$1.35	2481.0	2619.0	3194.1	3812.4	4374.0	5360.0	5099.0	-	-	-	-
Group A	United Kingdom 1 £ = \$1.720	345.9	338.7	257.2	205.4	234.4	555.6	418.0	385.2	344.0	357.7	344.0
Group A	Netherlands 1 Hfl = \$0.404	-	-	-	-	-	350.2	-	-	-	-	-
Group A	Canada 1 Can \$ = 0.944	-	-	-	-	-	521.3	-	-	-	-	-
	Subtotal	3327.3	3690.9	4296.0	5017.9	5675.7	8054.0	6853.2	1876.7	1968.9	2276.0	2402.9
Group B	Austria 1 Ös = \$0.0601	24.1	23.3	64.2	156.2	89.3	81.8	73.4	42.1	33.5	45.6	31.0
Group B	Belgium 100 FrB = \$2.77	114.7	136.0	112.9	78.8	95.3	143.6	131.7	91.2	82.7	30.0	29.3
Group B	Czechoslovakia 1 KrC = \$0.175	99.9	71.9	79.5	118.0	135.3	99.9	98.5	186.7	168.5	166.0	153.1
Group B	Yugoslavia 1 Din = \$0.0548	5.2	0.2	0.3	6.0	40.3	61.7	83.5	58.9	56.8	48.6	48.4
Group B	Hungary 100 Frt = \$2.425	23.6	24.9	20.5	29.2	35.6	63.9	68.7	47.8	63.5	49.4	46.4
Group B	Poland 1 Zlt = \$0.0502	45.1	49.2	66.3	110.1	97.6	76.4	111.6	145.6	172.7	163.9	156.0
Group B	Switzerland 1 SF = \$0.405	14.3	16.7	41.0	168.6	120.4	41.5	18.3	32.9	34.6	42.2	48.7
Group B	Italy 100 Lire = \$0.113	30.8	35.3	36.7	37.3	57.6	67.6	55.7	60.5	61.8	62.2	64.7
	Subtotal	357.7	357.5	421.4	704.2	671.4	636.4	641.4	665.7	674.1	607.9	577.6

TABLE 4 CONT'D

	Country	1970	1971	1972	1973	1974	1975	1976	1977	1978	1979	1980
Group C	Denmark 100 DKr = $16.57	2.4	2.4	1.2	4.2	1.8	1.5	2.4	5.8	7.0	8.0	10.0
	Finland 1 FMK = $0.25	0.4	0.2	7.9	30.1	8.7	1.7	0.2	0.1	-	-	-
	Sweden 1 SKr = $0.227	-	-	-	-	-	1.6	-	-	-	-	-
	Ireland 1£Ir. = $1.72	1.4	1.5	1.2	1.4	2.4	2.1	2.0	0.3	0.1	0.1	0.7
	Portugal 1 Esc. = $0.026	0.4	0.5	0.6	0.6	0.5	0.7	0.3	0.8	0.8	0.8	0.6
	Malta 1 £Malt. = $2.35	0.2	-	-	-	-	-	-	-	-	-	-
	Turkey 1 £Turk = $0.060	1.4	0.9	0.4	0.4	0.7	2.0	2.5	1.1	1.6	1.6	0.8
	Greece 1 Drannre = $0.027	-	-	-	-	-	0.2	-	-	-	-	-
	Subtotal	6.2	5.5	11.3	36.7	14.1	9.8	7.4	8.1	9.5	10.5	12.1
	Total	3691.2	4053.9	4728.7	5758.8	6361.2	8700.2	7502.0	2550.5	2652.5	2894.4	2992.0

TABLE 4(a)

Development Trends Recorded for the Three Groups from 1970 to 1976 (in 10^6 \$US)

	1970	1971	1972	1973	1974	1975	1976	Index 1970 = 100
Group A	3327.3	3690.9	4296.0	5017.9	5675.7	8054.0	6853.2	206
Group B	357.7	357.5	421.4	704.2	671.4	636.4	641.4	179
Group C	6.2	5.5	11.3	36.7	14.1	9.8	7.4	127
Total	3691.2	4053.9	4728.7	5758.8	6361.2	8700.2	7502.0	203

TABLE 5

Development of Investments in the Gas Industry 1970-1980 in Relative Terms
Index 1970 = 100

	Country	1970	1971	1972	1973	1974	1975	1976	1977	1978	1979	1980
Group A	France	100	118	134	138	176	282	178	274	296	413	434
	Federal Rep. of Germany	100	160	185	229	231	239	252	310	338	367	401
	Poland	100	109	147	244	216	169	247	323	383	364	348
	Italy	100	114	119	121	187	219	181	196	200	201	211
	USSR	100	106	130	155	178	218	206	-	-	-	-
Group B	Austria	100	97	267	649	371	340	305	175	139	190	131
	Switzerland	100	117	287	118	843	290	128	230	242	295	341
Group C	Czechoslovakia	100	71	74	117	133	104	99	194	175	173	159
	Denmark	100	100	50	179	79	64	100	250	300	343	450
	Yugoslavia	100	32	48	114	774	1184	1605	1132	1092	932	935
	Hungary	100	105	87	123	150	271	291	202	268	209	198
	Portugal	100	138	150	144	138	175	82	188	194	200	206
Group D	Belgium	100	119	98	69	83	125	115	79	-	-	-
	Finland	100	44	1772	6678	1933	380	66	11	6	6	6
	Ireland	100	111	84	105	177	156	148	2	14	13	68
	United Kingdom	100	98	74	59	68	115	121	111	100	103	100
	Turkey	100	65	29	29	53	141	176	77	112	112	112
Group E	Canada	no useful data available										
	Greece											
	Malta											
	Netherlands											
	Sweden											

TABLE 6

Financing of Investments 1975/76

Group	Country	Year	Total investments (in 10^6 of national currency)	of which: by own resources of the undertaking		by resources other than those of the undertaking		Percentage of investments made by "own resources" over those made by "other resources"
				in 10^6 of national currency	as percentage of total investments	in 10^6 of national currency	as percentage of total investments	
Group 1	Austria Natural gas producer and importers (only)	1974	886.2[a]	49.6	5.6	836.6	94.4	5.9
Group 1	Czechoslovakia	1976	563.0	256	45.4	307	54.6	83.4
Group 1	Hungary	1976	2831.0	1064.5	37.6	1766.5	62.4	60.3
Group 1	Poland	1976	2223.8	513.9	23.1	1709.9	76.9	30.1
Group 1	Switzerland	1976	45.3	19.0	42.0	26.3	58.0	72.3
Group 1	USSR	1976	3777.0	873.0	23.1	2904.0	76.9	30.1
Group 1	Yugoslavia	1974	735.0	194.0	26.0	541.0	74.0	35.0
Group 2	Federal Rep. of Germany	1976	1993.0	1100.0	55.0	893.0	45.0	123.2
Group 2	Italy[b]	1975	59 800.0	15 000.0	25.1	44 800.0	74.9	33.5
Group 2	Belgium	1976	4754	2441	51.3	2313	48.7	105.5
Group 3	Finland	1976	1.18	0.1	8.5	1.08	91.5	9.3
Group 3	France[c]	1975	1428.0	1668.0	73.7 (70)	594.9	(30)	280.4
Group 3	Greece	1975	6.3	6.3	100.0			
Group 3	United Kingdom	1976	243.0	217.0	89.3	26.0	10.7	834.6
Group 3	Luxembourg	1976	7.9	7.9	100.0			

(a) Preliminary. (b) Estimates. (c) (1976).

TABLE 7

Ratio of Investments over Turnover (investments and turnover expressed in 10^6 of national currency)

Country	National currency	1970	1971	1972	1973	1974	1975	1976
Belgium	FRB							
Total investments		4141.0	4910.0	4075.0	2844.0	3442.0	5183.0	4734.0
Turnover		8809.0	10 262.0	12 410.0	14 493.0	19 496.0	27 280.0	37 637.0
Ratio of I/T (in %)		47	48	33	20	18	19	13
Czechoslovakia	KrC							
Total investments		571.0	411.0	454.0	674.0	773.0	571.0	563.0
Turnover		3633.0	3986.0	4433.0	4687.0	5381.0	5200.0	5710.0
Ratio of I/T (in %)		16	10	10	14	14	11	10
Denmark	dKr							
Total investments		9.0	6.0	5.0	13.0	2.0	-	
Turnover		56.0	57.0	58.0	70.0	100.0	-	
Ratio of I/T (in %)		16	11	9	19	2	-	
Finland (Helsinki Gasworks)	FMK							
Total investments		1.8	0.7	1.7	2.2	0.5	-	-
Turnover		15.5	22.0	18.0	16.0	11.0	-	-
Ratio of I/T (in %)		12	3	9	14	5	-	-
France	FF							
Total investments		803.6	950.5	1079.2	1112.8	1415.0	2262.9	1428.0
Turnover		2655.2	3198.1	3847.2	4541.2	5789.5	7948.0	10 680.0
Ratio of I/T (in %)		30	30	28	25	24	29	13
Federal Rep. of Germany	DM							
Total investments		790.0	1265.0	1465.0	1813.0	1827.0	1891.0	1993.0
Turnover		4565.0	5368.0	6699.0	7755.0	-	14 000.0	17 265.0
Ratio of I/T (in %)		17	24	22	23	-	14	12

Hungary	Frt							
Total investments		-	-	-	-	-	4088.2	2831.0
Turnover		-	-	-	-	-	7324.5	
Ratio of I/T (in %)		-	-	-	-	-	56	
Ireland	£Ir							
Total investments		0.7920	0.8768	0.6615	0.8298	1.4008	-	
Turnover		5.0460	5.4860	5.6380	5.9640	9.5090	-	
Ratio of I/T (in %)		16	16	12	14	15	-	
Italy[b]	Lire							
Total investments		13.300	15.200	16.500	15.500	24.200	21.460	
Turnover		34.800	39.700	42.400	55.100	70.700	104.751	
Ratio of I/T (in %)		38	38	38	28	34	20	
Netherlands	Hfl							
Total investments		-	-	-	-	-	886.8	
Turnover		-	-	-	-	-	11 700.0	
Ratio of I/T (in %)		-	-	-	-	-	7	
Poland[c]	Zlt							
Total investments		1157.0	1178.0	1536.0	2450.0	2313.0	1879.6	3377.6
Turnover		5737.0	6289.0	6526.0	6770.0	6921.0	12 941.0	24 393.3
Ratio of I/T (in %)		20	18	23	36	33	14	14
Portugal	Esc.							
Total investments		16.0	22.0	24.0	23.0	22.0	-	13.2
Turnover		198.0	212.0	220.0	223.0	235.0	-	302.5
Ratio of I/T (in %)		8	10	11	10	9	-	4
Switzerland	SF							
Total investments		35.27	41.25	101.11	416.38	297.42	102.33	45.3
Turnover		265.41	316.82	387.54	-	-	-	631.8
Ratio of I/T (in %)		13	13	26	-	-	-	7

TABLE 7 CONT'D

Country	National currency	Year 1970	1971	1972	1973	1974	1975	1976
USSR	Rub.							
Total investments		-	-	-	-	2521.0	-	
Turnover		-	-	-	-	3332.0	-	
Ratio of I/T (in %)		-	-	-	-	76	-	
United Kingdom	£							
Total investments		201.100	196.933	149.533	119.400	136.300	323.0	243.0
Gross sales income (GS)		498.000	536.975	629.641	739.600	812.500	1340.0	1957.0
Gross income (GI)		659.400	695.500	787.000	896.800	970.400	1566.0	12
Ratio of I (GS)		40	37	24	16	17	24	
Ratio of I (GI)		30	28	19	13	14	21	
Luxembourg								
Total investments								7.9
Turnover								951.0
Ratio of I/T (in %)								1.0

(a) Total investment in Table 5 differs from that in Table 4.
(b) As national data are not available, these data are given on the basis of a group of distribution undertakings, which represent *ca.* 20 per cent of national total.
(c) The total investment amount includes also the expenses for investment other than indicated in Tables 2 and 4, e.g. residential constructions, social facilities, etc. Turnover at the prices being obligations since beginning 1975.

TABLE 8

Exchange Rates between Various National Currencies and the $US (on 30 June 1976)

1. Austria	1 Sch = $ 0.0601
2. Belgium	100 FrB = $ 2.77
3. Canada	1 C$ = $ 0.944
4. Czechoslovakia	1 Kr = $ 0.175
5. Denmark	100 dKr = $ 16.57
6. Finland	1 FM = $ 0.25
7. France	1 FF = $ 0.203
8. Federal Republic of Germany	1 DM = $ 0.427
9. Greece	1 Drachme = $ 0.027
10. Hungary	100 Frt = $ 2.425
11. Ireland	1 £Ir = $ 1.72
12. Italy	100 Lire = $ 0.113
13. Malta	1 £Malt = $ 2.35
14. Netherlands	1 Gul = $ 0.404
15. Poland	1 Zl = $ 0.0502
16. Portugal	1 Esc. = $ 0.026
17. Sweden	1 Skr. = $ 0.227
18. Switzerland	1 SFr. = $ 0.405
19. Turkey	1 T£ = $ 0.060
20. United Kingdom	1 £ = $ 1.720
21. Union of Soviet Socialist Republics	1 Rub. = $ 1.35
22. Yugoslavia	1 Din. = $ 0.0548

The French Gas Industry

Gaz de France

GAS IN FRANCE

The French gas industry has undergone a complete transformation since 1956, the date when the first production of the Lacq field marked the beginning of the conquest of the national market by natural gas which covers now practically all the needs. The Saint Marcet field which was discovered just before the war had only a regional importance. The natural gas imports formerly fulfilled the progressively growing needs. Coal gas, which was the only one to be used in the past, is produced only in the framework of the coal mine and the iron industry where it is applied either to internal utilizations as in blast furnaces or to subsidiary companies (chemistry or metallurgy) or to thermal stations.

Besides natural gas — a leading proportion — the gas availability on the market is made up of manufactured gas obtained by the cracking of petroleum products, of propane or butane delivered by pipeline, either pure or mixed with air, as well as of liquefied butane or propane delivered in bottles. The liquefied petroleum gases delivered in bottles are marketed by the subsidiaries of petroleum groups and do not belong to the gas industry *stricto sensu*. The repartition of the gas availability according to its nature and its evolution during the last 12 years shows an enormous progress of natural gas used in its pure state which goes from 15.5% of the availability in 1962 to 66.5% in 1974 and 72.4% in 1976 for volumes which nearly increased tenfold (Table 1).

This spectacular progress has strengthened the place of gas on the whole of the French energy consumption. The share it takes in the covering of the national needs went from 3.5% to 10.8% between 1962 and 1976 as far as the primary energy is concerned and from 4.8% to 15.6% as far as the final energy utilizations are concerned (Table 2). In 1985 the "Plan" foresees that gas will cover 15.9% of the needs of the primary energy in France.

The mortal gases apart (coke-oven gas, blast-furnace gas, refinery gas) natural gas represents nearly 81.5% of the French gas consumption and nearly 94% of the industrial consumption.

It is interesting to place the French gas industry in an international background, by comparing its results with the ones of certain foreign countries (statistics of the gas for Europe of the Economic Commission for Europe of the United Nations). The gross gas availability is thus, for 1975, for the gas industry (natural gas + manufactured gas) in petajoules (1 PJ = 10^9 MJ = 238.92 × 10^6 thermies) (Table 4).

TABLE 1

Net Gas Availability in France
(in petajoules)[a]

	1962		1965		1970		1975		1976	
1. Natural gas	66	15.5%	160	33.7%	324	50.-%	665	70.9%	740	72.4%
2. Manufactured gas	154	36.4%	75	15.7%	63	9.8%	26	2.7%	18	1.8%
3. Liquefied petroleum gases (by pipeline)	n.d.		n.d.		n.d.		12	1.2%	12	1.2%
4. Gas industry *stricto sensu* (1 + 2 + 3)	220	51.9%	235	49.4%	387	59.8%	703	74.8%	770	75.5%
5. Liquefied petroleum gases (in containers)	n.d.		n.d.		n.d.		114	12.2%	118	11.6%
6. Total liquefied petroleum gases (3 + 5)	50	11.8%	72	15.2%	103	16.-%	126	13.4%	130	12.7%
7. Total on the market (1 + 2 + 3 + 5)	270	63.7%	307	64.6%	490	75.8%	817	87.-%	888	87.-%
8. Coke-oven gas	42	10.-%	45	9.5%	48	7.4%	48	5.1%	51	5.-%
9. Blast-furnace gas	105	24.7%	100	20.9%	89	13.8%	69	7.1%	75	7.3%
10. Refinery gas	7	1.6%	24	5.-%	20	3.-%	7	0.8%	7	0.7%
Total availability	424		476		647	100.-%	941	100.-%	1021	100.-%

(a) 1 Pj = 10^9 MJ = 238.92 × 10^6 thermies, 10^9 thermies = 4.1855 PJ.

The increase of the availability is essentially due to the natural gas imports which have come to make up a more and more important part of the resources (production + imports) (Table 5).

ORGANIZATION OF THE FRENCH GAS INDUSTRY

Following the war, the gas production was carried out in France in 546 gas plants scattered all over the territory, using coal as a raw material. The plants belonged to companies with a very variable importance; big companies notably ensured the supply of the Paris area, subsidiaries of the coal companies delivered gas in the coal basins, in the north in particular.

All the French companies involved in gas production, transmission, distribution, exportation and importation (that is to say 615 undertakings depending on 127 companies) were nationalized in France in 1946, apart from a small number of undertakings

TABLE 2

Evolution of the Energy Consumption in France
(10^6 tep)

	1962	1965	1970	1974	1975	1976
Primary energy						
Solid mineral fuels	46.8	45.7	38.1	31.6	27.5	32.3
%	54.6	41.1	25.7	18.-	16.7	18.5
Petroleum products	26.9	49.7	87.3	112.4	101.7	108.9
%	31.4	44.7	58.8	64.-	61.9	62.4
Gas	3.-	5.1	9.3	16.-	17.5	18.8
%	3.5	4.6	6.3	9.1	10.6	10.8
Total fuels	76.7	100.5	134.7	160.-	146.7	160.-
%	89.5	90.4	90.8	91.1	89.2	91.7
Primary electricity	9.-	10.7	13.6	15.6	17.7	14.5
%	10.5	9.6	9.2	8.9	10.8	8.3
Total	85.7	111.2	148.3	175.6	164.4	174.5
%	100.-	100.-	100.-	100.-	100.-	100.-
Final utilizations						
Solid mineral fuel	31.2	28.9	21.2	19.2	14.8	14.6
%	39.7	30.6	16.7	12.5	10.3	9.6
Petroleum products	27.6	40.7	70.8	87.2	80.7	84.7
%	35.2	43.1	55.7	56.9	56.3	55.5
Gas	4.8	6.1	9.-	12.9	13.5	15.6
%	6.1	6.4	7.1	8.4	9.4	10.2
Total fuels	63.6	75.7	101.-	119.3	109.-	114.9
%	81.-	80.1	79.5	77.8	76.-	75.3
Electricity	14.9	18.8	26.-	34.1	34.4	37.6
%	19.-	19.9	20.5	22.2	24.-	24.7
Total	78.5	94.5	127.-	153.4	143.4	152.5
%	100.-	100.-	100.-	100.-	100.-	100.-

TABLE 3

Repartition of the French Gas Consumption in 1976
(in petajoules)

	Domestic utilization	Commercial utilization	Industrial utilization	Power gas	Total
Natural gas	180.-	127.-	432.-	< 1	740.-
%	24.4	17.1	58.4	0.1	100.-
Manufactured gas	10.-	4.-	4.-	-	18.-
%	56.8	22.7	20.5	-	100.-
Liquefied petroleum gas					
in containers	81.-	15.-	22.-	-	118.-
by pipeline	9.-	2.-	1.-	-	12.-
Total	90.-	17.-	23.-	-	130.-
%	68.8	13.2	18.-	-	130.-
Total	280.-	148.-	459.-	< 1	888.-
%	31.6	16.7	51.7	-	100.-

TABLE 4

United States	24 172
USSR	10 906
Netherlands	3200
West Germany	1917
United Kingdom	1598
Romania (1974)	1184
Italy	921
France	842
Belgium	443
East Germany	315
Poland	418
Czechoslovakia	248
Hungary	214
Austria	185

(These figures, which take into account production, imports, exports and stock variations, represent the activity of the French industry characterized by the gas volumes it commercializes.)

TABLE 5

French Resources in Natural Gas
(in petajoules)

	Production	Imports	Total
1960	114.-	-	114.-
%	100.-	-	100.-
1965	200.-	8.-	208.-
%	96.2	3.8	100.-
1970	270.-	132.-	402.-
%	67.2	32.8	100.-
1971	281.-	186.-	467.-
%	60.2	39.8	100.-
1972	291.-	258.-	549.-
%	53.-	47.-	100.-
1973	293.-	355.-	648.-
%	45.2	54.8	100.-
1974	295.-	425.-	720.-
%	41.-	59.-	100.-
1975	287.-	442.-	729.-
%	39.4	60.6	100.-
1976	277.-	548.-	825.-
%	33.5	66.5	100.-

of a slight importance and the ones which were managed by local organizations ("Régies", state monopolies) 94% of the French gas industry were thus granted to Gaz de France, a national public undertaking with an industrial and commercial character, endowed with a financial autonomy and in charge of a mission of public utility.

Gaz de France is an independent undertaking, subject to rules which regulate the activity of industrial and commercial companies; in particular, it has its own budget. Its employees, subject to special regulations, are not public officers, but, as a state enterprise, it is subject to state guardianship which is exercised by the Ministry of Industry and Research and the Ministry of Finance. Besides, it is forbidden for it to build, to sell, to install gas-using appliances, which are activities reserved to the manufacturers and the installers with whom it keeps naturally close relationships, observing, however, the freedom of action of the partners.

On the other hand, a law in 1949 excluded from the nationalization the transportation of natural gas ensured by producing undertakings, the majority of shares is held by the State, between the fields and the metre of the distribution plant, in order to take into account the existence of the "Régie Autonome des Pétroles" (RAP) which operated the network used for the transportation of the Saint Marcet natural gas.

The actual structures of the French gas industry correspond to the following legal framework:

According to the nationalization law, Gaz de France holds the monopoly of the transportation and distribution of gas in France, subject to the exceptions just mentioned. Thus, it possesses the greater part of public gas distribution which manages the majority of the customers, and besides, it supplies very big customers directly from the transmission network.

The "Régies" and "Sociétés d'économie mixte", which are excluded from the nationalization, together with the private companies, the sales of which are inferior to the limit foreseen for the nationalization, take a smaller role as many of them are transferred to Gaz de France when they no longer have the financial capacities required to modernize their network, in particular when converting to natural gas.

Natural gas from the Lacq field was discovered by SNPA (Société Nationale des Pétroles d'Aquitaine), which became SNEA (P) Société Nationale Elf Aquitaine (Production) when the group Elf Aquitaine was reorganized in 1976. In application of the law of 1949, the marketing of this gas, an undertaking with public capital, was made according to the following restrictive provisions, by which means subsidiaries of gas producers and also Gaz de France, holder of the distribution licence, were granted the right to transport. The principal requirement was that these companies must ensure the transportation of the gas they deliver to public distributions of Gaz de France or "Régies" and non-nationalized undertakings, while having at the same time their own customers who are the very large consumers directly connected to the transportation network.

In the southwest, the transportation of natural gas between the fields and the public distributions, together with the large consumers directly supplied by the transportation network, is ensured by the Société Nationale des Gaz du Sud Ouest (SNGSO). Its capital belongs 30% to Gaz de France, 35% to the Société Nationale Elf Aquitaine (Production), 34.9% to SOGERAP and 0.1% to ERAP, the latter three companies belonging to the group Elf ERAP. SNGSO operates a network of 3100 km in the southwest.

Outside the southwest, the Lacq gas is transported and sold by the Compagnie Français du Méthane (CeFeM). When this company was created it was a 50% subsidiary of SNEA (P) and Gaz de France. Since June 1976 the repartition of its capital was modified: the participation of SNEA(P) was reduced to 40% in order to enable the Companie Française des Pétroles (CFP) to acquire 10% of the capital of CeFeM. The share of Gaz de France (50%) remained unchanged.

These relationships are set out here to indicate the part which will be played by the capital invested by the licences of French petroleum companies in the North Sea in the supplies made to the customers of CeFeM. CeFeM buys gas from SNEA(P) and transports and sells it to public distributions and large consumers directly supplied by the transportation network. The network is built by Gaz de France which leases it to CeFeM. The activity of the latter extends to the network which covers the west of France, the centre and the Lyon and Dijon areas. The Paris area is excluded from this system and is directly supplied by Gaz de France which had built up over a long period of time a very dense network.

Outside these areas, gas is transported and commercialized only by Gaz de France. This is in particular the case of natural gases imported from Algeria via Le Havre and Fos sur Mer, from the Netherlands, from the North Sea and the USSR, and later from Iran.

The quantities of natural gas commercialized by Gaz de France represented in 1976 three-quarters of the gas sales of Gaz de France, its subsidiaries CeFeM and SNGSO and the "Régies" and non-nationalized undertakings.

The activities related to the marketing of the liquefied petroleum gases (butane and propane) are free and ensured by subsidiaries of these petroleum companies. These

TABLE 6

Production and Transportation of Natural Gas in France

Zone	Production	Transport	Sale to customer
Southwest	SNEA (P) →	SNGSO →	Gaz de France, "Régies" and non-nationalized undertakings (public distributions)
		(GDF 30%) (SNEA (P) 35%) (ERAP 35%) →	Large consumers directly connected to the transportation network, thermal stations (direct customers of SNGSO)
		owner of the network	
West Centre	SNEA (P) →	CeFeM →	Gaz de France, "Régies" and non-nationalized undertakings (public distributions)
Lyon area		(GDF 50%) (SNEA (P) 40%) (CFP 10%) →	Large consumers directly connected to the transportation network, thermal stations (direct customers of CeFeM)
		network leased by GDF	
Paris area	Lacq gas SNEA(P) →	→	Gaz de France, "Régies" and non-nationalized undertakings (public distributions)
North	Imports (the Netherlands, the USSR, North Sea,	Gaz de France	Large consumers directly connected to the transportation network, thermal stations (direct customers of Gaz de France)
East		(Production and Transportation Department)	
Southeast	Algeria, Iran)		

ones produce gases which are delivered to distributor companies which put them into bottles and market them by means of a network of concessionaries and depositories. A part of the production of these gases is also sold to Gaz de France which uses them to supply public distributions not connected to the transportation network, either in a pure state or mixed with air (propane air or butane air mixtures) (Table 6).

GAS RESOURCES IN FRANCE

National Production

Except two small gas fields in Alsace and in the Jura area, the national production is essentially concentrated in the southwestern part of France.

On 1 January 1975 the available gas reserves (recoverable and after purification) were the following:*

Deep Lacq	90×10^9 m^3
Meillon concession	35×10^9 m^3
Other	3×10^9 m^3

The working conditions give hope for the future for the production given in Table 7 (in 10^9 m^3).

TABLE 7

1978	7.5
1981	7.5
1983	6.1
1985	5.7
1987	4.6
1989	3.7
1990	3.2
1993	2.5
1994	2.2
1996	1.7
1997 to 2006	1.7/year

Imports

The importation prospects, on the base of the contracts approved on 31 July 1978, are as shown in Table 8.

Thus it is obvious that the level of the foreseen sales in 1985 may be obtained with the contracts already signed when taking the maximum quantity of gas allowed by the contract with the Netherlands; however, other contracts may be approved before this date.

*Source: CEDIGAZ.

TABLE 8

Countries of origin	Start-up date	Duration of the contract	Average annual quantities in billion m^3 (gas) in plateau period
Algeria I (Arzew)	1965	25 years	0.5
Algeria II (Skikda)	1973	25 years	3.5
Algeria III (Arzew)	1980	20 years	5.1
The Netherlands	1967	20 years	10[a]
Norway (Ekofisk - Eldfisk Amoco and Albuskjell)	1976 (en fait 1977)	20 years	3.2
URSS	1976	20 years[b]	4[b]
Iran	1981	22 years	3.7

(a) Average quantities taken per year.
(b) From 1 January 1980.